Chemical Pollution and Waste Management

Chemical Pollution and Waste Management

Edited by **Giselle Tang**

SYRAWOOD
PUBLISHING HOUSE

New York

Published by Syrawood Publishing House,
750 Third Avenue, 9th Floor,
New York, NY 10017, USA
www.syrawoodpublishinghouse.com

Chemical Pollution and Waste Management
Edited by Giselle Tang

International Standard Book Number: 978-1-68286-029-8 (Hardback)

Contents

Preface   IX

Chapter 1   **Effect of Goldcrew and Corexit on selected blood parameter of the African cat-fish**
             ***Clarias gariepinus* following sublethal exposures**   1
             K. J. Alagoa

Chapter 2   **Lead and coliform contaminants in potable groundwater sources in Ibadan,**
             South-West Nigeria   7
             Olusegun Peter Abiola

Chapter 3   **Environmental mobility and geochemical partitioning of Fe, Mn, Co, Ni and Mo in**
             sediments of an urban river   12
             H. M. Zakir and N. Shikazono

Chapter 4   **Acute toxic effects of Endosulfan and Diazinon pesticides on adult amphibians**
             (*Bufo regularis*)   23
             Ezemonye Lawrence and Tongo Isioma

Chapter 5   **An overview of the treatment of print ink wastewaters**   29
             Lichao Ding, Yunnen Chen and Jingbiao Fan

Chapter 6   **"Polyamine levels and pigment contents in rapeseed regenerated *in vitro* in the**
             presence of zinc"   34
             Asma Ben Ghnaya, Annick Hourmant, Michel Couderchet, Michel Branchard and
             Gilbert Charles

Chapter 7   **Acute toxicity and behavioral response of freshwater fish, *Mystus vittatus* exposed to**
             pulp mill effluent   42
             A. Mishra, C. P. M. Tripathi, A. K. Dwivedi and V. K. Dubey

Chapter 8   **Evaluation of the levels of organochlorine pesticide residues in water samples of**
             Lagos Lagoon using solid phase extraction method   48
             David Adeyemi, Chimezie Anyakora, Grace Ukpo, Adeleye Adedayo and
             Godfred Darko

Chapter 9   ***Ruditapes decussatus* embryo-larval toxicity bioassay for assessment of Tunisian coastal**
             water contamination   55
             Salem Fathallah, Mohamed Néjib Medhioub, Amel Medhioub and
             Mohamed Mejdeddine Kraiem

Chapter 10   **Metal concentrations in sediments and water from Rivers Doma, Farinruwa and**
             Mada in Nasarawa State, Nigeria   64
             Aremu, M. O., Atolaiye, B. O., Gav, B. L., Opaluwa, O. D., Sangari, D. U. and Madu, P. C.

Chapter 11   **Effect of cadmium on germination, growth, redox and oxidative properties in**
Pisum sativum seeds   72
Moêz Smiri

Chapter 12   **Pesticide residue analysis of fruits and vegetables**   80
Rohan Dasika, Siddharth Tangirala and Padmaja Naishadham

Chapter 13   **Toxicological effects of burrow pit effluent from a waste dump on periwinkle**
(_Tympanotonus fuscatus linne_)   90
Ogeleka D. F. And Tudararo-Aherobo L. E.

Chapter 14   **Toxicological effects of methomyl and remediation technologies of its residues in an**
aquatic system   97
Ismail I. El-Fakharany, Ahmed H. Massoud, Aly S. Derbalah and Mostafa S. Saad Allah

Chapter 15   **Potential climate effects on nitrogen eco-toxicology of freshwater Lake, Victoria**   105
Opio Alfonse

Chapter 16   **Geoelectric investigation of groundwater resources and aquifer characteristics in**
Utagba-Ogbe kingdom Ndokwa land area of Delta State, Nigeria   113
Julius Otutu Oseji

Chapter 17   **Influence of pre-oxidation with potassium permanganate on the efficiency of iron and**
manganese removal from surface water by coagulation-flocculation using
aluminium sulphate: Case of the Okpara dam in the Republic of Benin   122
Zogo D., Bawa L. M., Soclo H. H. and Atchekpe D.

Chapter 18   **Monitoring of basic and acid radicals load in main canal, Giza governorate:**
A risk to health of consumers   130
Abd El-Moneim M. R. Afify, Sayed A. Fayed and Emad A. Shalaby

Chapter 19   **Trace element concentrations of soils of Ife-Ijesa area Southwestern Nigeria**   138
Aderonke A. Okoya, Olabode I. Asubiojo and Adeagbo A. Amusan

Chapter 20   **A study on removal characteristics of para-nitrophenol from aqueous solution by fly ash**   145
Alinnor I. J. and Nwachukwu M. A.

Chapter 21   **Chlorpyrifos induces hypertension in rats**   150
Alvarez A. Anthon and Campaña-Salcido A. D.

Chapter 22   **Kinetic and equilibrium studies of the adsorption of lead (II) ions from aqueous**
solution onto two Cameroon clays: Kaolinite and smectite   155
Joseph Ketcha Mbadcam, Solomon Gabche Anagho, Julius Ndi Nsami and
Adélaïde Maguie Kammegne

Chapter 23   **Use of Mediterranean plant as potential adsorbent for municipal and**
industrial wastewater treatment   163
Amina Soudani, Mohamed Chiban, Mohamed Zerbet and Fouad Sinan

Chapter 24   **Occurrence of paraquat residues in some Nigerian crops, vegetables and fruits**   170
Akinloye, O. A., Adamson, I., Ademuyiwa, O. and Arowolo, T. A.

Chapter 25   **Essentials of ecotoxicology in the tanning industry**   174
Mwinyikione Mwinyihija

Chapter 26   **Physicochemical dynamics of the impact of paper mill effluents on Owerrinta River, eastern Nigeria**   **183**
Ihejirika Chinedu Emeka, Emereibeole Enos Ihediohamma, Nwaogu Linus,
Uzoka Christopher Ndubuisi and Amaku Grace Ebele

Permissions

List of Contributors

Preface

Chemical pollution is the result of hazardous industrial chemicals which are not properly disposed and treated. Some industrial chemicals are highly poisonous as they contaminate not only air, water and soil but food chains as well. This book is a valuable compilation of topics, ranging from the basic to the most complex advancements in the field of chemical waste management. Different approaches, evaluations, methodologies as well as advanced studies on controlling the adverse effects of chemicals and management practices to reduce the hazardous impacts have been included in the book. This text is a vital tool for all researching and studying this field.

The information contained in this book is the result of intensive hard work done by researchers in this field. All due efforts have been made to make this book serve as a complete guiding source for students and researchers. The topics in this book have been comprehensively explained to help readers understand the growing trends in the field.

I would like to thank the entire group of writers who made sincere efforts in this book and my family who supported me in my efforts of working on this book. I take this opportunity to thank all those who have been a guiding force throughout my life.

Editor

Effect of Goldcrew and Corexit on selected blood parameter of the African cat-fish *Clarias gariepinus* following sublethal exposures

K. J. Alagoa

Department of Agric-Technology, Bayelsa State College of Arts and Science, P. M. B. 74, Agudama-Epie, Yenagoa, Bayelsa State, Nigeria. E-mail: mrkjalagoa@yahoo.com.

Sublethal effect of the dispersants Goldcrew and Corexit on selected blood parameters (White blood cell count- WBCC; Red blood cell count- RBCC; Platelet; Haemoglobin-Hb; and Packed Cell Volume - PCV) in *Clarias gariepinus* (Burchell) were studied. This was done in other to gauge the usefulness of blood parameters as good indicators of pollution and simulate the possible effect of clean-up operations on fish. Fish were exposed for four consecutive weeks in a static renewal condition and blood parameters measured weekly for the entire period at different levels of inclusion of dispersants. Results from the investigation revealed that dispersant effect on all blood parameters were not significantly different for the two dispersants ($P > 0.05$). Concentration of dispersants had a marked significance on blood parameters ($P < 0.05$); RBCC, PCV, Hb, platelet counts all decreased with increased concentration, while WBCC increased with increase in dispersant concentration. Duration of exposure had a marked significance on all blood parameters ($P < 0.05$) except in WBCC which was not significant at all exposure times. This study inferred that the two dispersants (Goldcrew and Corexit) are equally deleterious to fish health as they impact negatively on fish blood on the same scale.

Key words: *Clarias gariepinus*, corexit, blood, goldcrew, sublethal.

INTRODUCTION

Clarias gariepinus are outstanding members of the family claridae. They contribute greatly to the commercial catch of fishermen in coastal and fresh waters of Nigeria and an important source of protein in these times when beef availability has continued to show signs of steady decline (Borode, 1998). Regrettably, fresh water fishes are often subjected to the problem of pollution especially near industrial or populated area (Annune et al., 1994). The presence of crude oil in the aquatic environment through industrial accidents is one of such perennial problems in Nigeria (Alagoa, 2005).

Crude oil spills are sometimes allowed to disperse and degrade naturally, but clean up efforts are initiated when biological and economically important areas are threatened (GECL, 1991). The use of dispersants is one of the classical options for dealing with crude oil spills (Oyewo, 1986). However the addition of these chemicals to floating oils for the purpose of dispersing the oil can modify the effects of the oil as the dispersants them-selves have some toxic properties that could independently affect the ecosystem (Oyewo, 1986; Akintonwa

and Ebere, 1990).

The use of haematological values as indices of the state of fish health is receiving research efforts (Blaxhall, 1972; Musa and Omoregie, 1990). Studies of fish blood might reveal conditions within the fish long before there is any outward manifestation of disease (Sampath et al, 1993).

The objective of this investigation is to determine haematological responses of *C. gariepinus* to the dispersants Goldcrew and Corexit in a bid to simulate their possible effects on fish and the implication within the ecosystem. This will serve useful purpose for the management of the environment and protection of our fisheries.

MATERIALS AND METHODS

Goldcrew and Corexit were obtained from the Shell Petroleum Development Company of Nigeria Limited (SPDC) Forcados Terminal Pollution Control Store in Warri, Delta State, Nigeria. Live Juveniles of *C. gariepinus* average length 15.60 ± 0.2 cm were obtained from Ellah Lakes Obrikom in Ogba Egbema Ndoni Local

Table 1. Mean blood parameters of C. gariepinus exposed to different concentrations of Goldcrew and Corexit over four weeks.

Concentration (ml/l)/dispersant	WBCC ($\times 10^{9}$/L)	Platelet ($\times 10^{9}$)	RBCC ($\times 10^{2}$/L)	PCV (%)	Hb (2/dl)
Control	5.70 ± 0.24	395.25 ± 4.11	5.18 ± 0.28	42.03 ± 1.38	15.80 ± 1.07*
0.0625					
G	6.90 ± 0.14	324.00 ± 57.90	4.88 ± 0.70	38.95 ± 1.40	14.78 ± 0.90
C	6.22 ± 0.56	329.50 ± 52.80	4.90 ± 0.53	38.74 ± 0.95	15.01 ± 0.92
0.125					
G	7.12 ± 0.15	327.75 ± 37.77	4.80 ± 0.82	38.07 ± 1.91	14.05 ± 0.42
C	6.75 ± 0.53	374.25 ± 17.74	4.60 ± 0.37	38.68 ± 2.75	14.30 ± 0.42
0.1875					
G	6.70 ± 0.48	333.50 ± 53.20	4.35 ± 0.72	38.36 ± 3.93	13.85 ± 0.87
C	6.15 ± 0.64	276.25 ± 26.89	4.42 ± 0.27	38.07 ± 3.40	13.75 ± 0.42
0.25					
G	6.85 ± 1.07	227.25 ± 102.55	4.00 ± 0.59	35.52 ± 5.64	13.28 ± 1.45
C	7.12 ± 0.22	222.25 ± 28.76	4.27 ± 0.19	36.98 ± 3.60	12.92 ± 0.31

G - Goldcrew
C - Corexit
*Standard Deviation (SD).

Government Area of Rivers State Nigeria. Acclimatization of fish to experimental conditions was done in plastic basins containing borehole water in groups of six (6) for seven (7) days in twenty seven (27) plastic basins. A total of 162 fish were acclimatized. The mean weekly water characteristic and standard deviations were temperature 16.8 ± 0.60°C, pH 7.12 ± 0.02 and dissolved oxygen 6.8 ± 0.22, alkalinity 22.7 ± 2.0 mg/l.

The fish were fed *ad-libitum* twice daily during and after acclimation using African Regional Centre (ARAC) compounded experimental feed. A hunzu[R] air pump was used to provide aeration during acclimation. Mortality during this period was 2.5% of total fish population. Therefore fish were considered healthy for experimentation.

Test concentrations of 0.0625, 0.125, 0.1875 and 0.25 m/l of Goldcrew and Corexit in dilution water were prepared by dissolving 0.1, 0.2, 0.3 and 0.4 ml of Goldcrew and Corexit respectively in 16 L of water. Homogenous mixing was achieved by measuring pre-determined volumes of the dispersants into a bottle to which dilution water was measured and added into the bottle from the plastic basin to be prepared. The bottle was corked then vigorous shaking carried out to give a homogenous mixture before being poured into the plastic basin. The process was repeated severally for each test concentration and each replicate basin using one dispersant at a time. These ranges were obtained after conducting a range finder test to establish safe sublethal limits. This was done by exposing test fish to 0.025, 0.05 and 0.075 m/l of Goldcrew and Corexit respectively and measuring mortality of fish over 96 h. Test concentrations of 0.05 and 0.075 m/l resulted in total mortality after only 48 h. Test concentration of 0.025 m/l did not lead to mortality after 96 h for the two dispersants. Test concentrations and water in plastic basins were renewed daily by siphoning using rubber hoses.

The control basins (0 m/l) had no dispersants and were prepared simply by adding 16 L of water into the pre-washed plastic basins. Introduction of fish into plastic basins containing various treatment levels of the dispersants and control was done randomly. Each plastic basin contained 5 fish. All treatment levels and control had 3 replicates. There was no sexual consideration.

A 21 "gauge hypodermic needles and syringes (Innoson shanchuan®) were used to collect blood from only one fish at a time in each plastic basin weekly for four consecutive weeks. Each sampled fish was removed from the test medium after sampling. The method of cardiac puncture using physical restraint was employed because of its relative ease, and the probability of collecting 'frank' blood is highest (Snieszko and Axelrod, 1989). The site chosen for the puncture was about half an inch behind the apex of the 'V' formed by the gill cover and isthmus as described by Klontz and Smith (1968).

Collected blood samples were put in previously tagged anticoagulant bottles containing potassium salt of ethylene diamine tetra acetic acid (EDTA) and sent for laboratory analysis at the University of Port Harcourt Teaching Hospital, Port Harcourt. Cyanment-haemoglobin method was used to obtain haemoglobin values (Larsen and Sniezko, 1961). Packed Cell Volume (PCV) was determined by the method of Snieszko (1960).

In this method micro-haematocrit tubes were ¾ filled with the blood samples and then the end of the tubes sealed. The capillary tubes were centrifuged at a speed of 12,000 rev/min. for 10 min. The centrifuged tubes were then placed on a PCV reader and PCV values read for each of the blood samples. Red blood cell count was determined by the use of formal citrate, formal-dehyde and a counting chamber. White blood cell count (WBCC), using Turks solution and an improved Neubauer counting chamber. Platelet was determined by using 1% ammonium oxalate solution and a counting chamber.

Statistical analysis

Data were subjected to analysis of variance (ANOVA) at the 95% probability level. Duncan multiple range tests were employed to compare means according to standard procedures using the general linear model (GLM) of statistical analysis system software (SAS, 1999). Pearson correlation coefficient was employed to determine the relationship between the two dispersant effects,

Table 2. Mean weekly blood parameters of *C. gariepinus* to cumulative treatment levels of Goldcrew and Corexit and control.

Weeks	Dispersant	WBCC ($\times 10^9$/L)	Platelet ($\times 10^9$)	RBCC ($\times 10^{2/}$L)	PCV (%)	Hb (g/dl)
	G	6.22 ± 0.69	363.60 ± 47.63	5.50 ± 0.46	42.34 ± 1.53	15.40 ± 1.09
	C	6.62 ± 0.83	340.20 ± 67.75	5.02 ± 0.53	42.07 ± 1.24	14.96 ± 1.68
	Control	6.00 ± 1.56	390.00 ± 36.55	5.50 ± 1.13	43.53 ± 1.28	17.00 ± 2.64
	G	6.90 ± 0.67	335.40 ± 83.23	4.62 ± 0.54	38.87 ± 1.99	14.62 ± 1.33
	C	6.60 ± 0.58	327.20 ± 69.83	4.82 ± 0.37	38.90 ± 1.96	14.75 ± 1.24
	Control	5.80 ± 0.20	396.00 ± 2.64	5.30 ± 0.26	42.30 ± 0.81	16.40 ± 0.52
	G	6.88 ± 0.79	304.20 ± 81.51	4.28 ± 0.43	36.75 ± 3.01	13.80 ± 1.04
	C	6.02 ± 0.57	301.60 ± 85.27	4.48 ± 0.29	36.82 ± 2.63	14.10 ± 0.74
	Control	5.50 ± 0.10	395.00 ± 2.60	4.90 ± 0.30	40.46 ± 0.25	15.00 ± 1.73
	G	6.62 ± 0.64	283.00 ± 87.15	4.16 ± 0.61	36.38 ± 4.38	13.58 ± 0.95
	C	6.32 ± 0.63	309.00 ± 76.03	4.38 ± 0.38	37.71 ± 2.70	13.62 ± 0.86
	Control	5.50 ± 0.55	400.00 ± 32.78	5.00 ± 1.32	42.20 ± 1.85	14.80 ± 0.72

Table 3. Chemical properties of the dispersants.

Chemicals	Lead (Pb) (mg/l)	Copper (Cu) (mg/l)	Zinc (Zn) (mg/l)	Iron (Fe) (mg/l)	Cadmium (Cd) mg/l	Nickel (Ni) (mg/l)	Cyanide	Arsenic (ppm)	Chlorinated hydrocarbon	Chromium (ppm)
Goldcrew	0.089	0.149	0.096	0.092	0.003	0.055	ND	<1.0	ND	<1.0
Corexit	0.011	0.204	0.062	0.128	0.005	0.039	*ND	0.16	ND	<1.0

Adapted: Global environmental consultant limited, (1991)
ND = Not detected.

blood parameters and other variables.

RESULTS

The mean physico-chemical parameters of the experimental water and the mean lengths of the experimental fish are highlighted in the introduction. The results of the investigation are presented in Tables 1 - 5. Table 1 shows the mean blood parameters of *C. gariepinus* with varying concentrations of dispersants (Goldcrew and Corexit) and control over a four week period. The result reveals a significant difference ($p < 0.05$) in all blood parameters measured with concentration changes and control over the four week period. Apart from the white blood cell count (WBCC) which increased from control to higher treatment levels; all other measured blood parameters declined in value from control to higher treatment levels.

Table 2 shows the mean weekly changes of blood parameters to summative treatment levels of dispersant and control. The result shows a significant weekly reduction in all measured blood parameters except in WBCC which did not exhibit any significant reduction in weekly values.

Table 3 shows the chemical composition of the dispersants (Goldcrew and Corexit). The Table shows little but negligible variations in the chemical constituents of the two dispersants.

Table 4 shows the analysis of variance (ANOVA) revealing degrees of significance of measured blood parameters to sources of variability. It shows that there are no significant differences in toxicant effects on blood parameters of the two dispersants. The Table also reveals that treatment levels of the dispersants have a significant effect on all measured blood parameters. Except for WBCC which was not significant at all exposure times, duration of exposure has a marked effect on all blood parameters. The interaction of toxicant and concentration reveal that apart from platelet counts, no other blood parameter showed any form of significance.

The interaction of Toxicant and Exposure period reveal that only platelet counts was significant ($p < 0.005$) of all blood parameters. Also, the complex interactions of concentration and duration of exposure (weeks) and the interaction of toxicant, concentration and duration of exposure (weeks) reveal that only PCV and platelet counts to be highly significant.

Table 5 shows the Pearson's correlation coefficient

Table 4. Analysis of variance showing levels of significance of measured blood parameters to sources of variability.

Source of variation	Df	White blood cell count	Red blood cell count	Packed cell volume	Haemoglobin	Platelet count
*Toxciant	1	1.98^{ns}	0.55^{ns}	5.47^{ns}	0.0009^{ns}	104.53^{ns}
Concentration	4	6.51***	2.91***	107.55***	25.82***	9475.14***
Duration of exposure (week)	3	0.69^{ns}	5.04***	208.20***	15.3***	19897.0***
Toxicant × concentration	4	0.90^{ns}	0.61^{ns}	4.23^{ns}	0.38^{ns}	7439.1***
Toxicant × week	3	1.90^{ns}	1.55^{ns}	2.91^{ns}	0.75^{ns}	3124.60*
Concentration × week	12	0.77^{ns}	0.38^{ns}	15.82***	0.72^{ns}	8113.1***
Toxicant × concentration × week	12	0.47^{ns}	0.37^{ns}	4.90^{ns}	1.16^{ns}	4051.16***
Error	80	0.96	0.67	4.79	2.31	1118.38
R^2		0.42	0.45	0.78	0.49	0.87
CV (%)		14.99	17.45	5.64	10.58	10.44

ns = not significant
* $p = 0.05$
** $p = 0.01$
*** $p = 0.001$

Table 5. Pearson correlation coefficients showing the level of inter-relationship between all variables.

	Treat	Conc	Week	WBCC	Platelet	RBCC	PCV	HB
Treat	1.00000	0.00000 1.0000	0.00000 1.0000	0.12287 0.1812	0.01211 0.8956	- 0.06593 0.4743	- 0.04273 0.6431	- 0.00159 0.9863
Conc	0.00000 1.0000	1.00000	0.00000 1.0000	0.32892 0.0002	- 0.67238 <.0001	- 0.34013 0.0001	- 0.47430 <.0001	- 0.53091 <.0001
Week	0.00000 1.0000	0.00000 1.0000	1.00000	- 0.01499 0.8709	- 0.28276 0.0018	- 0.35008 <.0001	- 0.53596 <.0001	- 0.35424 <.001
WBCC	0.12287 0.8956	- 0.67238 <.0001	- 0.28276 0.0018	- 0.34294 0.0001	0.0001	0.31648 0.0004	0.57687 <.0001	0.54781 <.0001
Platelet	0.01211 0.8956	- 0.67238 <.0001	- 0.28276 0.0018	- 0.34294 0.0001	0.0001	0.31648 0.0004	0.57687 <.0001	0.54781 <.0001
RBCC	- 0.06593 0.4743	- 0.34013 0.0001	- 0.35008 <.0001	- 0.19360 0.0341	0.31648 0.0004	1.00000	0.42785 <.0001	0.30794 0.0006
PCV	- 0.04273 0.4731	- 0.47430 <.0001	- 0.53596 <.0001	- 0.38912 <.0001	0.57687 <.0001	0.42785 <.0001	1.00000	0.46736 <.0001
HB	- 0.00159 0.9863	- 0.53091 <.0001	- 0.35424 <.0001	- 0.34400 0.0001	0.54781 <.0001	0.30794 0.0006	0.46736 <.0001	1.00000

revealing the inter-relationship between variables.

DISCUSSION

The mean values of the physico-chemical parameters of the bore-hole water used for the analysis revealed that the water parameters measured were within the limits recommended by the United States Environmental Protection Agency (1973) for optional health of warm and cold water fishes. This implied that there was no water quality mediated stress and therefore any stress factor observed in the study was due only to the addition of the dispersants at various concentration levels.

Haematological parameters relating to oxygen transport, red blood cell count [RBCC], haemoglobin (Hb), packed cell volume (PCV) and platelet declined in values with increases in concentration of the dispersants over a four week period (Table 1). This reduction is in agreement with Sharma and Gupta (1984) whom noticed a dose dependent reduction of erythrocytes following the exposure of C. batracus to carbon-tetrachloride. Similar reductions in haemoglobin, packed cell volume and erythrocytes of fish have been reported by Omoregie et al, (1994). The decline in RBCC, Hb, platelet and PCV may be due to the swelling of red blood cells, haemo-dilution or the distortion of cellular components of fish by invading toxin (Musa and Omoregie, 1990).

The decrease in Hb, PCV, platelets and RBCC with increasing concentration of dispersants may also suggest a progressive macrocytic anaemia. This is an indication of bone marrow erythropoietin response to an ensuing anaemia resulting from possible metabolic dysfunction in organs like the liver and kidney (Olukunle et al., 2002). This occurs if the diet is deficient in iron, vitamin B_{12} and folic acid the digestive system will be unable to absorb them, the marrow therefore fails to produce enough red blood cells or produces defective ones.

It may also be a signal to the unset of a hemolytic anaemia. An anaemia of this type results when if red blood cells are produced in adequate numbers but are destroyed faster than new ones can be produced. Anaemia of this type is produced by severe infectious diseases such as scarlet fever, malaria and puerperal fever. This can also be caused by poisons and certain drugs especially the sulfonamides.

In contrast, the parameter relating to defense mechanism (white blood cell count, (WBCC) showed a gradual but fluctuant pattern of increase with increase concentration of dispersants (Table1). This is an indication of increased production of leucocytes into peripheral circulation from marginated pool, the bone marrow or spleen probably in response to bacterial infection, invading toxin in the already compromised and anaemic fish (Olukunle et al., 2002). This may also be due to the stress-mediated release of WBCC as noted by Sampath, et al (1993). Who reported increase in lymphocytes of the Nile tilapia *Oreochromis niloticus* exposed to toxic condition.

However WBCC did not show any significant decline during the four weeks of exposure (Table 2). This may be due to a time related homeostasis of WBC or a remote indication of actualization of threshold concentrations of dispersants as suggested by Davids et al. (2002). They noticed a time related stability in white blood cell counts in the fishes *Sarotherodon melanotheron* and *Tilapia guiniensis* exposed to varying treatment levels of industrial effluents.

The dispersants Goldcrew and Corexit contain similar contents of heavy metals and hydrocarbons (Table 3). This may be due to the fact that both dispersants are sulfactants of the same group and are used routinely together by multi-national oil company's operating in Nigeria.

Goldcrew and Corexit dispersants respectively did not exhibit any significant difference in measured blood parameter presentation between the two dispersant as shown in the analysis of variance (ANOVA) Table 4. This may be due to the fact that both dispersants presented similar blood parameter presentations.

The findings in this study indicate that all haematological parameters assessed except WBCC were affected by concentrations of dispersants and exposure periods. This agreed with the findings of Annune et al. (1994) that the most susceptible parameters PCV, Hb, WBCC, RBCC and platelet counts of fresh water fish were affected by the stress agents.

Although Goldcrew and Corexit are routinely used oil spill dispersants in Nigeria, this study indicate their deleterious effect to our environment and our fisheries.

REFERENCES

Akintonwa C, Ebere E (1990). Toxicity of crude oil (Asabo, 16c) and two Dispersants (Conco-K and Teepol), to *Barbus* sp and *Clarias* sp. Environ. Pollut. (Series A) 41: 31-33.

Alagoa KJ (2005). Sublethal effect of the dispersants Goldcrew and Corexit on selected blood parameters of *Clarias gariepinus*. M. Phil Thesis. Rivers State University of Science and Technology, Port Harcourt, Nigeria. 68 pp.

Annune PA, Lyanikura FT, Hbele SO, Oladimeji AA (1994). Effect of sublethal concentration of zinc on haematological parameters of fresh water fishes *Clarias gariepinus* (Burch) and *Oreochromis niloticus* (Trewavas). J. Aquat. Sci. 9: 1-6.

Blaxhall PC (1972). Haematological assessment of the health of fresh water fish. J. Biol. 4: 593-604.

Borode AO (1998) Comparative study of the embryology and hatchability of *Clarias gariepinus* and *Heterobranchus bidersalis* hybrid and their parental crosses. Appl. Trop. Agric. 3(1): 52-57.

Davids CBD, Eweozor KKE, Daka ER, Dambo WB, Bartimaeus EAS (2002). Effect of industrial effluents on some haematological parameters of *Sarotherodon melanotheron* and *Tilapia guiniensis*. Global J.Pure Appl. Sci. (3): 305-309.

GECL (1991) Dispersant Ecotoxicological studies report presented to Shell Petroleum Development Company of Nigeria Limited. 45 pp.

Klontz GW, Smith LS (1968). Methods of using fish as biological research subjects'. In Gay WR (Ed) Methods of Animal experimentation. Academic press. Inc. New York. 3: 323-385

Larsen HN, Snieszko SF (1961) Comparison of various methods of determination of haemoglobin in trout blood. Prague Fish Cult., 23: 6-7.

Musa SO, Omoregie E (1990). Haematological changes in mud-fish

Clarias gariepinus (Burchell) exposed to malachite green. J. Aquat. Sci. 14: 37-42.

Olukunle OA, Ogunsanmi AO, Taiwo VO, Samuel AA (2002). The nutritional value of cow blood meal and its effects on growth performance, haematology and plasma enzymes of hybrid cat fish. Trop. J. Anim. Sci. 5(1): 75-85.

Omoregie E, Ufodike EBC, Keke IR (1994). Tissue Chemistry of *Oreoclromis niloticus* exposed to sub lethal concentrations of Gammalim 20 and Aetellic 25 E.C. J. Aquat. Sci. 5: 33-36.

Oyewo FO (1986).The acute toxicity of three oil dispersants. Environ. Pollut. (Series A) 41: 23-31.

Sampath K, Valamnial S, Kennedy IJ, Jane R (1993). Haematological changes and their recovery in *Oreochromis mossambicus* as a function of exposure period and sub lethal levels of concentration ACTA Hydrobiol. 35: 73-83.

Sharma RC, Gupta N (1984). Carbon-tetra-chloride induced haematological alterations in *C. batrachus* (L). J. Environ. Biol. 3: 127-131.

SAS (1999). Statistics Analysis System user guide: SAS Institute Inc. carry N.C; USA. pp. 18-20.

Snieszko ES, Axelrod RH (1989). Diseases of fishes TH publication Inc. N.J 07755. pp. 130-136.

Snieszko SF (1960). Microhaematocrit as a tool in fisheries research. U.S fish and wildlife service special scientific report. pp. 341-15.

United States Environmental Protection Agency (1973). Water quality criteria. EPA. 73 033. 594 pp.

2

Lead and coliform contaminants in potable groundwater sources in Ibadan, South-West Nigeria

Olusegun Peter Abiola

Department of Science Laboratory Technology, Faculty of Science, The Polytechnic, Ibadan, Oyo State, Nigeria.

The present study investigates possible contamination of untreated and treated groundwater by lead, faecal coliform and *Escherichia coli* in a hundred randomly selected boreholes from different parts of Ibadan metropolis, located in South-West Nigeria. Total lead contents in the water samples were measured by atomic absorption spectrophotometer while coliform count was undertaken using the multiple tube fermentation technique. Detection of *E. coli* in the water was by the presumptive and confirmative tests. Data obtained showed that all the untreated water samples contained lead concentrations in the range of 1.0 - 12.0 ppb, with a mean value (X) of 4.9 ± 0.18 ppb. Seventy-three percent of the borehole water samples had coliform, with 18% of these borehole samples having detectable *E. coli*. All the sachet "water" samples that were supposedly treated having lead concentrations that were in the range of 2.0 - 9.0 ppb with one of them having coliform bacteria. The results obtained supported previous findings that severe environmental degradation, which is readily observable in most parts of Ibadan city, could possibly contribute to pollution of ground water source like boreholes. Supply of adequately treated water from public waterworks to the teeming population in Ibadan city is an important problem that must be solved by the government.

Key words: Groundwater, sachet water, Ibadan, lead, coliform, environmental pollution.

INTRODUCTION

Boreholes and wells are groundwater types that form an integral part of water supply systems in urban and rural communities of Nigeria (Pickering and Owen, 1995), and so can be described as indispensable because of inadequate public water supply systems in most communities in Nigeria (Sayjad et al., 1998). According to Egwari and Aboaba (2002), natural processes and anthropogenic activities of man can contaminate groundwater, and such activities could be domestic, agricultural or industrial in nature. Uncontrolled discharge of toxic effluents to the soil, stream and rivers by industries and indiscriminate dumping of garbage and faeces have been reported to heavily contaminate groundwater in Nigeria (Erah et al., 2002).

Anaele (2004) reported extensive contamination of residential wells and boreholes by sewage from the numerous septic tanks, latrines and soak away pits often sited near them. Majority of residents drink water from these groundwater sources without any form of treatment. Reasons adduced for this unhealthy practice ranged from lack of access to basic methods of water treatment to simply ignorance of hazards associated with the ingestion of contaminated water (Anaele, 2004). This is particularly true of borehole water since its sparkling look gives a false impression of absolute purity to unsuspecting consumers (Anaele, 2004).

Several research findings such as those of Egwari and Aboaba (2002) and Erah et al., (2002) showed that water from several boreholes and wells in some urban centers in Nigeria were heavily contaminated with lead, toxic organic wastes, faecal coliform and *Salmonella typhii*. Coliform in water, though harmless to human health had been shown to portend the most probable presence of other pathogenic microorganisms like *S. typhii, E. coli, Pseudomonas, Vibrio* spp., *Shigella* spp., *Aeromonas hydrophilia* among others (USEPA, 2009; Egwari and Aboaba, 2002; Shelton, 2002).

Indiscriminate dumping of materials laden with lead on land and use of leaded gasoline had been shown to contribute to the lead load of underground water sources of many Nigerian cities (Ademoroti, 1986; Ademoroti, 1986;

*Corresponding author. E-mail: olupabiola@yahoo.com.

Ayoola, 1979; Erah et al., 2002). A very recent assessment of soils for heavy metals' concentrations of Ibadan metropolis by the duo of Adewara and Akinlolu (2008) using petrographic studies and X-ray diffractograms, confirmed a heavy contamination of the industrial and densely populated residential areas of Ibadan metropolis by lead, copper and zinc.

Although, Shelton (2002), suggested that the permissible limit for lead in drinking water is 0.05 µg/dl, but the current limit by WHO standards is 10 ppb (or 0.5 ppb), it has been established that long-term accumulation of lead in body tissues has neurotoxic, nephrotoxic, fetotoxic, and teratogenic effects on man just like most heavy metal contaminants (Asogwa, 1979; Hoekman, 2005; The Washington Times, 2005). The situation in Nigeria is disturbing as recent reports also suggested that 80% of the hospital patients on admission in Nigeria had water related problems while kidney diseases are on the rise in the last two decades (Anaele, 2004).

In view of a recent study by Adewara and Akinlolu (2009), and survey findings from urban centres like Lagos and Benin as reported by Anaele (2004) and Erah (2002), this present study intends to investigate the current trend of contamination of groundwater sources especially boreholes in Ibadan. The present study analyzed water samples from one hundred carefully selected boreholes in Ibadan for assessment of total lead and total microbial contamination.

Ibadan is the largest indigenous city located within the coordinates; 7°23′47″N 3°55′0″E / 7.39639°N 3.916667°E in Nigeria, West Africa has an area of 828 km^2 and a population of approximately 2.6 million by the 2006 census (Wikipedia, 2010). Ibadan cannot be described as an industrial city because, it has a few small and medium manufacturing industries, but its economic landscape is heavily dotted with hundreds of artisans that handle or process numerous articles containing metals, which they often dispose carelessly to their immediate environment. A recent assessment of soils for heavy metals' concentrations of Ibadan metropolis using petrographic studies and X-ray diffractograms confirmed a heavy contamination of the industrial and densely populated residential areas of Ibadan metropolis by lead, copper and zinc (Adewara and Akinlolu, 2008). The city just like most urban centres in Nigeria (Anaele, 2004), has an almost non-existent public water treatment facility for the supply of adequately treated water to the residents of this city. This situation had forced the majority of Ibadan residents to source water for drinking and other domestic needs from several other water source especially wells and boreholes.

MATERIALS AND METHODS

All the samples for this study were drawn from the city's residential sectors, with none of them having any industry using lead or its compounds. The sectors were identified and demarcated from the

approved geographical map of the Ibadan city. The main city, exclusive of the expansive suburban areas was divided into twenty sectors for the purpose of sampling, with five boreholes per sector. None of the boreholes were located near any in battery producing industry or battery repair and maintenance artisans. The five functional boreholes which were randomly selected were chosen if a criterion of a separation from each other by a minimum of 500 meters distance was met and were collected per city sector for analysis using standard procedures for the collection of the water samples (APHA, 1998).

Twenty out of the hundred boreholes sampled were located in small-scale factories involved in the production of sachet packaged water (popularly called "pure water"). Water from each of the boreholes had earlier been pumped into 3000 to 5000 L plastic storage tanks. To ensure that the holding tanks of the water to be sampled is contaminant free, each of these tanks were thoroughly washed, with deionized water, sterilized with chlorinated water, and then rinsed again with excess of deionized water until all traces of chlorine had been eliminated prior to sampling. Each water sample was collected from a tap that had its mouth sterilized appropriately with cotton wool soaked in methylated spirit and made to pass through sterilized filters, after which each was collected in new 1L PVC bottles, which had previously been washed, soaked in 10% nitric acid and finally rinsed in deionized water. Each bottle was labeled and stored in the freezer immediately after collection. Faecal coliform counts and lead concentrations were measured and used as indicators of possible contamination of water sampled in the survey. Presence of E. coli was also assessed in all the samples analysed. Actual analysis for total coliform was done within twenty-four hours of collection.

METHODS

Lead determination

Complete digestion of the raw borehole and treated sachet water samples was done with nitric, perchloric, and hydrofluoric acids in a fume chamber. Metals in water samples were extracted and analyzed in accordance with the standard method of analysis raw water samples (APHA, 2002). Concentration of total lead was determined in each of the water samples using a Buck Scientific Atomic Absorption Spectrophotometer Model 200A at appropriate wavelengths (Erah et al., 2002). The digestion and analytical procedures were checked by analysis of DOLT-3 Matrix Certified Reference Material with known concentration for heavy metals (Cantillo and Calder, 1990)

Assessment of coliform contamination

Total coliform counts in the samples were determined using the multiple tube fermentation technique (APHA, 2002). This involved inoculating multiple fermentation tubes containing MacConkey broth with 1 ml of water sample at 37°C for 24 h, after which the count was done with a Suwtex 560 colony counter. Detection of E. coli in the water was carried using the presumptive and confirmative tests (APHA, 1998).

RESULT AND DISCUSSION

The results of mineral lead analysis are as shown in Table 1 and presented in Figure 1. The mean lead content of 4.91 ppb was about ten times the WHO permissible limit of 0.05 µg/dl (0.5 ppb) of lead in drinking water (WHO, 1998),

Table 1. Frequency distribution of total lead concentrations in borehole water samples.

Lead concentration (x) ppb	Frequency (f)	Cumulative frequency	Fx
1.0	08	08	08
2.0	13	21	26
3.0	09	30	27
4.0	22	52	88
5.0	13	65	65
6.0	07	72	42
7.0	13	85	91
8.0	06	91	48
9.0	04	95	36
12.0	05	100	60
-	$\Sigma f = 100$	$\Sigma f = 100$	$\Sigma fx = 100$

(ppb = parts per billion = µg/L) (WHO permissible limit for lead in drinking water 0.05 µg/dl or 0.5 ppb).

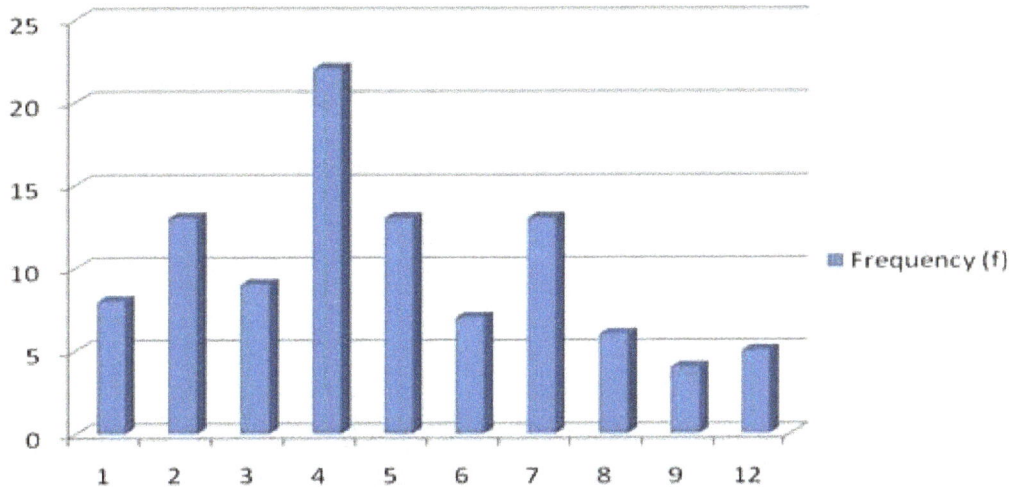

Figure 1. Histogram presentation of the distribution of total lead in borehole water samples.

confirming possible contamination of underground water by lead pollutants. The degree of contamination ranged from 1.0 -12.0 ppb (24 times the permissible limit). Since the residents totally depend on such water for all their daily domestic needs, long-term accumulation of lead in the body via oral ingestion and dermal absorption is a possibility (Florence et al., 2003). Lead as a cumulative body poison (Bernard and Becker, 1998) may be contributory to the onset of chronic or sub clinical symptomatic lead poisoning (HMT files, 2005). Clinical symptoms associated with such poisonings in adults that often occur at blood lead levels greater than 80 micrograms per dl (3.9 micromole per litre) had been well documented by several sources (Sofoluwe et al., 1971; Williams et al., 1999). Such symptoms include abdominal pains, head-

ache, irritability, joints' pain, fatigue, anaemia, peripheral motor neuropathy, deficits in short –term memory, and hypertension (Burns and Baghurst, 1999). Since the boreholes sampled were not located near any important source of industrial battery wastes and yet had significant levels of lead contaminants, data obtained confirmed earlier research findings that several other less recognized sources, apart from occupational or industrial release of lead wastes could contribute significantly to the problem of pollution of underground water by lead (Erah et al., 2002; HMT files, 2005). The results buttressed the findings of Adewara and Akinlolu (2008) that confirmed significant concentrations of metals like lead, copper and zinc in the soils of residential and industrial areas of Ibadan metropolis. Microbiological examination of the

Table 2. Frequency distribution of coliform counts in borehole water samples.

Coliform counts (c)	Frequency (f)	Fc
0	27	00
1	23	23
2	11	22
3	19	57
4	07	28
5	08	40
6	05	30
-	$\Sigma f = 100$	$\Sigma fc = 200$

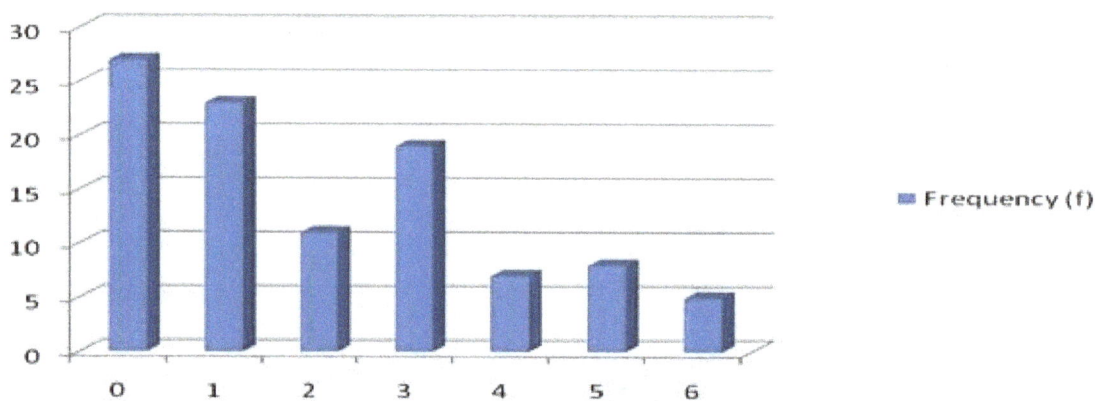

Figure 2. Histogram presentation of the distribution of total coliform counts in borehole water samples.

water samples clearly showed that 73% of the boreholes had coliform bacteria with 18 having *E. coli* (Table 2 and Figure 2).

Data as obtained was disturbing since the presence of any type of coliform bacteria in drinking water is not acceptable by WHO water quality standards. This observation supports earlier reports by Anaele (2004), Erah et al. (2002) and Egwari and Aboaba (2002) that most underground water sources in urban centres of Nigeria may contain substantial microbial contaminants including very harmful ones like *E. coli, S. typhii* among others. According to Anaele (2004), indiscriminate sinking of boreholes near toilet soak-away, pit latrines, dirty gutters and poor drainages had been known to contribute to the presence of contaminants in water from such boreholes. This is of particular significance considering the implicit trust, a substantial proportion of the Nigeria population has in the "absolute purity" of water from boreholes (Anaele, 2004). The population is often igno-rant of possible chemical and microbial contaminants in untreated water and is often deceived by the sparkling nature of borehole water.

Mean total lead concentration in the twenty borehole water sample from the "pure water" factories was 5.7 ppb while the mean lead level in the processed sachet packaged water (pure water) was 4.8 ± 0.2 ppb. Nineteen

(19) of the twenty samples were coliform free confirming that majority disinfected the raw borehole water properly. Analysis also showed that all the treated "pure water" samples had high lead concentrations that were in the range of 2.0 - 9.0 ppb. This is an indictment against the quality assurance control measures of these so called 'pure water' industries since their products had excessive lead levels. Since many unsuspecting consumers in Nigeria depend on such water to quench their thirst especially when away from home, it is obvious that very few can hardly escape from lead contaminated water

CONCLUSION AND RECOMMENDATIONS

Data obtained in this study agreed with several other surveys earlier carried out in other Nigerian urban centers like Benin City and Lagos (Erah et al, 2002; Anaele, 2004; Egwari and Aboaba, 2002). It can therefore be concluded from the study that Ibadan residents con-suming untreated borehole water are potentially exposed to possible acute, sub chronic or even chronic plumbism and water borne diseases like typhoid fever, dysentery, diarrhea etc. It is recommended here that chlorinating agents be provided by the government at heavily sub-sidized prices to all and sundry to assist in the elimination of pathogenic microorganisms in the untreated water supplies (Schute, 1995). Intensive education of the Nigerian population on correct treatment procedures of water for domestic use should be done on the electronic media especially TV and radio on continual basis. This measure, though might reduce the incidence of diseases attributable to microbiologically unsafe water, it will not remove chemical contaminants such as lead (Anon, 1980).

To reduce lead contamination, the local mass media should disseminate information on the need for a more careful handling and disposal of materials that contain lead especially lead paints. Enactment of appropriate legislations to regulate the handling and disposal of lead accumulators by battery chargers in order to control pollution of the environment by lead wastes is likewise recommended. Practical measures to ensure this is the formulation and implementation of policies that will promote comprehensive planning, and redesign of our urban centers to allow for construction of sewage and drainage systems.

Legal framework should be put in place at the national level to put stringent laws to regulate the citing of wells and boreholes, as findings from this study corroborate observations of Erah et al. (2002) that indiscriminate sinking of boreholes and wells without proper geological surveys contributes to the presence of faecal coliform in underground water. And more importantly, the government at all levels in Nigeria should also be admonished and take the issue of supply of adequately treated water to the public as an essential public service.

REFERENCES

Anaele A (2004) Boreholes: Harbinger of Death. The Punch of 27th October, pp. 42

Ademoroti CMA (1986). Levels of heavy metals in barks and fruits in Benin City. Nigerian Environmental Pollution Series B. (11): 241 – 253.

Adewara AO, Akinlolu FA (2008) Contamination indices and heavy metal concentrations in urban soil of Ibadan metropolis, south-western Nigeria. Environ. Geochem. Health., 30(3): 243-254. Published by SpringerLink Journal (accessed on linefrom http://www.springeronline.com on 22nd January 2010).

Anon (1980). Corrective Action for water that does not meet the recommended guideline (http://www.lookd.com).

APHA (2002). Standard Methods for Examination of Water and Wastewater. (20th edition). American Waterworks Association. And Water Pollution Control Federation. American Public Health Association.pp. 213-256

APHA (1998). Standard Methods for Examination of Water and Wastewater. (16th edition). American Waterworks Association. And Water Pollution Control Federation. American Public Health Association. pp. 80-115

Asogwa SE (1979). The Risk of Lead Poisoning in Battery Chargers and possible hazard of their Occupation on the Environment. Nig. Med. J., 9(2): 189-193.

Ayoola EA (1979). Lead Poisoning In Adults. Nig. Med. J., (2) 185-188.

Benard B, Becker CE (1998). Environmental Lead poisoning. J. Clin. Toxicol., 26(4): 223-226

Burns PA, Baghurst MG (1999). Lifetime low level exposure to Environmental lead and children's emotion and behavioural development. Am. J. Epidemiol. 149: 740-749.

Cantillo A, Calder J (1990). Reference materials for marine science. Fresenius J. Anal. Chem., 338: 380-382.

Egwari L Aboaba OO (2002). Bacteriological quality of domestic waters. Rev. Saude Publica, 36(4): 513-520.

Erah OP, Akujieze CN, Oteze GE (2002). The Quality of Groundwater in Benin City:A baseline study on inorganic chemicals and microbial contaminants of health importance in boreholes and open wells. Trop. J. Pharmaceu. R., 1(2): 75-82.

Florence TM, Stauder JL, Dale LS, Henderson D, Izard BE, Belbin K (2003). Skin Absorption of Ionic Lead Compounds. J. Aust. College Nutr. Environ. Med. 15(2): 11-12.

Hoekman T (2005) Heavy Metals Toxicology.http://www.luminet-hydro/heavymet;htm Heavy Metal Toxicity (HMT) Files. Heavy Metal Toxicity. http:// tuberose .com/.html

Pickering KT, Owen LA (1995). Water Resources and Pollution. In; Introduction to Global Environmental Issues. Routledge, 11 New Fetter Lane London EC 4P4EE pp. 133-151.

Sayjad MM, Rahim Sand Tahir SS (1998). Chemical Quality of Groundwater of Rawalpindi/Islamabad, Pakistan 24th WEDC Conference Islamabad Pakistan. pp.101-105

Schute CF (1995). Surface water treatment disinfection:In New World Water. Routledge, 11 New Fetter Lane London EC4.P4.EE. pp. 63-65.

Shelton TB (2002). Interpreting drinking water quality London EC4P4EE. pp. 78-81.

The Washington Times (2005). Heavy Metals. http://washingtontimes.com/ metro0530-94712-9017. Htm.

Sofoluwe GO, Adegbola A, Akinyanju PA (1971) Urinary delta-aminolaevulinic acid determination among workers charging lead accumulator batteries in Lagos, Nigeria. Archives of Environmental Health 23:18-20US-EPA. Drinking water contaminants. EPA paper 816-F-09-004 of May 2009. (Accessed on line on 22nd January 25, 2010 from http://www.epa.gov/safewater/contaminants.

Williams F, Robertson R, Roworth M (1999). Scottish Center for Infection and Environmental Health: Detailed Profile of 25 Major organic and inorganic substances. 1st ed. Glasgow: SCEIH. pp. 223-235

WHO (1998). Guidelines for Drinking Water. 2nd edition. Volume 2. Health criteria and other supporting information. Geneva, Switzerland; World Health Organization (WHO), pp. 940-949.

3

Environmental mobility and geochemical partitioning of Fe, Mn, Co, Ni and Mo in sediments of an urban river

H. M. Zakir[1]* and N. Shikazono[2]

[1]Department of Agricultural Chemistry, Faculty of Agriculture, Bangladesh Agricultural University, Mymensingh- 2202, Bangladesh.
[2]Laboratory of Geochemistry, School of Science for Open and Environmental Systems, Faculty of Science and Technology, Keio University, Hiyoshi 3-14-1, Yokohama 223-8522, Japan.

Environmental mobility and geochemical partitioning of iron (Fe), manganese (Mn), cobalt (Co), nickel (Ni) and molybdenum (Mo) were examined in sediments collected from the whole old Nakagawa river (NR), Tokyo, Japan. A combined 6- step sequential extraction procedure was employed for the partitioning of the metals and the concentrations were measured in the liquid extracts by atomic absorption spectrophotometry (AAS) and inductively coupled plasma mass spectrometry (ICP-MS). The observed order of potential mobility of metals in the aquatic system of NR was: Ni > Co > Mo > Mn > Fe. The association of Co, Ni and Mo (26.6-30.5; 23.2-38.5 and 19.0-40.1% of total, respectively) were found highest with amorphous Fe oxyhydroxide phase, and the maximum fraction of Fe and Mn were in silicates and residual phase. The normalization of the concentrations of metals using Al as a conservative element confirmed that most of the sampling stations of NR were enriched with Ni, Mo and Mn. Geoaccumulation index (I_{geo}) values for most of the sites have higher for Ni and Mo. Environmental risk of metals were also evaluated by using risk assessment code (RAC) and found medium risk for Ni, Mo, Co and Mn.

Key words: Mobility, partitioning, pollution, metals, sediment.

INTRODUCTION

Urbanization is of considerable importance for socio-economic growth and is continuously modifying the physical, chemical and biological composition of our living environment. As a result, millions of people living in and around urban centres are exposed to an unnatural and unhealthy environment. Sediments are important carriers of metals in the environment and reflect the current quality of the system. As in natural environments, urban river sediments have a high potential for storage of different metals. Unlike natural rivers, however, a large proportion of the element load contained in urban sediments is not associated with the original geologic parent material, but with the steady supply of those elements, both dissolved and in particulate form. Thus, a river close to an urban centre has the opportunity to be polluted by both naturally occurring and anthropogenically originated metals. Undoubtedly, natural sources come from physical and chemical weathering of parent materials (rocks and minerals) of the river area and the anthropogenic sources include industrialization, wastes and sewage effluents from urban centres, underground deposition of industrial wastes and others.

The banks of NR in Tokyo, Japan is a location where a significant amount of different types of industrial units were available at the time of second world war (USAMS,

*Corresponding author. E-mail: zakirhm.ac.bau@gmail.com.

Abbreviations: Fe, Iron; **Mn,** manganese; **Co,** cobalt; **Ni,** nickel; **Mo,** molybdenum; **NR,** Nakagawa river; **AAS,** atomic absorption spectrophotometry; **ICP-MS,** inductively coupled plasma mass spectrometry; **RAC,** risk assessment code; **ICP-MS,** inductively coupled plasma- mass spectrometry; **AR,** Arakawa river; **OR,** Onaki river.

1945-46) and a huge amount of industrial metal (mainly Cr) contained waste was deposited about 30 years ago (Zakir, 2008; Zakir et al., 2008).

Metals are usually distributed throughout the sediment components and associated with them in various geochemical forms including ion exchange, adsorption, precipitation, and complexation. Due to changes in environmental conditions, such as acidification, appearance of strong current, dredgings, redox petential or organic ligand concentrations, oxidation state and others can cause mobilization of metals from solid to liquid phase. Then metals accumulated in sediments can cause contamination of surrounding waters and represent a potential risk for an aquatic environment. Heavy metals are most important among the frequently observed contaminants in the environment. Both industrial activities and urbanization have greatly increased the metal burden in the environment. In Japan, there has been variable human damage caused by heavy metals, such as Cu poisoning in the Ashio Cu mine area and the Itai-itai disease induced by Cd in the Kamioka mine area.

In sediments, metals can be present in a number of chemical forms, and generally exhibit different physical and chemical behaviour in terms of chemical interaction, mobility, biological availability and potential toxicity, which has been demonstrated by several authors (Amiard et. al., 2007; Zakir et. al., 2008; Wang et al., 2010). Therefore, it is necessary to identify and quantify the forms in which a metal is present in sediment to gain a more precise understanding of the potential and actual impacts of elevated levels of metals in sediments and to evaluate processes of downstream transport, deposition and release under changing environmental conditions. Methods for the determination of different forms of metals in sediments include sequential extraction, whereby a series of reagents is used to extract operationally defined discrete phases from sediments in an outlined sequence. The overall behaviour of metals in an aquatic environment is strongly influenced by the associations of metals with various geochemical phases in sediments (Horowitz, 1991).

Sequential extraction techniques have been applied to study the geochemical partitioning of metals in contaminated soils (Hickey and Kittrick, 1984; Basta and Gradwohl, 2000) and sediments (Pardo et al., 1990; Jain, 2004; Morillo et al., 2004; Zakir and Shikazono, 2008). Geochemical distribution results have also been used as an aid in predicting potential contaminant mobility and bioavailability (Pueyo et al., 2003; Zakir et al., 2008). This could help us to understand the geochemical processes governing metal mobilization and potential risks induced. According to Singh et al. (2005), there are five major mechanisms of accumulation of metals in sediment, namely- exchangeable, bound to carbonate, bound to reducible phases (amorphous Fe oxyhydroxide and crystalline Fe-oxide), bound to organic matter (oxidizable fraction) and residual. These categories have different

behaviours with respect to remobilization under changing environmental conditions. The fractions introduced by anthropogenic activity include the adsorptive, exchangeable and bound to carbonate fractions. These are considered to be weakly bounded metals, which may equilibrate with the aqueous phase and thus become more rapidly bioavailable (Pardo et al., 1990).

In this study, sediment samples from a river in an urbanized area of Japan were analyzed by sequential extraction scheme to measure the geochemical distribution of Fe, Mn, Co, Ni and Mo; to assess the environmental mobility of aforementioned metals and to determine the pollution levels in sediments due to different anthropogenic activities.

MATERIALS AND METHODS

General description of the study area

NR is located in the eastern part of Tokyo (Edogawa ward) with a length of about 5 km. Both side of the river is connected with Arakawa river (AR) by sluice gates and with Onaki river (OR) by a narrow channel. Depth, width, flow of current and water level (about 3 m) in AR is higher than the NR and OR. There were several types of small and large industrial units on both banks of NR during 40th century (USAMA, 1945-46), such as refinery and mineral oil company, several chemical and steel company, paper and pulp mills, soap industry, hide and tanning factory, paints and dyeing industry and others, and still the river is received a significant amount of various industrial waste and house hold discharge from residential areas (Zakir, 2008).

The area mainly consists of tertiary and quaternary sedimentary rocks (shale and sandstone) overlain by quaternary volcanic materials and weathered soils of volcanic origin. Alluvium sediments consisting of various kinds of rock fragments (granite, basalt, chert, limestone, shale, sandstone) were derived from upper stream region where Paleozoic rocks are distributed (Omori et al., 1986). Among the clay minerals, vermiculite was dominant, and chlorite and illite were also observed. These crystalline clay minerals were deduced to be derived from aeolian mica that was contained in a coast deposit distributed along coasts of this region, and might be mixed with volcanic ash soils (Takesako et al., 2002).

Collection and preparation of samples

A total of 21 sediment samples were collected, among those 17 were from NR, 2 from OR and the other 2 from AR, as described in Figure 1. The sampling distance from one station to another was at least about 100 m. The surface sediment samples were taken from 0-10 cm and quickly packed in airtight polythene bags. The sample mass collected in each case was about 500 g. Sub-samples of the material were oven dried at 50°C for 24 h and sieved (aperture 125 μm). The lower particle size fraction was homogenized by grinding in an agate mortar and stored in glass bottles for chemical analyses.

Reagents

All chemicals and reagents were of analytical reagent grade quality (Sigma-Aldrich, USA and Wako, Japan). Millipore water (18 MΩ) was used throughout all the experiments. Before use, all glass and plastic ware were soaked in 14% HNO_3 for 24 h. The washing was completed with Millipore water rinse.

Apparatus and instrumentations

Atomic absorption measurements for Fe and Mn were performed with a Hitachi Z-6100 polarized Zeeman spectrophotometer, equipped with single element hollow-cathode lamps. The instrument was operated at maximum sensitivity with an air-acetylene flame.

Lamp intensity and band pass were used according to the manufacturer's recommendations. A centrifuge (Kokusan H-27F, Tokyo, Japan) for complete separation of the extract from the residue, a Yamato SA-320 (Tokyo, Japan) shaker and a TOA-DKK HM-20P (Tokyo, Japan) pH meter were used throughout all the experiments. Inductively coupled plasma-mass spectrometry (ICP-MS, Hewlett-Packard 4500, USA) was applied for determination of Co, Mo and Ni in this study. A BrukerAXS: D-8 Advance (Berlin, Germany) X-ray diffractometer was employed for XRD analysis.

Sequential extraction experiment

The sequential extraction scheme adopted in this study was a combination of Tessier et al. (1979) and Hall et al. (1996) schemes. The first two (2) steps were adopted from the procedure proposed by Tessier et al. (1979), which will provide more information about the availability of the metals studied. On the other hand, Zakir et al. (2008) reported that the oxide phase of the sediments of old Nakagawa river is dominant for most of the trace metals, and Hall et al. (1996) scheme has the opportunity to get much more knowledge about the oxide phase as amorphous Fe oxyhydroxide and crystalline Fe oxide. Due to these reasons, the last four (4) steps were adopted from the scheme proposed by Hall et al. (1996). However, the successive chemical extraction steps involved

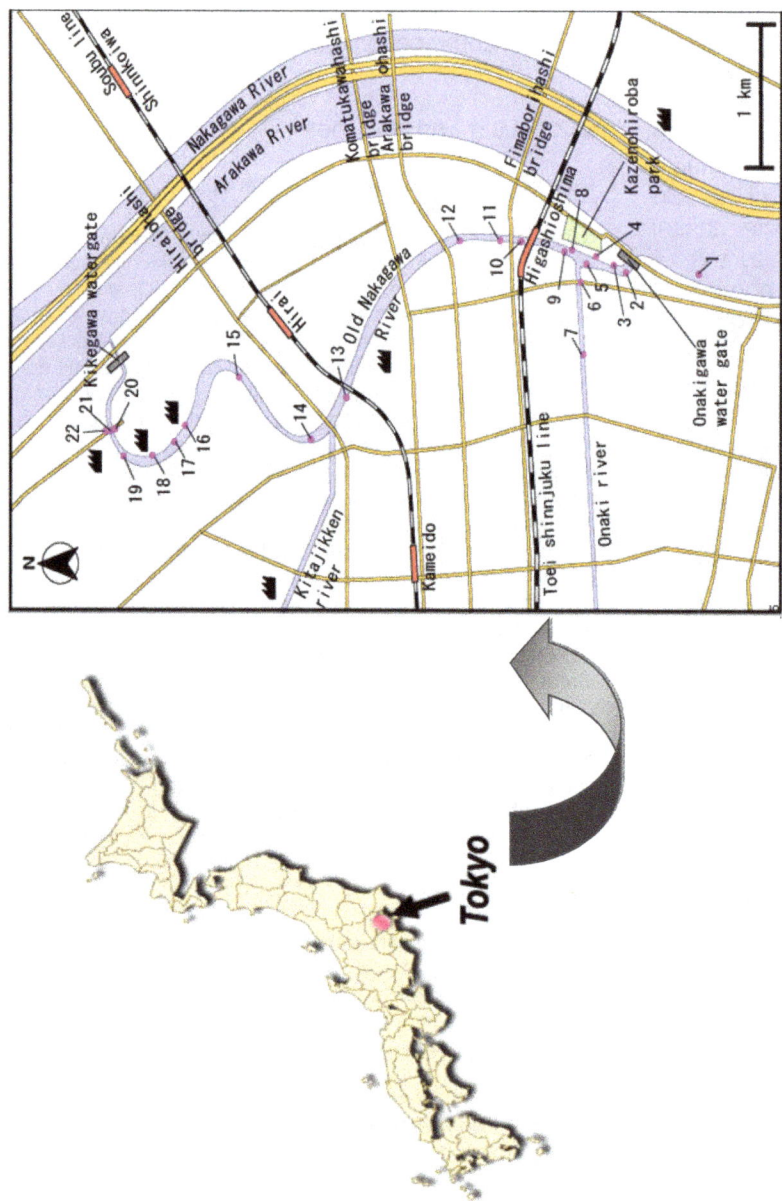

Figure 1. Locations of the sampling stations in Old Nakagawa, Onaki and Arakawa rivers.

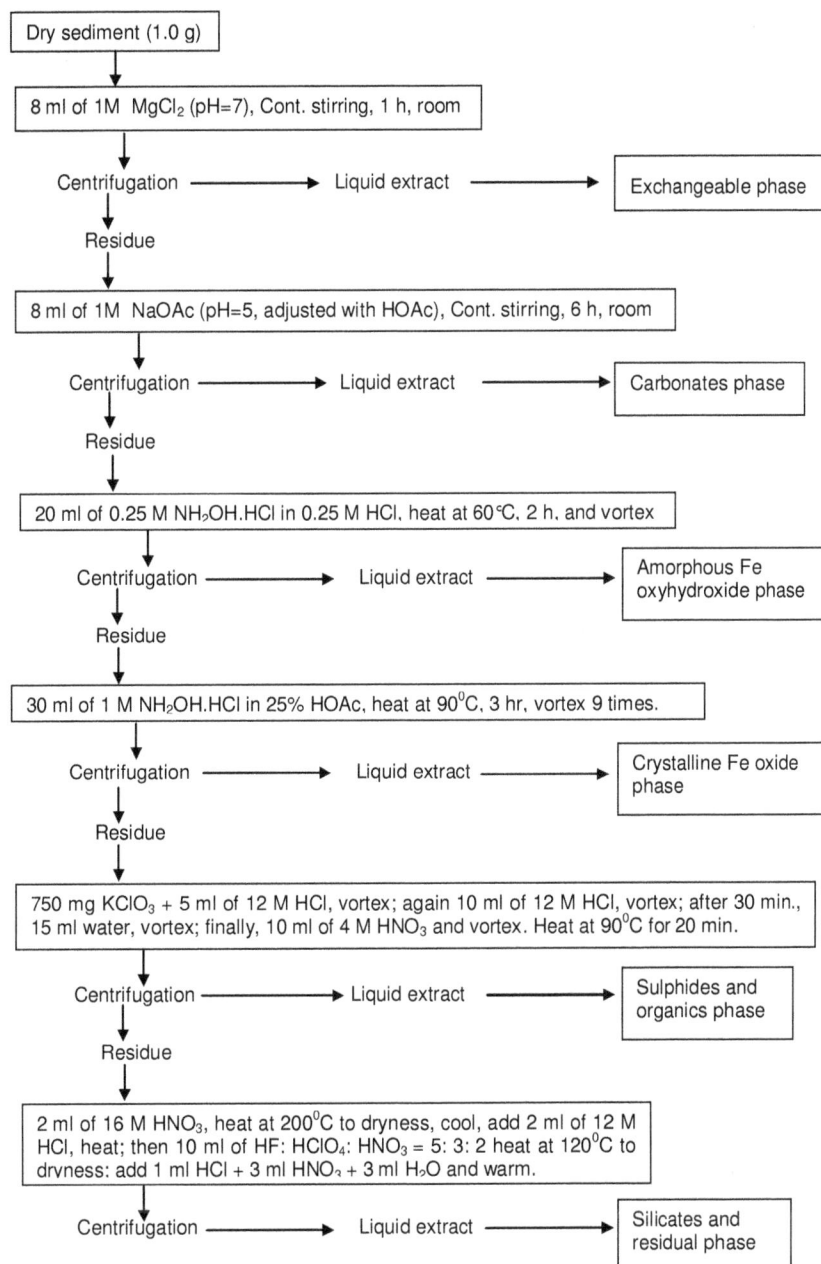

Figure 2. Flow-chart of the sequential extraction scheme used in this study.

in the sequential procedure employed are summarized in the flow-chart presented in Figure 2.

All the operations were carried out in 50 ml polypropylene centrifuge tubes (Nalgene, New York) and Teflon (PTFE) containers provided with screw stoppers. The initial mass of sediment was 1.0 ± 0.0001 g. The extracts were separated from the solid residue by centrifugation at 3000 rpm for 30 min. After each step, the solution was filtered by suction through a 0.45 µm Millipore filter, and the filtrate collected in a polyester container. Then the solutions for each step were prepared accordingly for ICP-MS measurement following the manufacturer's recommendations. During extraction, extractant quality (especially the required pH) maintained carefully. A certified reference stream sediment JSd-2, provided by the

geological survey of Japan was also analyzed using the same procedure as check for total concentration and reached 89-107% recovery for the studied metals. As a quality assurance measure, each sediment sample was subjected to triplicate analyses and the measurements are given as mean, unless noted.

Sediment quality assessment

Normalization of metals

In order to account for geochemical variations along the NR of Tokyo, Japan, normalization by a conservative element Al was

Table 1. Classification of risk assessment code (RAC).

RAC	The sum of exchangeable and carbonate bound fractions (in % of total)
No risk	< 1
Low risk	1 - 10
Medium risk	11 - 30
High risk	31 - 50
Very high risk	> 50

used, as geochemical normalization can compensate for both the granulometric and mineralogical variability of metal concentration in sediments (Liu et al., 2003). Consequently, in order to assess the possible anthropogenic impact, several authors (Datta and Subramanian, 1998; Badr et al., 2009; Milacic et al., 2010) have successfully used Al to normalize the metal contaminants. In this study, we have also used Al as a conservative tracer to differentiate natural from anthropogenic components and Earth's crust average (Huheey, 1983) data have been considered as baseline values, as Al is abundant in the earth's crust and is scarcely influenced by anthropogenic inputs.

Determination of geoaccumulation index (I_{geo})

The geoaccumulation index (I_{geo}) values were calculated for Ni, Mo, Co, Mn and Fe as introduced by Muller (1969) is as follows:
$I_{geo} = \log_2 (C_n / 1.5 \times B_n)$
Where C_n is the measured concentration of metal in the sediment, and B_n is the geochemical background for the same element which is either directly measured in precivilization sediments of the area or taken from the literature (average shale value described by Turekian and Wedepohl, 1961). The factor 1.5 is introduced to include possible variations of the background values that are due to lithologic variations.

According to Muller (1969), there are seven grades or classes of the geoaccumulation index. Class 0 (practically uncontaminated/unpolluted): $I_{geo} < 0$; Class 1 (Uncontaminated to moderately contaminated): $0 < I_{geo} < 1$; Class 2 (moderately contaminated): $1 < I_{geo} < 2$; Class 3 (moderately to strongly contaminated): $2 < I_{geo} < 3$; Class 4 (strongly contaminated): $3 < I_{geo} < 4$; Class 5 (strongly to extremely contaminated): $4 < I_{geo} < 5$; Class 6 (extremely contaminated): $I_{geo} > 5$, which is an open class and comprises all values of the index higher than Class 5.

Risk assessment code (RAC)

The metals in the sediments are bound with different strengths to the different fractions. The risk assessment code (RAC) as proposed by Perin et al. (1985), mainly applies to the sum of exchangeable and carbonate bound fractions for assessing the availability of metals in sediments (Table 1). If a sediment sample can release in these fractions less than 1% of the total metal will be considered safe for the environment. On the contrary, sediment releasing in the same fractions more than 50% of the total metal has to be considered highly dangerous and can easily enter into the food chain.

RESULTS AND DISCUSSION

Geochemical fractionation profile of Fe, Mn, Co, Ni and Mo

The following sections describe geochemical distribution of Fe, Mn, Co, Ni and Mo according to the sequential extraction procedure applied in this study.

Exchangeable fraction

The fraction of exchangeable metals included the portion, which is held by the electrostatic adsorption as well as those specially adsorbed. The amount of metals in this phase indicated the environmental conditions of the overlying water bodies. Metals in this fraction are the most mobile and readily available for biological uptake in the environment (Singh et al., 2005; Zakir et al., 2008). The exchangeable fraction recovered for Co and Fe in sediments of the study area were comparatively low (0 and 0- 0.98% of total, respectively). On the other hand, the range of Ni, Mo and Mn for this fraction varied from 0-10.3 (0 - 4.1% of total), 0 - 1.0 (0- 7% of total) and 0-108.1 (0- 8% of total) $\mu g\ g^{-1}$, respectively in sediments of the study area (Figure 3). The adsorbed/ exchangeable/ carbonate (AEC) fractions recovered for Fe, Mn and Ni in sediments of Nomi river, Tokyo, Japan were also reported low (0.57 - 1.57, 0.66 - 4.11 and 0.60 - 20.06% of total, respectively) (Sharmin et al., 2010). Metals accumulated in this fraction can return to waters in the suspended or dissolved form and represent a potential risk for an aquatic ecology as well as the environment. However, the affinity order of different metals for exchangeable fraction was Ni > Mo > Mn > Fe > Co.

Fraction bound to carbonate

Preferential association of carbonate bound Ni, Co and Mo (on an average 22.1, 18.1 and 16.5% of total, respectively) were found in sediments of the study area (Figure 3), which may readily available following a slight lowering of pH (Jain, 2004; Zakir et al., 2008). Metal recovered by the use of 1 M NaOAc adjusted to pH 5 is associated with carbonate minerals likely to be bioavailable. The XRD data of our previous study also indicate presence of higher amount of carbonate and several clay minerals in sediments of NR (Zakir et al., 2008), which support these findings. However, metals associated with this fraction are not strongly bound to the sediment solids and can be released to the sediment pore water in acidic conditions (pH< 5). Metals in this fraction are adsorbed on sediments or on their essential components namely clays, Fe and Mn, hydrooxides or oxyhydroxides and humic acids. Turner and Olsen (2000) determined extractability of metals in contaminated

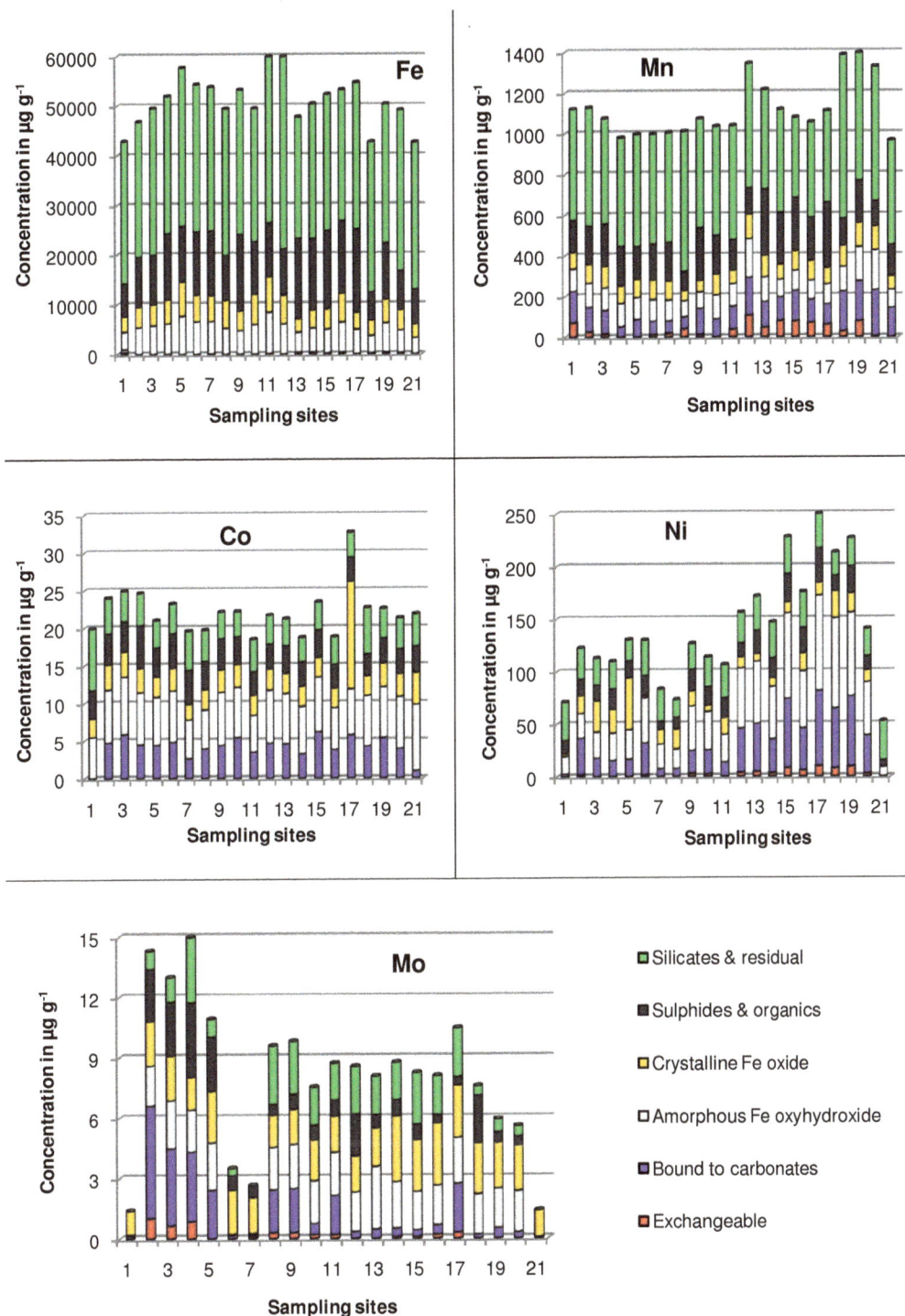

Figure 3. Geochemical distribution of Fe, Mn, Co, Ni and Mo concentration (μg g^{-1}) in sediment samples at different sampling sites of the study area.

estuarine sediments by chemical and enzymetic reactions. Among the chemical reagents, acetic acid best represented the fraction that was likely to be bioavailable to sediment ingesting and benthic organisms. Finally association of metals with this fraction is probably the best example of human-induced influence in the sediments of the study area. On the other hand, fraction bound to carbonate recovered for Fe and Mn in sediments of the study area were comparatively low (on an average 0.14 and 11.1% of total, respectively), suggesting lower pollution risk (Figure 3). However, the affinity order of different metals for the fraction bound to carbonate was: Ni > Co > Mo > Mn > Fe.

Amorphous Fe oxyhydroxide fraction

Due to the large surface area and surface site, amorphous Fe oxyhydroxide are one of the most important geochemical phases impacting the mobility and behaviour of metals in sediments (Bilinski et al., 1991; Turner, 2000; Wang et. al., 2010). The metal content in this phase has already been proved to be sensitive to anthropogenic inputs (Modak et al., 1992). Metals bound to amorphous Fe oxyhydroxide fraction are unstable under reducing conditions. These conditions result in the release of metal ions to the dissolve fraction. Amorphous Fe oxyhydroxide phase is well recognized for its scavenging properties of metals in the surficial environment and is defined by the extraction with 20 mL of 0.25 M $NH_2OH.HCL$ in 0.25 M HCl and heated for 2 h at 60 °C (Hall et al., 1996).

The geochemical fractionation results from the present study are consistent with the high affinity of metals for amorphous oxide minerals. This phase accumulates metals from the aqueous system by the mechanism of adsorption and co-precipitation (Bordas and Bourg, 2001). The relatively higher concentrations of metals such as Ni, Mo, Co and others associated with this fraction are caused by the adsorption of these metals by the Fe-Mn colloids (Purushothaman and Chakrapani, 2007). Industrial discharge may be one of the factors for the increased concentrations of metals in sediments of the study area.

The geochemical fractionation result from the present study revealed the order of affinity of metals as Co > Mo > Ni > Mn > Fe (26.6- 30.5, 23.2- 38.5, 19.0- 40.1, 7.7-14.7 and 8.3- 13.8% of total, respectively) for amorphous Fe oxyhydroxides of NR sediments (Figure 3). The observed trend in the association of Co, Mo, Ni, Mn and Fe with amorphous Fe oxyhydroxide minerals were moderately well explained and the presence of Fe oxyhydroxide minerals, such as goethite in the samples of NR sediments also detected by the XRD data, which already been published in our previous report (Zakir et al., 2008). The results obtained for the sediments of Nomi river of Tokyo (Sharmin et al., 2010) are also at par with the present study.

Crystalline Fe-oxide fraction

In contrast to amorphous Fe oxyhydroxide, the mean percentage for crystalline Fe oxide fraction was lower for Co and Ni (Figure 3). This is probably due to the much greater surface area of amorphous minerals in comparison with crystalline materials (Kampf et al., 2000;

Zakir et al., 2008). On the other hand, considering the mean percentage only 26.4, 15.7, 10.8, 9.0 and 8.2% of total Mo, Co, Ni, Fe and Mn, respectively were associated with the operationally defined crystalline Fe oxide fraction. It is apparent from the Figure 3 that the association of Fe, Mn and Mo in this fraction was almost similar to the amorphous Fe oxyhydroxide phase. Metals associated with oxide (both amorphous and crystalline) minerals are likely to be released in reducing condition.

However, the affinity order of different metals for crystalline Fe oxide was: Mo > Co > Ni > Fe > Mn. According to Sharmin et al. (2010) the affinity order of trace metals for the same fraction of sediment samples of Nomi river was Cu > Fe > Ni > Cr > Mn.

Fraction bound to organic matters and sulfides

Organic matter and sulphides are important factors controlling the mobility and bioavailability of metals. Current opinions reflects on the toxicity of metals influenced by organic matters (Hoss et al., 2001; Besser et al., 2003). Degradation of organic matter under oxidizing conditions can lead to the release of soluble metals bound to this fraction (Purushothaman and Chakrapani, 2007). The affinity of metals for organic substances and their decomposition products are of great importance for the release of the metals into water. Metal bound to this fraction is assumed to reflect strong association with sediment organic material. The fractionation result from the present study revealed that on an average 22.5, 17.4, 16.9, 12.5 and 14.3% of total Fe, Mn, Co, Ni and Mo, respectively were associated with the sulfides and organic (oxidizable) fraction (Figure 3).

On the other hand, after residual fraction, major association with this fraction was found for Fe and Mn, may be for high affinity with organic matter and also scavenging properties of this fraction. This indicates that Fe and Mn occurred in the form of stable organic complexes and metal sulphides. This result is in good agreement with the literature reported by Prasad et al. (2006) and Sharmin et al. (2010).

Silicate and residual fraction

Silicate and residual fraction of metals are generally much less toxic for organisms in aquatic environment because this fraction is chemically stable and biologically inactive. In the present study residual form was the dominant for Fe and Mn. Fractionation profile of present study indicating average affinities of Fe, Mn, Co, Ni and Mo were 57.5, 49.3, 19.4, 20.9 and 18.2 % of total, respectively for silicate and residual fraction (Figure 3).

The metal present in the residual fraction can be used as a baseline data for the assessment of the degree of contamination of the system. The association between metals and the residual fraction of uncontaminated soils is so strong that metal association with non-residual fraction has been used as an indicator of anthropogenic enrichment (Sutherland et al., 2000; Zakir, 2008).

Environmental mobility and geochemical partitioning trend of metals

Here, a summary of geochemical partitioning results is described and environmental mobility pattern is assessed for a particular metal. Overall, the order of importance of different geochemical fractions of metals in NR sediment samples obtained for the study were:

1) Co: amorphous Fe oxyhydroxide > bound to carbonate > silicates and residual ≥ sulphides and organics > crystaline Fe oxide > exchangeable
2) Fe: silicates and residual > sulphide and organics > amorphous Fe oxyhydroxide ≥ crystaline Fe oxide > bound to carbonate > exchangeable
3) Mn: silicates and residual > sulphide and organics > bound to carbonate > amorphous Fe oxyhydroxide ≥ crystaline Fe oxide > exchangeable
4) Mo: amorphous Fe oxyhydroxide > crystalline Fe oxide > silicates and residual > bound to carbonate > sulphide and organics > exchangeable
5) Ni: amorphous Fe oxyhydroxide > bound to carbonate > silicates and residual > sulphide and organics > crystalline Fe oxide > exchangeable

These findings suggest that the order of potential mobility of metals in the aquatic environment of the study area is: Ni ≥ Co > Mo > Mn > Fe

Assessment of metal pollution in sediments

Normalization of metals

The normalization of the concentrations of metals using Al as a conservative element confirmed that with little exception all the sampling sites are the most polluted by Ni, Mo and Mn (Figure 4). Furthermore, the normalization revealed that all the sampling stations (2- 20) of NR and OR were enriched with Ni, Mo and Mn when the ratios are compared with the baseline values. It is also evident from Figure 4 that the calculated ratios for Fe and Co at all the sampling sites were less or very close to the ratios obtained from the Earth's crust average (Huheey, 1983) data, which have been considered as baseline value indicates the study area is not polluted by Fe and Co.

Risk assessment code (RAC)

The distribution of metal speciation associated with different geochemical fraction is a critical parameter to assess the potential mobility and bioavailability of metals in sediments. RAC as applied to the present study revealed that on an average 24.4, 19.1, 18.1 and 14.5% of total Ni, Mo, Co and Mn, respectively of the study sites either is adsorb, exchangeable or carbonate bound and therefore comes under the medium risk category to local environment and can easily enter into the food chain (Tables 1 and 2).

Because of the toxicity and availability of metals, they can pose serious problem to the ecosystem, and can be remobilized by changes in environmental conditions such as pH, redox potential, salinity etc. On the other hand, only 0- 0.95% of total Fe was found in the exchangeable and carbonate bound fraction with an average value of 0.32% and therefore comes under the no risk category indicating lower availability from which Fe cannot be easily leached out for the aquatic environment (Tables 1 and 2). So, the potential hazard of Ni, Mo, Co and Mn were larger than those of Fe which occurred mostly in the inert residual fraction.

Geoaccumulation Index (I_{geo})

The calculated index of geoaccumulation (I_{geo}) of the metals in the sediments of NR and their corresponding contamination intensity are shown in Figure 5. The I_{geo} values for Ni ranged from 0.3 to 1.4 corresponded with class 0 to 2, which exhibited unpolluted/ uncontaminated to moderately contaminated sediment quality. Similarly, the I_{geo} values for Mo ranged from 0.2 to 2.8 corresponded with class 0 to 3, which exhibited unpolluted/ uncontaminated and moderately to strongly polluted sediment quality. It is evident from Figure 5 that all the sampling stations of NR and OR are moderately to strongly polluted (class 1-3) for Mo and Ni, but the sampling sites (nos. 1 and 21) of AR were unpolluted (class 0) may be due to higher flow of current and water level.

The results revealed that Ni and Mo contamination of the sediments of NR and OR could have significant impact on aquatic ecology. On the other hand, it is also evident from the Figure 5 that the sediments of the study area are unpolluted with respect to Fe, Mn and Co as the I_{geo} values for those metals were negative in the study sites.

Conclusion

Identification and quantification of metal sources in soil and sediment, as well as the fate of those metals, are important environmental scientific issues. This study has been focused mainly on environmental mobility and geochemical partitioning of Fe, Mn, Co, Ni and Mo in sediment profiles of NR, Tokyo, Japan. The present study also used several tools, methods and indices for the

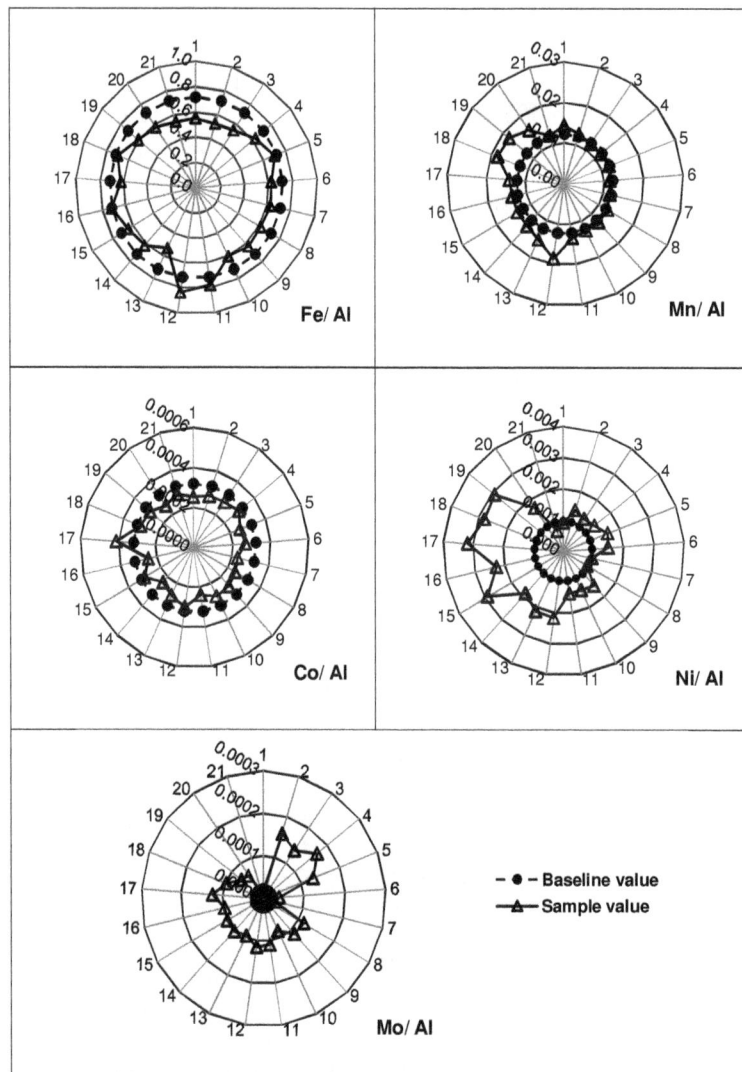

Figure 4. Normalized ratio of Fe, Mn, Co, Ni and Mo among the sampling sites using Al as conservative element. Broken line indicates the Earth's crust average as baseline value.

Table 2. Speciation pattern up to carbonate fraction for Fe, Mn, Co, Ni and Mo in sediments of the sampling sites and risk assessment code (RAC).

	Average percentage in each fraction (in % of total)			Level of risk on the basis of RAC
Metal	Exchangeable	Bound to carbonates	Total in both fraction	
Fe	0.18	0.14	0.32	No risk
Mn	3.4	11.1	14.5	Medium risk
Co	0.0	18.1	18.1	Medium risk
Ni	2.3	22.1	24.4	Medium risk
Mo	2.6	16.5	19.1	Medium risk

evaluation of sediment quality, which provides valuable information for decision-making processes and management involving natural resources. However, a detailed examination on geochemical partitioning and normalization results indicate that Ni and Mo areanthropogenically enriched in sediments of NR,which

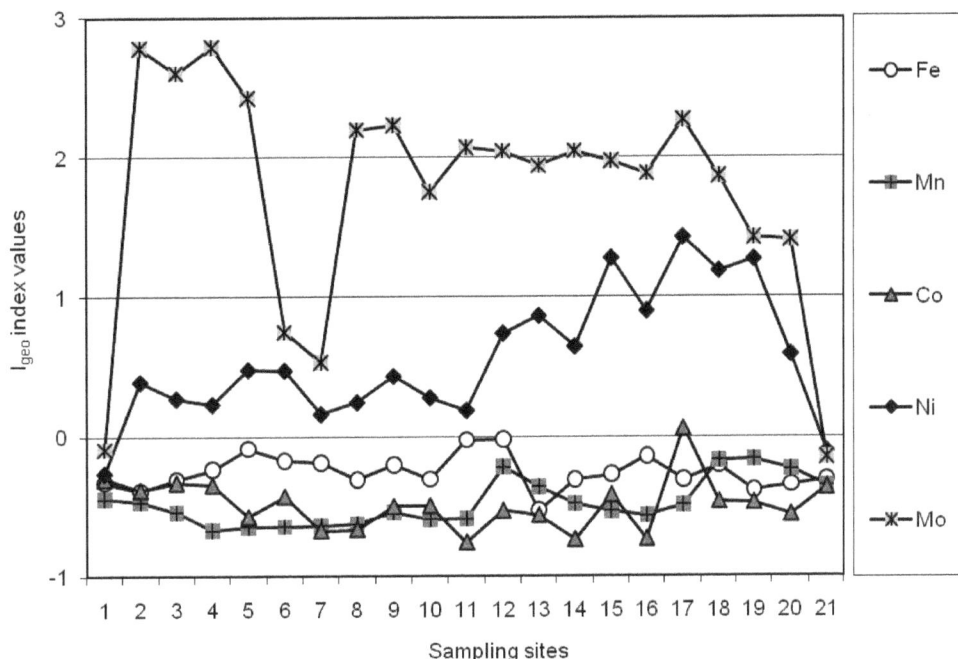

Figure 5. Geoaccumulation index (I_{geo}) of Fe, Mn, Co, Ni and Mo in sediment samples at different sampling sites of the study area.

has much effect on environment and humans.

According to geoaccumulation index, Ni and Mo also exhibited uncontaminated to strongly polluted sediment quality. On the other hand, residual and silicate fraction was the main geochemical association for Fe, Mn and Co, which has little effect on environment suggesting weak pollution risk. RAC as applied to the present study revealed that except Fe, all other metals comes under the medium risk category to local environment. However, the lower mean percentage of active fractions is an advantage to control the environmental pollution in the study sites.

REFERENCES

Amiard JC, Geffard A, Amiard-Triquet C, Crouzet C (2007). Relationship between the lability of sediment-bound metals (Cd, Cu, Zn) and their bioaccumulation in benthic invertebrates. Estuar. Coast. Shelf Sci., 72: 511-521.

Badr NBE, El-Fiky AA, Mostafa AR, Al-Mur BA (2009). Metal pollution records in core sediments of some Red Sea coastal areas, Kingdom of Saudi Arabia. Environ. Monit. Assess., 155: 509-526.

Basta N, Gradwohl R (2000). Estimation of Cd, Pb and Zn bioavailability in smelter-contaminated soils by a sequential extraction procedure. J. Soil Contam., 9: 149-164.

Besser JM, Brumbaugh WG, May TW, Ingersoll CG (2003). Effects of organic amendments on the toxicity and bioavailability of cadmium and copper in spiked formulated sediments. Environ. Toxicol. Chem., 22: 805-815.

Bilinski H, Kozar M, Plavsic M, Kwokal Z, Branica M (1991). Trace metal adsorption on inorganic solid phases under estuarine conditions. Mar. Chem., 32: 225-233.

Bordas F, Bourg A (2001). Effect of solid/liquid ratio on the remobilization of Cu, Pb, Cd and Zn from polluted river sediment.

Water Air Soil Pollut., 128: 391-400.

Datta DK, Subramanian V (1998). Distribution and fractionation of heavy metals in the surface sediments of the Ganges-Brahmaputra-Meghna river system in the Bengal basin. Environ. Geol., 36(1-2): 93-101.

Hall GEM, Vaive JE, Beer R, Hoashi M (1996). Selective leaches revisited, with emphasis on the amorphous Fe oxyhydroxide phase extraction. J. Geochem. Explor., 56: 59-78.

Hickey MG, Kittrick JA (1984). Chemical partitioning of cadmium, copper, nickel and zinc in soils and sediments containing high levels of heavy metals. J. Environ. Qual., 13: 372-376.

Horowitz AJ (1991). A Primer on Sediment- Trace Element Chemistry. 2nd edition, Lewis Publishers, Chelsea (Michigan), p.136

Hoss S, Henschel T, Haitzer M, Traunspurger W, Steinberg CEW (2001). Toxicity of cadmium to Caenorhabditis elegans (Nematoda) in whole sediment and pore water – The ambiguous role of organic matter. Environ. Toxicol. Chem., 20: 2794-2801.

Huheey JE (1983). Inorganic Chemistry: Principle of Structure and Reactivity. Harper and Row Publisher, New York, p.912

Jain CK (2004). Metal fractionation study on bed sediments of River Yamuna, India. Water Res., 38: 569-578.

Kampf N, Scheinost AL, Schulze DG (2000). Oxide Minerals. In: Sumner ME (ed.) Handbook of Soil Science. CRC Press, Boca Raton, FL. pp. 125-168.

Liu WX, Li XD, Shen ZG, Wang DC, Wai OWH, Li YS (2003). Multivariate statistical study of heavy metal enrichment in sediment of the Pearl River Estuary. Environ. Pollut., 121: 377-388.

Milacic R, Scancar J, Murko S, Kocman D, Horvat M (2010). A complex investigation of the extent of pollution in sediments of the Sava River. Part 1: Selected elements. Environ. Monit. Assess., 163: 263-275.

Modak DP, Singh KP, Chandra H, Ray PK (1992). Mobile and bound form of trace metals in sediments of the lower Ganges. Water Res., 26(11): 1541-1548.

Morillo J, Usero J, Gracia I (2004). Heavy metal distribution in marine sediments from the southwest coast of Spain. Chemosphere, 55(3): 431-442.

Muller G (1969). Index of geoaccumulation in sediments of the Rhine River. Geo. J., 2(3): 108-118.

Omori M, Hatayama Y, Horiguchi M (Eds.) (1986). Geology of Japan, Kanto Districts (in Japanese). 1st ed., Kyooritsu Publishing Co., Tokyo, Japan, p.350

Pardo R, Barrado E, Perez L, Vega M (1990). Determination and association of heavy metals in sediments of the Pisucraga, river. Water Res., 24(3): 373-379.

Perin G, Craboledda L, Lucchese M, Cirillo R, Dotta L, Zanetta ML, Oro AA (1985). Heavy metal speciation in the sediments of northern Adriatic sea. A new approach for environmental toxicity determination. In: Lakkas TD (Ed.). Heavy Metals in the Environment, CEP Consultants, Edinburgh. 2: 454-456.

Prasad MBK, Ramanathan AL, Shrivastav SKR, Anshumali, Saxena R (2006). Metal fractionation studies in surfacial and core sediments in the Achankovil river basin in India. Environ. Monit. Assess., 121: 77-102.

Pueyo M, Sastre J, Hernandez E, Vidal M, Lopez-Sanchez JF, Rauret G (2003). Prediction of trace element mobility in contaminated soils by sequential extraction. J. Environ. Qual., 32: 2054-2066.

Purushothaman P, Chakrapani GJ (2007). Heavy metals fractionation in Ganga river sediments, India. Environ. Monit. Assess., 132: 475-489.

Sharmin Shaila, Zakir HM, Shikazono Naotatsu (2010). Fractionation profile and mobility pattern of trace metals in sediments of Nomi River, Tokyo, Japan. J. Soil Sci. Environ. Manag., 1: 1-14.

Singh KV, Singh PK, Mohan D (2005). Status of heavy metals in water and bed sediments of river Gomti- a tributary of the Ganga River, India. Environ. Monit. Assess., 105: 43-67.

Sutherland RA, Tack FMG, Tolosa CA, Verloo MG (2000). Operationally defined metal fractions in road deposited sediment, Honolulu, Hawaii. J. Environ. Qual., 29: 1431-1439.

Takesako H, Wada N, Sumida H, Kawahigashi M, Miyamoto J, Suzuki S, Tanaka H, Yokotagawa T, Inubushi K (2002). Properties of soil in Chiba University's Atagawa Farm. III. Mineralogical properties and soil classification of volcanic ash soils (in Japanese). Chiba University,Tech. Bull. Fac. Hort., 56: 27-37.

Tessier A, Campbell PGC, Bisson M (1979). Sequential extraction procedure for the speciation of particulate trace metals. Anal. Chem., 51(7): 844-851.

Turekian KK, Wedepohl KH (1961). Distribution of the elements in some major units of the earth's crust. Geol. Soc. Am. Bull., 72: 175-192.

Turner A (2000). Trace metal contamination in sediments from U.K. estuaries: an empirical evaluation of the role of hydrous iron and manganese oxides. Estuar. Coast. Shelf Sci., 50: 355-371.

Turner A, Olsen YS (2000). Bioavailability of trace metal in contaminated estuarine sediments: A comparison of chemical and enzymatic extractants. Estuar. Coast. Shelf Sci., 51: 717-728.

USAMS (US Army Map Service) (1945-46). US Army map service city plans for the Tokyo area. Sheet 9- Honjo. Reproduced from compilation manuscript by overprints. UT Library online, http://www.lib.utexas.edu/maps/ams/japan_city_plans/index_tokyo.html

Wang S, Jia Y, Wang S, Wang X, Wang H, Zhao Z, Liu B (2010). Fractionation of heavy metals in shallow marine sediments from Jinzhou Bay, China. J. Environ. Sci., 22(1): 23-31.

Zakir HM (2008). Geochemical partitioning of trace metals: an evaluation of different fractionation methods and assessment of anthropogenic pollution in river sediments. Ph D thesis, Keio Univ. Yokohama, Japan. pp. 1-156

Zakir HM, Shikazono N (2008). Metal fractionation in sediments: a comparative assessment of four sequential extraction schemes. J. Environ. Sci. Sustainable Soc., 2: 1-12.

Zakir HM, Shikazono N, Otomo K (2008). Geochemical distribution of trace metals and assessment of anthropogenic pollution in sediments of Old Nakagawa River, Tokyo, Japan. Am. J. Environ. Sci., 4(6): 661-672.

4

Acute toxic effects of Endosulfan and Diazinon pesticides on adult amphibians (*Bufo regularis*)

Ezemonye Lawrence and Tongo Isioma*

Department of Animal and Environmental Biology (AEB) University of Benin, P. M. B. 1154, Benin City, Edo State, Nigeria.

The acute toxicity of Endosulfan (organochlorine) and Diazinon (organophosphate) pesticides to adult amphibians, *Bufo regularis* was evaluated to determine uptake and effect of environmentally relevant concentrations on survival, morphology and behaviour. Toxicity characterizations were also assessed using standard indices. Toads were exposed for 96 h to varying concentrations of the pesticides; 0.25, 0.50, 0.75 and 1 mg/l. Mean percentage mortality increased significantly ($p < 0.05$) with concentrations and exposure duration for Endosulfan and Diazinon pesticides and was significantly ($p < 0.05$) different from the control, indicating that pesticide induced lethality. The results showed that Diazinon (LC_{50} = 0.44 mg/l) was more toxic than Endosulfan (LC_{50} = 0.73 mg/l). Derived safe concentrations were 0.07 and 0.04 mg/l for Endosulfan and Diazinon, respectively. Estimated Toxicity index values (TIV) and Hazard Quotients (HQ) for all the concentrations were above one (1) indicating potential risk of the pesticides to the toad. Bioconcentration of the pesticides after 96 h increased with increasing concentrations indicating that uptake was concentration dependent. There was a significant positive correlation between tissue concentration and mortality ($p < 0.01$) for both pesticides. The pesticides also caused dose-dependent deformities and behavioural abnormalities. More pronounced poisoning symptoms were observed in Diazinon and at higher concentrations.

Key words: Acute toxicity, Endosulfan, Diazinon, adult amphibian.

INTRODUCTION

Agricultural pesticides are indispensable in contemporary agriculture. They are beneficial by providing reliable, persistent and relatively complete control against harmful pests with less cost and effort. They have, no doubt, increased crop yields by killing different types of pests, which are known to cause substantial or total crop damage. However, their effects are less than desirable when they leave the target compartment of the agricultural ecosystem. Up to 90% of the pesticides applied never reach the intended targets (Sparling et al., 2001); as a result, many other organisms sharing the same environment as pests are accidentally poisoned. One of the non-target biological groups mostly affected by pesticides is amphibians (Fulton and Chambers, 1985; Berrill et al., 1994; Sparling et al., 2001).

Amphibians are exposed to pesticides by many routes but perhaps the most common route is agricultural runoffs. Agricultural practices affect natural habitat in several ways such as through land conservation, increased fragmentation and agrochemical contamination (Hecnar, 1995; Davidson et al., 2002). Much of the interest on amphibian's declines is currently focused on the role of pesticides on the observed global declines (Houlahan et al., 2001). A diversity of pesticides and their residues are present in a wide variety of aquatic habitats (McConnell et al., 1998; LeNoir et al., 1999; Kolpin et al., 2002). Declining amphibian populations have been correlated with greater amounts of upwind agriculture, where pesticide use is common (Davidson et al., 2001; 2002). While these correlative studies suggest that pesticides may affect amphibian communities, there are

---

*Corresponding author. E-mail: isquared27@yahoo.com.

few rigorous experiments to confirm that pesticides are altering amphibian communities.

This study therefore, concurrently evaluates the effects of Endosulfan, an organochlorine insecticide and Diazinon, an organophosphate insecticide at acute lethal doses on survival, morphology and behaviour of the adult toad *Bufo regularis*. The study was designed to identify the acute toxic effects of these pesticides widely used in Nigeria, and to contribute to the knowledge of the effects of pesticides on amphibians.

MATERIALS AND METHODS

Collection of test organisms

Adult toads, B. regularis of both sexes were collected by hand net from their spawning ponds in unpolluted and non agricultural sites (Harri et al., 1979; Allran and Karasov, 2001; Vogiatzis and Loumbourdis, 1997; Khan et al., 2003).

Toads samples were collected by hand net from their spawning ponds and were transported to the laboratory in a covered basket. Adult toads of the same size and almost the same weight (32.87 ± 0.03 g) were acclimatized in glass tanks (51 × 32 × 33 cm³) containing 2 L of dechlorinated tap water, for 7 days prior to the experiments (Vogiatzis and Loumbourdis, 1997). Tanks were placed on a slant to provide the option of both aqueous and dry environment (Allran and Karasov, 2001). Water was changed every two (2) days and the tank cleaned thoroughly. Toads were fed with earthworms twice weekly. Uneaten earthworms and faecal wastes were removed and water replenished regularly (Allran and Karasov, 2001).

Test chemicals

Endosulfan and Diazinon pesticides were purchased from Coromandel Fertilizers Limited (North Arcot District, Ranipet 632 401 Tamilnadu, India).

Pesticide toxicity tests (bioassay procedure)

Test water

Water for toxicity tests was dechlorinated tap water. The water was dechlorinated by allowing it to stand and exposed for 36 h (Ezemonye and Enuneku, 2005). This water was used for acclimatization, control tests and for making the various concentrations of the test chemicals. Water qualities assessed during the exposure periods for the bioassay were, temperature, pH, dissolved oxygen, conductivity, turbidity and alkalinity. They were determined using standard methods (APHA, 1998).

Test concentrations

The bioassay procedures started with a range finding test (ASTM, 1996). For the definitive test, stock solutions of the required concentrations were prepared for both pesticides. The stock was then diluted serially into environmental relevant treatment concentrations of 0.25, 0.50, 0.75 and 1 mg/l, for Endosulfan and Diazinon pesticides. Two replicates per test concentration were used to avoid test repetition and to provide a stronger statistical baseline. Each test chamber contained an equal volume of test solution (2 L) and equal numbers of toads (10). Replicate test chambers were physically separated.

Toxicity tests

Acute toxicity tests were conducted according to standard procedures (ASTM, 1996). Adult amphibians were exposed for 96 h to each selected concentration of Endosulfan and Diazinon pesticide solutions. Ten (10) toads each were assigned to individual experimental units containing one of the treatments of Endosulfan and Diazinon pesticides (0.25, 0.50, 0.75 and 1 mg/L). The control contained only dechlorinated tap water. Feeding was discontinued 24 h before the commencement of the experiment and during the test. Observations were made on a daily basis and mortality recorded. Behavioural and morphological changes were also observed.

Mortality rate

Mortality was recorded at an interval of 24 h over a period of four days. Toads were assumed dead when they turned upside down and sank to the bottom of the tank or when they showed no form of movement even when prodded with a glass rod (Harri et al., 1979). The behaviour and morphological changes in exposed toads were assessed at 24 h interval.

Bioconcentration

Concentrations were achieved according to the method described by Harri et al. (1979) and Steinwandter (1990). For the whole-body tissue concentration studies, toads (Dead samples after 96 h) were frozen to -18°C. Deep frozen tissues were grounded while still partially frozen, as this makes the tissue more brittle.

Statistical analysis

The susceptibility of the Adult toads to Endosulfan and Diazinon pesticides were determined using the Probit (Probit Software) method of analysis (Finney, 1971) for median lethal concentration at 96 h. Computation of confidence interval of mortality rate was also obtained from the Probit analysis used to determine the LC_{50}. Student's t-test, Pearson correlation and one-way analysis of variance SPSS (14.0 version), SPSS Inc, Chicago, USA was employed to test the variable at $p < 0.05$ level of significance. Multiple bar graphs and line graphs were also used in this study for the pictorial representation of assessment endpoints.

Risk characterization using standard indices

Safe concentrations at 96 h were obtained by multiplying the lethal concentration by a factor of 0.1 (EIFAC, 1998). Bioconcentration factor was calculated as the concentration of the pesticide in the tissue per concentration of the pesticide in water (Falandysz and Chwir, 1997). Hazard Quotient (HQ) was calculated as the ratio of the concentrations of chemicals in tissues to the Toxicity Reference Value (TRV) [Endosulfan = 0.056 (EPA, 1992), Diazinon = 0.043 (EPA, 1996)] (Moloche, 2008). Toxicity index value was calculated as the concentration of the pesticide in the tissue divide by LC_{50} (Battaglin and Fairchild, 2002).

RESULTS

Behavioural and morphological changes

Observations of the behavioral responses of toads

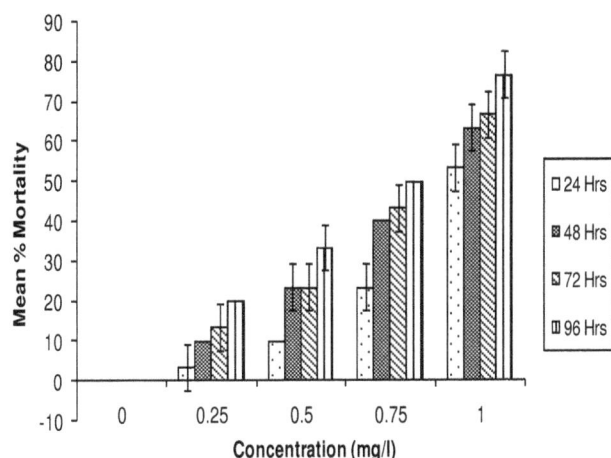

Figure 1. Mean percentage mortality of toads exposed to different concentrations of Endosulfan pesticide.

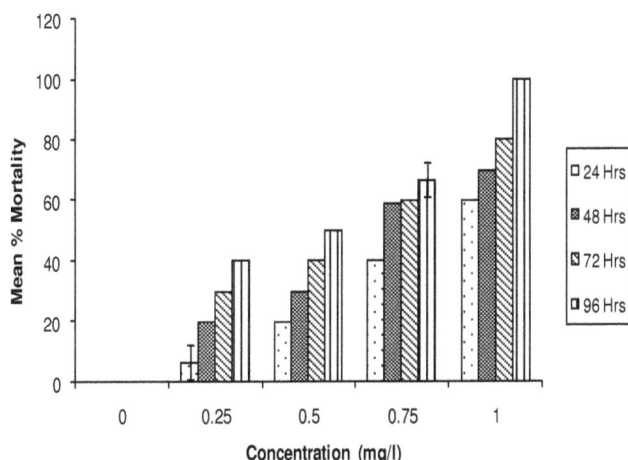

Figure 2. Mean percentage mortality of toads exposed to different concentrations of Diazinon pesticide.

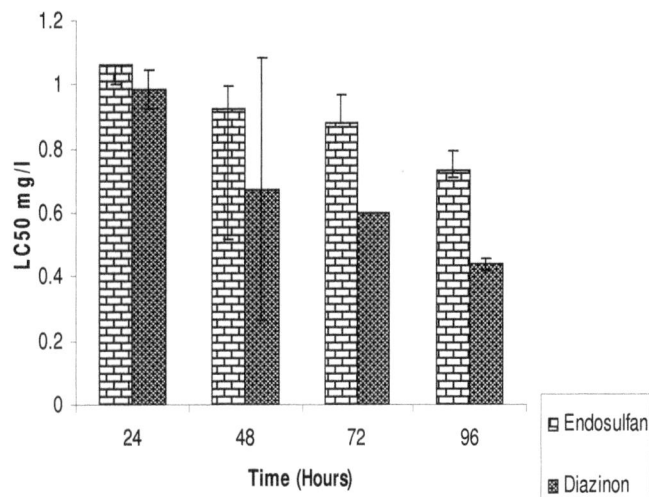

Figure 3. Mean 24, 48, 72 and 96 LC_{50} in adult B. regularis exposed to Endosulfan and Diazinon.

exposed to Endosulfan pesticide were conducted during the 96 h acute toxicity test. The control group showed normal behavior during the test period. Toads exposed to the lowest concentration of 0.25 mg/l had close to normal behavior. At 0.50 mg/l and at all concentrations above 0.50 mg/l, less general activity was recorded when compared with the control group. The highest concentration (1.0 mg/l) showed all responses at high levels. Behavioural changes observed at all concentrations above 0.50 mg/l were characterized by initial hyperactivity (thrashing and leaping) followed by loss of coordination in both front and hind limbs, erratic swimming, unusual retention of water, prolonged and motionless laying down on the aquarium bottom. Finally death was defined as lack of response to mechanical stimuli.

Similarly, toads exposed to Diazinon pesticide also showed the aforementioned poisoning symptoms,

however, it was more severe with skin discoloration, reddening of the snout and the protrusion of the intestine from the anus (heamorrhoid).

Mortality

No mortality or morphological changes were observed in the controls for the 96-h acute toxicity test. Toads in the control experiment appeared active and healthy throughout the test period.

Adult B. regularis exposed to varying Endosulfan and Diazinon concentrations recorded mortality in all the concentrations (0.25, 0.50, 0.75, 1.0 mg/l). The mean percentage mortality increased with increase in concentration and exposure duration for both pesticides (Figures 1 and 2). One hundred percent (100%) mortality was observed in Diazinon exposed toads in 1 mg/l at 96 h. This indicated that mortality was concentration dependent. The exposure showed a clear significant positive correlation between dose and mortality (p < 0.01, r = 1) at all the exposure durations.

Derived 96-h LC_{50} values observed for 24, 48, 72 and 96 h for Endosulan and Diazinon pesticides were 1.06, 0.93, 0.88, and 0.73 mg/l (Endosulfan) and 0.98, 0.67, 0.60, and 0.44 mg/l (Diazinon), respectively (Figure 3). The Probit analysis showed that 96-h LC_{50} values decreased with increase in exposure time. This indicates an increase in toxicity with exposure duration (Figure 1). Safe concentrations at 96 h for Edosulfan and Diazinon pesticides were 0.07 and 0.04 mg/l, respectively (Table 1).

Bioconcentration

The 96 h bioconcentration values obtained increased with

Table 1. Bioconcentration, bioconcentration factors, toxicity index values (TIV) and toxicity quotient (TQ) for *B. regularis* exposed to Endosulfan and Diazinon for 96 h (Mean ± SD).

Test pesticides	Bioconcentration (mg g⁻¹)					Bioconcentration Factors (BCF)					Toxicity Index Value (TIV)					Hazard Quotient (HQ)					96 h-LC$_{50}$	Safe Conc.
	0	0.25	0.50	0.75	1.0	0	0.25	0.50	0.75	1.0	0	0.25	0.50	0.75	1.0	0	0.25	0.50	0.75	1.0		
Endosulfan	0	0.40 ±0.03	1.29 ±0.11	2.44 ±0.01	2.70 ±0	0	1.61 ±0.12	2.57 ±0.21	3.26 ±0.02	27.00 ±0	0	0.56 ±0.04	1.76 ±0.15	3.35 ±0.02	3.70 ±0	0	7.20	22.98	43.63	48.21	0.730 mg/l	0.073 mg/l
Diazinon	0	0.63 ±0	2.19 ±0.08	4.15 ±0.13	4.71 ±0.01	0	2.52 ±0	4.38 ±0.21	5.53 ±0.18	47.07 ±0.12	0	1.47 ±0	5.09 ±0.19	9.65 ±0.30	10.95 ±0.03	0	14.65	50.93	96.51	109.53	0.438 mg/L	0.044 mg/L

increase in concentration of test pesticides (Table 1). There was also a significant positive correlation between accumulated residues and mortality (p<0.01) for both pesticides.

Risk characterization

Bioconcentration factors (BCF), toxicity index values (TIV) and hazard quotient (HQ) of Endosulfan and Diazinon pesticides in *Bufo regularis* are shown in Table I. Toxicity index values increased with concentration. The highest TIV of 10.95 was recoded in 1 mg/l of Diazinon. Toxicity index values increased with concentration. The highest TIV of 10.95 was recoded in 1 mg/L of Diazinon.

The bioconcentration factors ranged from 1.61 - 27 for toads exposed to Endosulfan pesticide and 2.52 - 47.07 for toads exposed to Diazinon pesticide. In toads exposed to the highest concentration (1 mg/L) of Endosulfan, BCF was approximately 25 times more than the lowest concentration of 0.25 mg/l, while for Diazinon, BCF was 45 times more.

All the estimated HQs in all the concentrations for both pesticides were above 1 (Table 1).

Diazinon pesticide had the highest TQ of 109.53 in 1 mg/l.

DISCUSSION

Behavioural and morphological observations

Toads exposed to different lethal concentrations of Endosulfan and Diazinon pesticides, undoubtedly experienced stress due to their irritating and neurotoxic effects. The increased erratic movements exhibited by the toads may be an attempt to be relieved from such stressful environment. The changes in the animal's behaviour after exposure to toxicants may be related to the consequent alteration in physiological process (Marler et al., 1966). Certain signs of toxicity, such as hyperactive symptoms at the beginning, then loss of balance, followed by motionlessness and finally death were observed in the behaviour of *B. regularis* with higher symptoms occurring for Diazinon exposed toads. Similar behavioural effects in response to laboratory exposure to carbaryl, carbofuran, malathion, dimethoate atrazine and basudin in larvae amphibians have been reported (Bishop, 1992; Saym and

Akyurtlaki, 1999; Saym and Kaya, 2006; Ezemonye and Tongo, 2009; Ezemonye and Ilechie, 2007). These behavioural effects are not surprising as most of these pesticides are neurotoxins. It is also known that behavioural changes of these types increase the chances of amphibian predation (Bishop and Pettit, 1992). Morphological changes of skin discoloration observed in Diazinon exposed toads is comparable with the findings of Kaplan and Overpeck (1964) who reported skin discoloration in *Rana pippiens* exposed to near lethal concentrations of aldrin and chloradane pesticides.

Mortality

No mortality was reported in all the control tests in this study, while varying degrees of mortality were reported in the tests concentrations. This is a clear indication that the effects of the pesticides could be regarded as possible cause of death of the test organisms. The results clearly indicate that both pesticides, Endosulfan and Diazinon varied greatly in their effects on survival of *B. regularis*. The highest mortality was found at the highest concentrations, suggesting dose-dependent survival and concentration graded lethality. The 96 h LC_{50} for *B. regularis* exposed to Endosulfan in this study was found to be 0.730mg/l while the 96-h LC_{50} values for Diazinon was 0.438mg/l. The safe concentrations were 0.07 and 0.04 mg/l for Endosulfan and Diazinon, respectively. Using hazard ratings and 96 h LC_{50} estimates (Table 1) from the present study, Endosulfan and Diazinon would be classified as very highly toxic (that is, $LC_{50} > 0.1$ to 1). The current pesticide water quality criterion of > 0.01 mg/L (FEPA, 1991) however, appears protective of these toad species.

Mortality patterns were pesticide specific. Endosulfan and Diazinon pesticides significantly induced mortality in the toad *B. regularis*, with Diazinon recording complete mortality (100%) at the highest treatment concentration within 96 h. Although, no data on the toxicity of Endosulfan and Diazinon on adult amphibians are available for comparison with the results of this study, mortality rates were however, comparable to those reported for larvae amphibians (Broomhall, 2002; Sparling and Fellers, 2009; Sumanadasa et al., 2008; Harris et al., 1998; Relyea, 2004; Ezemonye and Ilechie, 2007) in which lethality was dose and time dependent.

In comparison with fish for endosulfan pesticide, Nowak and Sunderam (1991) reported LC_{50} values of 2.0 µg/l at 30°C and 4.6 µg/L at 35°C when mosquito fish was exposed to technical grade Endosulfan. Smith (1991), reported LC_{50} for rainbow trout to be 1.4 µg/l. For Diazinon pesticide, Hoque et al. (1993) observed the 96-h LC_{50} values for *Puntius gonionotus* exposed to Diazinon to be 3.67 mg/l, Al-Arabic et al. (1992) reported the 96-h LC_{50} value of *Labio calbasu* fingerlines exposed to Diazinon to be 1.54 mg/l, Svobodova et al. (2001) reported 96-h LC_{50} value of 26.7 mg/l for common carp exposed to Diazinon.

The recorded 96-h LC_{50} values observed for fishes exposed to Diazinon pesticides were higher than the 96-h LC_{50} values of 0.44 mg/l for *B. regularis* observed in this study; this implies that *B. regularis* is more sensitive to diazinon compared to the fish species. Standard toxicity tests show that for some contaminants, amphibians are more sensitive than fish (Birge et al., 2000). However, *B. regularis* was less sensitive to Endosulfan than in the reported fish species.

Higher mortality values recorded for Diazinon may be mainly due to the metabolite Diazoxon that is formed in animals. Diazoxon is a potent enzyme inhibitor capable of killing organisms directly by inhibiting acetylecholinesterase and numerous other important enzymes with molecular structures that are similar to it (Eisler, 1986). Most disorders in animals exposed to organophosphate compound like Diazinon have been linked to their toxic effects in the central nervous system (Berrill et al., 1994; Saglio et al., 1998). The pesticides have been found to concentrate in tissue of frogs with depressed cholinesterase activity (Sparling et al., 2001) and have induced hyperactivity in frogs followed by paralysis (Berrill et al., 1998).

Bioconcentration

The high mortality recorded could also be explained by bioconcentration of these agrochemicals in the tissues. There was a clear significant positive correlation between accumulated residues and mortality ($p < 0.01$). Each of the two focal compounds in this study has been reported to bioconcentrate (Sparling et al., 2001) in the tissues of test organisms and the estimates of tissue concentrations may be more valuable for the assessment of situations in the natural environment (Sparling et al., 2001). Bioconcentration in tissues has serious ecological consequence because these pesticides are retained in the amphibian's body tissue which when fed on by a predator can lead to the concentration of the chemical from one trophic level to the next (ASTM, 1998; Suter, 1993). The high BCF values observed for Diazinon pesticide may be due to persistence of the chemical in the system (Hall and Swineford, 1980). Results of the measured indices of toxicity (TIV and TQ) indicate that the pesticide Diazinon was highly toxic to the toad *B. regularis* than Endosulfan.

REFERENCES

Al- Arabi SSM, Mazid MA, Alam MGA (1992). Toxicity of Diazinon to three species of Indian major carps. Banglasesh J. Tran. Dev. 5(1): 77-86.

Allran JW, Karasov WH (2001). Effects of atrazine on embryos, larvae and adults of anuran amphibians. Environ. Toxicol. Chem. 20: 761-775.

American Public Health Association (APHA) (1998). Standard methods for the examination of water and waste water 20th Edn American Public Health Association, New York, USA, p.1976.

American Society for Testing and Materials (1996). Standard practices for conducting acute toxicity test with fishes, macro invertebrates, and amphibians In Annual Book of ASTM standards. 11 (5): 1 - 29.

American Society for Testing and Materials (ASTM) (1998). Standard guide for conducting the Frog Embryo Teratogenessis Assay - Xenopus (FFTAX) E1439 volume 1105 E47, Annual book of ASTM standards West Conshohocken PA: Committee on Biological Effects and Environmental fate. E1439-1498

Battaglin W, Fairchild J (2002). Potential toxicity of pesticides measured in midwestern streams to aquatic organisms Water Sci. Technol. 45(9): 95-103.

Berrill M, Coulson D, McGillivray L, Pauli B (1998). Toxicity of endosulfan to aquatic stages of anuran amphibians. Environ. Toxicol. Chem. 17: 1738-1744.

Berrill M, Bertram S, McGillivary L, Kolohan M, Paul B (1994). Effects of low concentrations of forest use pesticides on frogs' embryo and tadpoles Environ. Toxicol. Chem. 18: 657 -664.

Birge WJ, Westerman AG, Spromberg JA (2000). Comparative toxicology and risk assessment of amphibians, In Sparling DW, Linder G, and Bishop CA (Eds)Ecotoxicology of amphibians and reptiles SETAC Press, Pensacola, FL, USA p. 877.

Bishop CA (1992) The effects of pesticides on amphibians and the implications for determining the causes of decline in amphibian populations In: Bishop CA, Pettit KE, editors Declines in Canadian amphibian populations designing a national monitoring strategy Ottawa ON: Canadian wide life service p. 76.

Bishop CA, Pettit KE (1992). Declines in Canadian amphibian populations: designing a national monitoring strategy, Minister of Supply and Services Canada, Ontario. 61: 243-250.

Broomhall S (2002). The effects of endosulfan and variable water temperature   on   survivorship   and   subsequent   vulnerability   in predation in *Litoria citropa* tadpoles. Aqua. Toxicol., 61: 243-250.

Davidson CH, Shafer B, Jennings MR (2002). Spatial tests of the pesticide drift, habitat destruction, UVB and Climate change hypothesis for California amphibian declines. Conservat. Biol. 16: 1588-1601.

Davidson CH, Shafer B, Jennings MR (2001). Decline of the California red-legged frog: climate, UVB, habitat and pesticides hypothesis Ecol. Appl. 11: 464- 479.

Eisler R (1986). Diazinon Hazards to Fish Wildlife and Invertebrates, a synoptic viewRep; 85 (1-9) US Fish and wildlife service, US Department of interior Washington DC p. 37.

Environmental Protection Agency (EPA) (1992). National Recommended Water Quality Criteria Federal Register. 57- 60848.

Environmental Protection Agency (EPA) (1996). Ecotox Thresholds Eco update Office of Solid Waste and Emergency Response. EPA 540/D-95/038.

European Inland Fisheries Advisory Commission (EIFAC) (1998). Revise report on fish toxicology testing procedures: EIFAC Tech paper, 24, Rev 1: FAO Rome p. 37.

Ezemonye LIN, Tongo I (2009). Lethal and Sublethal Effects of Atrazine to Amphibian Larvae. Jordan J. Biol. Sci. 2(1): 29-36.

Ezemonye LIN, Enuneku A (2005). Acute toxicity of cadmium to tadpoles of Bufo maculatus and Ptychedena bibroni. Pollut. Health 4(1): 13 - 20.

Ezemonye LIN, Ilechie I (2007). Acute and chronic effects of organophosphate pesticides (Basudin) to amphibian tadpoles (Ptychadena bibroni). Afr. J. Biotech. 6(11): 1554 - 1568.

Falandysz J, Chwir A (1997). The concentrations and bioconcentration factors of mercury in mushrooms from the Mierzeja Wislana sand-bar, northern Poland. Sci. Total Environ. 203(3): 221-228.

FEPA (1991). Guidelines and Standards for Environmental Pollution Control in Nigeria Federal Environmental Protection Agency, Lagos.

Finney DJ (1971). Probit Analysis Cambridge, England, Cambridge University Press pp. 333.

Fulton MH, Chambers JE (1985).The toxic and teratogenic effects of selected organophosphorus compounds on the embryos of three species of amphibians Toxicol. lett. 26: 175-180.

Hall RJ, Swineford D (1980). Toxic effects of edrin and toxaphene on the southern leopard frog Rana sphenocephala. Environ. Pollut. 23(A): 53-56.

Harri MNE, Laitinen J, Valkama EL (1979). Toxicity and retention of DDT in adult frogs, Rana temporaria L. Environ Poll. 20(1): 45-55.

Harris ML, Bishop CA, Struger J, Ripley B, Bogart JB (1998). The functional integrity of northern leopard frog (*Rana pipiens*) and green frog (*Rana clamitans*) populations in Orchard wetlands II Genetics, physiology and biochemistry of breeding adults and young-of-the year. Environ. Toxicol. chem. 17: 1338-1350.

Hecnar SD (1995). Acute and chronic toxicity of ammonium nitrate fertilizer to amphibians from southern Onatara. Environ. Toxicol. chem. 141: 2131-2157.

Hoque MM, Mirja MJA, Miah MS (1993). Toxicity of Diazinon and Sumithion to Puntus gonionotus.. Bangladesh J. Trans. Dev. 6(1): 19-26.

Houlahan JE, Fridlay CS, Schmidt BR, Mayers AH, Kuzmin SL (2001). Quantitative evidence for global amphibian population declines. Nature 404: 752-755.

Kaplan HM, Overpeck JG (1964). Toxicity of halogenated hydrocarbon insecticides for the frog. Rana pipiens Herpetologica 20: 163-169.

Khan MZ, Tabassum R, Shah EZ, Tabassum F, Ahmad I, Fatima F, Khan MF (2003). Effect of Cypermethrin and Permethrin on Cholinesterase Activity and Protein Contents in Rana tigrina (Amphibia). Turk. J. Zool. 27: 43-246.

Kolpin DW, Furlong ET, Meyer MT, Thuranan EM, Zaugg SD, Barber LB, Buxton HT (2002). Pharmaceuticals, hormones, and other organic wastewater contaminant in US Streams 1999–2000, a national reconnaissance. Environ. Sci. Technol. 36: 1202- 1211.

LeNoir JS, McConnell LL, Fellers GM, Cahill TM, Serber JN (1999). Summertime transport of current use pesticides from California's central valley of the Sierra Nevada Mountain range, USA Environ. Toxicol. Chem. 18: 2715 - 2722.

Marler PR, Hamilton WJ (1966). Mechanism of Animal Behaviour New York: John Wiley p. 345.

McConnell LL, LeNoir JS, Datta S, Seibor JN (1998). Wet deposition of current-use pesticides in the Sierra Nevada Mountain range, California, USA Environ. Toxicol. Chem. 17: 1908-1916.

Moloche L (2008). Risk assessment of contaminated soil.http://wwwprsssca/pdf.

Nowak B, Sunderam RIM (1991). Toxicity and bioaccumulation of endosulfan to Mosquito fish Gambusia affinis (Baird and Girard). Verh Internat Verein Limnol. 24: 23-29.

Relyea AR (2004). Growth and survival of five amphibian species exposed to combinations of pesticides. Environ. Toxicol. Chem. 23(7): 1737-1742.

Saglio P, Trijasse S (1998). Behavioural responses to Atrazine and diuron in goldfish. Arch Environ contain Toxicol. 35: 484-491.

Sayim F, Akyurtlakli N (1999). Acute toxicity of malathion on the 25th stage larvae of Rana ridibunda. J. Fisheries. Aquatic Sci. 16: 19-29.

Sayim F, Kaya U (2006). Effects of Dimethoate on Tree Frog, Hyla arborea Larvae. Turk. J. Zool. 30: 261-266.

Svoboda M, Luskova V, Drastichova J, Zlabek V (2001). The effect of diazinon on haematological indices of common carp (Cyprinus carpio L). Acta Vet. Brno. 70: 457-465.

Smith AG (1991). Chlorinated hydrocarbon insecticides In: Handbook of Pesticide Toxicology, Volume 2, Classes of Pesticides, Hayes, WJ and Laws, ER (Eds) Publ: Academic Press, Inc pp. 731-915.

Sparling DW, Fellers GM (2009). Toxicity of two insecticides to California, USA, anurans and its relevance to declining amphibian populations. Environ. Toxicol. Chem. 28(8): 1696-1703.

Sparling DW, Fellers GM, McConnell LS (2001). Pesticides and amphibian population declines in California, USA Environ. Toxicol. Chem. 20: 1581-1595.

Steinwandter H (1990). Contributions to the on-line method For the extraction and isolating of pesticide residues and environmental chemicals II Miniaturization of the on-line method. Fresenius J. Anal. Chem. 336: 8-11.

Sumanadasa DM, Mayuri R, Wijesinghe Ratnasooriya WD (2008). Effects of diazinon on survival and growth of two amphibian larvae. J. Natn. Sci. Foundation Sri Lanka 36(2): 165-169.

Suter GW (1993). Ecological risk assessment. Boca Raron FL: Lewis p. 538.

Volgiatzis AK, Loumbourdis N (1997). Uptake, tissue distribution and depuration of cadmium (Cd) in the frog Rana ridibunda. Bull. Environ. Contamination Toxicol. 59: 770-776.

5

An overview of the treatment of print ink wastewaters

Lichao DING, Yunnen CHEN* and Jingbiao FAN

School of Resource and Environmental Engineering, Jiangxi University of Science and Technology, Ganzhou, China.

Based on a large number of domestic and international literatures, the characteristics of print ink wastewaters such as high concentration and chroma, complex of the wastewater composition, poor biodegradabilitiy, and the present treatment of it were summarized. It has made a comment on treatment for ink wastewaters.

Key words: Print ink wastewaters, treatment, COD_{cr}, decolor.

INTRODUCTION

Ink are mainly applied in books, magazines and low-grade prints, corrugated paper and low-required printing industries, the packing of cigarette factory, brewery, pharmaceutical factory, cosmetic, children toy, milk and drink, and other carton packing industries. According to the United States (U.S.) census of printing inks, the quantity of lithographic and offset inks sold in 1992 amounted to a total of 378.6 million kg, including 48.9 million kg of sheet fed inks. The inks for engraving may be of drying oil type or of solvent type formulated with a resin and suitable solvents. Solvent inks are made by dissolving resins, such as nitrocellulose, maleic resins, vinyl acetate resin, Gilsonite, or natural resins in solvents, as xylol, toluol, and high-boiling mineral thinners.

Typically, printing ink is a complex, multi-component compound composed principally of dyes and pigments, resins, binders, solvents and optional additives. Obviously, the wastewater generated from the printing process is highly colored, and contaminated with organic minerals because of the afore-referred compounds. Hence, the wastewaters from such printing installation cannot be directly discharged into receiving streams without any treatment, not only due to its deleterious effect on human health and the environment, but aesthetically due to visibility of color even at low concentration.

Characteristics of ink wastewaters

High concentration and chroma

Ink wastewaters are kind of high concentration organic wastewaters (Cai and Zhang, 2006; Li et al., 2007). The COD_{cr} of that is generally above 20000 mg/L and sometimes more than 100000 mg/L, SS about 800-1200 mg/L, higher pH commonly, the color dark blue with high chroma which is above 100000 tmes.

Great differences of water quality

Water quality varies with the different kinds of ink, which have different connected makings, pigments and additives. So the wastewaters have great differences correspondingly. The quality and quantity of wastewaters also varies with the different process section of the ink production.

Complex of the wastewaters composition

The main pollutants in wastewater of ink production are soluble resin of acrylic, organic with color group, alcohol bases of high molecular and phenyl dispersant. Moreover, acrylic resin is main component of COD_{cr} in wastewater, which is more than 80%. And it also contains tens of additives such as stabilizers, defoaming agent,

*Corresponding author. E-mail: cyn70yellow@yahoo.com.cn.

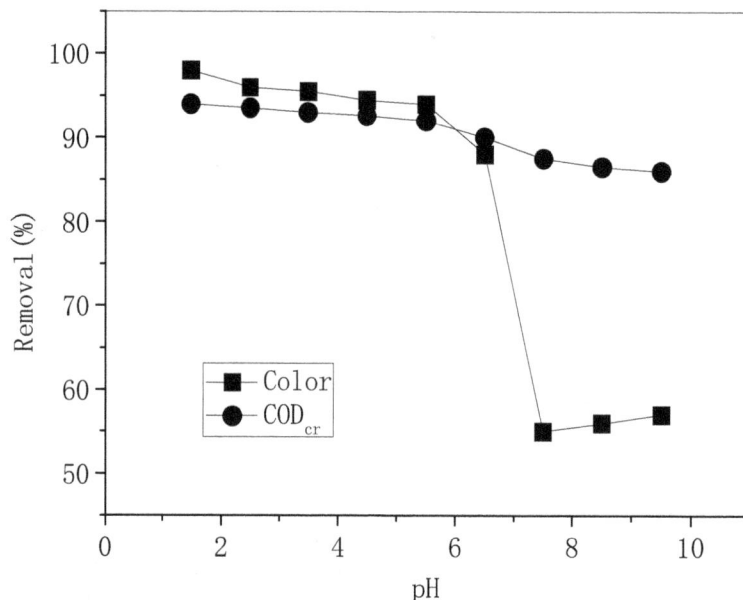

Figure 1. pH effect on the removal of ink wastewaters by PAMAM.

blockers, surfactants and preservative.

Poor biodegradabilitiy

Most of the components of wastewater are synthetic organic polymer with strong stability, which bring on the BOD_5/COD_{cr} of wastewaters is often lower than 0.4. Some wastewater has plenty of substances of inhibiting biological reaction and heavy metal ions, making for difficulty to be microbial decomposed and transformated.

Treatment methods of print ink wastewaters

Coagulation

Coagulation is a common method in wastewater treatments. Coagulation can reduce turbidity and chromaticity of ink wastewaters. It's important to choose coagulant in wastewater treatments. Considering about sediment time, decoloring rate and other factors, Wu et al., (2002) showed that COD_{cr} removal achieved 92.1%, decoloring rate achieved 97.4% after coagulation by polyferric chloride (PFC). Others (Cai and Zhang, 2006) indicated that the removal of decoloring and COD_{cr} were both low using ferrous sulfate as coagulants, while the decoloring rate can achieve 99% and COD_{cr} removal 45 to 60% using polymerization aluminium chloride (PAC). Tapas et al. (2003) demonstrated PAC was the best coagulant, in which decoloring rate, SS removal, BOD_5 removal and COD_{cr} removal achieved 95.9 to 96.5%, 96.5 to 97%, 61.3 to 65.8% and 54.8 to 61.8%, respectively.

However, due to the complexity of the actual wastewater and the strict request of the treatment effect, it needs a variety of coagulants to be used corporately. Metes et al. (2000) regarded chemical coagulation as a potential method. They selected a series of coagulants and found out the comparative effective combination: $AlCl_3 \cdot 6H_2O$ and $FeCl_3 \cdot 6H_2O$ united. Selecting $FeCl_3$ as flocculant and coagulant to treat water-based ink wastewaters, some researchers (Zhao et al., 2005) attained COD_{cr} removal 88.75% and chroma removal 99%. United PAC with Polyscrylamide (PAM) to pretreat packing and printing wastewaters, Sun et al. (2005) attained the removal of COD_{cr} and chroma above 50 and 90%, respectively. But coagulation will generate a lot of sludge which are difficult to dewater and disposing. So it will cause secondary pollution.

Adsorption

Adsorption is a process of using porous solid phase material to adsorb pollutants. Taking on pore structure and huge surface area, activate carbon is widely used to reduce chroma and COD_{cr} concentration. The results from Zhou et al. (2007) showed that active carbon has good efficiency for removing chroma and COD_{cr}. The wastewaters become colorless and transparent after adsorption. But active carbon tends to be saturated and high cost.

Zhang et al. (2010) use polyamidoamine (PAMAM) to modify zeolite and then treat ink wastewaters. The pH effect on the removal of ink wastewaters by PAMAM was shown in Figure 1 when the contact time 90 min. The chroma removal achieved 98% and COD_{cr} 93%.

Combined   poly-dimethyl-diallylammonium   chloride

Table 1. Proportion of H_2O_2 to $FeSO_4 \cdot 7H_2O$ effect on the COD_{cr} removal and decolor.

$H_2O_2/FeSO_4 \cdot 7H_2O$	4/1	3/1	2/1	1/1	1/2	1/3	1/4
COD_{cr} removal (%)	56.3	62.5	67.5	80.5	87.5	76.7	70.1
Decolor (%)	20	30	50	70	80	90	90

(PDMDAAC) and fly ash to treat ink wastewaters, some authors (Xiao et al, 2005) found the removal of chroma and COD_{cr} achieved 94 and 74%, respectively. Metes et al. (2004) used zeolites as adsorbent to treat ink wastewaters. The removal of COD_{cr} was 88% as dosage of zeolites being 5 g/L. The maximal adsorptional capacity was 34.48 mg/g When the ink wastewaters treated by adsorbent electroplate-compost (Netpradit et al., 2004).

Electrolysis method

Electrolysis method to treat wastewater mainly includes oxidation, reduction, agglomeration, air-floating in the electrolytic process to make pollutants transfer, degradation, mineralization, and reduce BOD_5, COD_{cr}, NH_3-N. Wang (2007) used iron as anode and aluminum cathode to treat print ink wastewaters under the strong electric current. The removal mechanism was iron dissolving into Fe^{2+} gradually in electrolytic process, and hydrolysis to Fe $(OH)_2$ which has agglomeration. At the same time, the cathode produces hydrogen which has strong reduction ability and occurs redox reactions with the pollutants in the wastewaters. Meanwhile macromolecular pollutants decompose into small molecules. The removal of COD_{cr} achieved 47%, BOD_5 60% and decoloring 84% after being treated.

Zhang et al. (2005) used iron electrolysis method to treat ink wastewaters. It is the comprehensive action of electroplating, Fenton action, coagulation and adsorption. Results showed chroma removal can attain above 90% and COD_{cr} about 50%. Electrolysis method has lots of advantages such as no secondary pollution, simple equipment, high chroma removal, but high cost and low COD_{cr} degradation.

Oxidation

The common oxidation methods are chemical oxidation and advanced oxidation. Oxidants in chemical oxidation are often NaClO, $KMnO_4$, O_3, $C_2H_2O_4 \cdot H_2O$. They can reduce most organic to a specified concentration, but not complete and the cost is high. Advanced oxidations such as Fenton oxidation, ultrasonic wave radiation oxidation, photo-chemical catalytic oxidation are new and effective chemical oxidation process in treating organic wastewater. The reaction mechanism is generally considered as the free radical oxidation composite oxidant, illumination, electricity or catalyst to induce and produce

various forms of strong oxidation active substances. Especially oxyhydrogen free radicals can make most of the organic pollutants completely mineralization or partial decomposition.

Ma and Xia, (2009) used Fenton combining with coagulation to treat ink wastewaters. When pH was 4.5, H_2O_2 4.5 mg/L, $FeSO_4$ 25 mg/L, and PAC 700 mg/L after a stated contact time, the removal of chroma and COD_{cr} can be achieved 100% and 93.4%, respectively. Adopted H_2O_2 and $FeSO_4 \cdot 7H_2O$ as oxidant, coordinated UV light to treat ink wastewaters, it can reduce COD_{cr} and decolor at the same time (Liang and Liang, 2005). Table 1 showed the proportion of H_2O_2 to $FeSO_4 \cdot 7H_2O$ effect on the removal of COD_{cr} and decolor at pH 4. Si and Ma, (2009) introduced $FeCl_3$, CaO, ultraviolet radiation, TiO_2 at the same time to deal with the ink wastewaters. The removal of chroma and COD_{cr} can all be attained some 90%.

Chen (2010) used UV-Fenton to treat ink wastewaters from circuit board production which COD_{cr} removal achieved 92.3%. Using ultrasound (US) combining with Fenton oxidation to treat ink wastewaters (Chua and Loh, 2008), the experimental results (Figure 2) showed the treating effective for COD_{cr} from ink wastewaters by US-Fenton was higher than that of both simple sum, which COD_{cr} removal achieve 81.4% (He et al., 2009).

Biological method

Due to the poor biodegradabilitiy of ink wastewaters, biologiocal method is generally used as deeply treatment. Liao et al. (2009) adopted two-stage SBR process to treat the ink wastewater after pretreatment. The results showed that COD_{cr} removal remained above 93% and decoloring rate 80%. Using UASB reactor showed that the ink wastewaters after degradating in the sludge nitrification process was feasible and the decoloring rate can achieve 80% (Wu et al., 2005).

Thanks to the complexity of the actual ink wastewaters, it always combines several methods to treat ink wastewaters. Liang (2003) chose air-float-biochemical process to treat high concentration of printing ink wastewaters which made up the unsatisfactory effect of physic-chemical and biochemical. Some (Liu and Cui, 2001) adopted coagulation-air-float-micro-SBR process to treat carton packaging industry ink and adhesives wastewater to achieve National Level of Discharging Standard. Ceng et al. (2010) used sulfide precipitation – neutralization precipitation - coagulation precipitate – catalytic oxidation

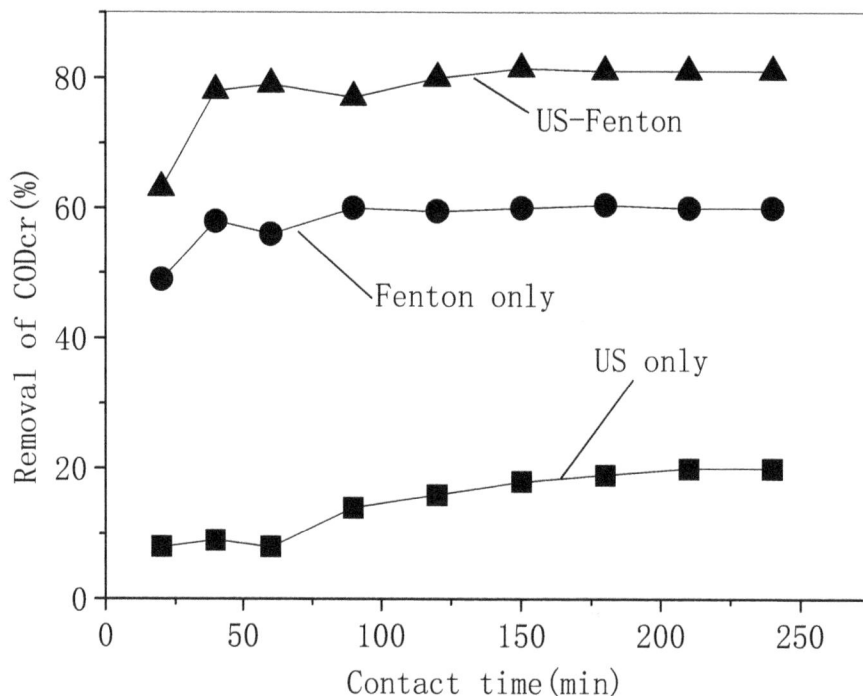

Figure 2. COD_{cr} removal treated by different methods.

combination process to treat high concentration ink wastewaters from print circuit board (PCB) manufacturer. The experimental results showed that the outlet of COD_{cr} was less than 90 mg/L, $\rho(Cu^{2+})$ less than 0.5 mg/L, SS less than 50 mg/L.

Some insufficiency on the treatment of ink wastewaters

Consulting larges of literatures from oversea and domestic can be known that most researchers saw about chroma and COD_{cr} as evaluating indexes for treating ink wastewaters. Few looked heavy metals and benzene as treating objects. Although the concentration of heavy metals and benzene is low, it is harmful to the water environment.

While there are a lot of reports about the technology of dealing with ink wastewaters, most of technologies (Cenek, 2004; Barredo et al., 2005; Zhang, 2004) are rest on the laboratory. It has a long way for industrialization.

Summary

The demand for ink will increases with the progress of social economy. In current, there are a lot of treatment methods for ink wastewaters. But deficiencies still exist for each kind of method. So it should continue to seek more effective and low cost treatment methods for print ink wastewaters.

REFERENCES

Barredo DS, Iborra-Clar MI, Bes-Piaeta A (2005). Study of preozonation influence on the physical-chemical Treatment of textile wastewater, Desalination, 182: 267-274.

Cai Y, Zhang Z (2006). Study on Water-Based Ink Wastewater Treatment with the Process of Flocculation Sedimentation and Biocontact Oxidation. Shanghai Chem. Ind., 31(4): 13-17.

Cai Y, Zhang Z (2006). Environmental Water-Based Ink and Its Wastewater Treatment. Shanghai Chem. Ind., 31(5): 23-26.

Cenek N (2004). Biodegeradation of synthetic dyes by Irpex Lacteus under various growth conditions. Int. Biodeterioration Biodegeradation, 54: 215-223.

Ceng X, Lu Z, Wang C, Wu R, Su Z (2010). Treatment of wastewater containing high-concentration printing ink from printed circuit board plant. Environ. Protection Chem. Ind., 30(1): 56-58.

Chen X (2010). Study on Treatment Technology of Ink Wastewater in Printed Circuit Board Manufacturing Process. Environ. Sci. Technol., 23(1): 30-32.

Chua CY, Loh KC (2008). Ultrasound-Facilitated Electro-Oxidation for Treating Cyan Ink Effluent. Canadian J. Chem. Eng., 86: 739-746.

He D, Qin Y, Wang W, Song D, Liang D, Du L (2009). Experimental research on treatment of ink wastewater by combination technology of ultrasonic irradiation and Fenton oxidation. J. Central South University. Sci. Technol., 40(6): 1482-1487.

Li Y, Cheng J, Shi W, Wang C (2007). Printing-ink and Environmental Protection. J. Xi'an University of Arts Sci. (Natural Science Edition), 10(2): 111-114.

Liang J (2003). Pispose printing ink waste water by air blow aeration-biochemstry techmics. Environ. Technol., pp. 52-54.

Liang W, Liang J (2005). Studies on the treatment of ink wastewater. Environment, z1: 83-85.

Liao S, Hu S, Shang J (2009). Treatment of Water- based Ink Wastewater by Using SBR. J. Dali Univ., 8(8): 46-49.

Liu L, Cui Y (2001). Treatment of wastewater of printing ink and adhesive by coagulative floatation-microelectrolysis-sbr process. Environ. Eng., 19(5): 16-18.

Ma XJ, Xia HL (2009). Treatment of water-based printing ink

wastewater by Fenton process combined with coagulation, J. Hazard. Mat., 162: 386-390.

Metes A, Koprivanac N, Glasnovic A (2000). Flocculation as a treatment method for printing ink wastewater. Water Environ Res., 72(6): 680-688.

Metes A, Kovacevic D, Vujevic D, Vujevic D (2004). The role of zeolites in wastewater treatment of printing inks, Water Res., 38: 3373-3381.

Netpradit S, Thiravetyan P, Nakbanpote W, Rattanakajhonsakul K, Tantarawong S, Jantarangsri P (2004). Waste metal hydroxide sludge as a new adsorbent, Environ. Eng. Sci., 21(5): 575-582.

Si AE, Ma EK (2009).Treatment of ink wastewater via heterogeneous photocatalytic oxidation. Desalination and Water Treatment. 7(3): 1-5.

Sun L, Miao Q, Liu Z (2005). Experimental study on pretreatment of package printing wastewater by coagulation precipitation technology. Water Resources Protection, 21(1): 75-78.

Tapas N, Sunita S, Pathe P, Kaul S (2003). Pre-treatment of currency printing ink wastewater through coagulation-flocculation process. Water, Air Soil Pollut., 148: 15-30.

Xiao J, Qian J, Lai X, Tong H, Jiang X (2005). Research on PDMDAAC-modified-fly ash in the Pre-treatment of Printing-ink Wastewater. Sichuan Environment. 24(4): 14-16.

Wang W (2007). Application of Electrolysis to Ink Sewage Treatment. J. Maoming Univ., 17(3):15-18.

Wu D, Luo Y, Fu W, Wu Z (2002). Study on Treatment of Printing Ink Wastewater and Its Effect on the Water's Chroma by Coagulation Method. Environ. Protection Sci., 28(113): 16-19.

Wu G, Yang R, Ceng J (2005). Test of sludge digestion process for treatment of printing ink wastewater. Ind. Water Wastewater, 36(4): 29-31.

Zhang GJ, Liu ZZ, Song L, Hu J, Ong S (2004). One-step cleaning method for flux recovery of an ultrafiltration membrane fouled by banknote printing works wastewater, Desalination, 170: 271-280.

Zhang G, Shen Y, Li W, Zhu X (2010). Preparation of PAMAM-modified Zeolite and Application in Wastewater Treatment of Printing Ink. Leather Chem., 27(1): 36-39.

Zhao T, Hu S, Zhou D (2005). Coagulation Process for Treating Wastewater Containing Water-based Ink. Environ. Sci. Technol., 28(3): 93-95.

Zhou D, Zhao T, Hu S (2007). Adsorbing Properties of Activated carbon for Treating Water-based Ink-Manufacturing Wastewater. Environ. Sci. Technol., 30(3): 85-86.

6

"Polyamine levels and pigment contents in rapeseed regenerated *in vitro* in the presence of zinc"

Asma Ben Ghnaya[1,2*], Annick Hourmant[1,2,3], Michel Couderchet[4], Michel Branchard[1,2] and Gilbert Charles[1,2,5]

[1]Université Européenne de Bretagne, France.
[2]Laboratoire de Biotechnologie et Physiologie Végétales, ESMISAB, Université de Bretagne Occidentale Technopôle Brest-Iroise, 29280 Plouzané, France.
[3]Laboratoire de Toxicologie Alimentaire et Cellulaire, Université de Bretagne Occidentale, UFR Sciences et Techniques, C.S. 93 837, 29238 Brest Cedex 3, France.
[4]Laboratoire Plantes Pesticides et Développement Durable, Université de Reims Champagne-Ardenne, URVVC-SE, BP 1039, 51687 Reims, France.
[5]Laboratoire d'Ecophysiologie et Biotechnologie des Halophytes et Algues Marines (LEBHAM), Université de Bretagne Occidentale, IUEM, Technopole Brest Iroise, Place Nicolas Copernic, 29280 Plouzané, France.

The present work was designed to select hyperaccumulator plants by *in vitro* breeding from transversal thin cell layers. The thin layer explants of *Brassica napus* cv. Jumbo were cultivated directly in the presence of $ZnSO_4$ at different concentrations (0.1 to 1 mM). Regenerated shoots were transferred into pots, acclimatized, and cultured in the greenhouse. After 3 weeks, the regenerants were treated with 2 mM of $ZnSO_4$ during 7 days. Growth parameters, guaiacol peroxidase activity (GPOX), malondialdehyde (MDA), polyamines (PAs) and pigment contents were evaluated. $ZnSO_4$ applied during the regeneration process influenced significantly both physiological and biochemical characteristics of regenerated plants. Particularly, an increase of Zn level in the tissue culture induced an increase in MDA content and GPOX activity in the leaves and a decrease in pigment contents. Exposure to 2 mM Zn induced 112% increase in free polyamines in the leaves and roots, while this increase was as high as 399% for plants regenerated in the presence of the metal.

Key words: Phytoremediation, hyperaccumulator, *in vitro*, polyamines, malondialdehyde, guaiacol peroxidase, zinc stress.

INTRODUCTION

Soil contamination by heavy metals is a consequence of industrial activities and phytoremediation appears as cheap and environmentally friendly method to restore soil quality (Pilon-Smits, 2005). However, the efficacy of this technique could be improved if plants were better

adapted to grow on sites rich in pollutants. A number of research works are being carried out aiming at increasing the ability of plants to satisfy this application, including biotechnology and genetic engineering (Rugh et al., 1996; Pilon-Smits, 2005). In this respect, *in vitro* culture and selective pressure were proposed to develop more efficient phytoremediation system through the selection of hyperaccumulator plants to remediate polluted soils (Ghnaya et al., 2007).

In an earlier study, plants regenerated on metal rich media were selected and their ability to accumulate metals and some morphological and physiological characteristics were described. However, mechanisms by which these regenerated plants could overcome metal excess effects were not studied. Heavy metals such as

---

*Corresponding author. E-mail: benghnaya_asma@yahoo.fr.

Abbreviations: BAP, 6-benzylamino-purine; **GPOX,** guaiacol peroxidase; **MDA,** malondialdehyde; **MS,** Murashige and Skoog's medium (1962); **NAA,** α-naphthaleneacetic acid; **PAs,** polyamines; **Put,** putrescine; **ROS,** reactive oxygen species; **Spd,** spermidine; **Spm,** spermine; **tTCL(s),** transverse thin cell layer(s).

Zn are essential for normal plant growth and development since they are constituents of many enzymes and other proteins. However, high concentrations of both essential and non essential heavy metals in soils can induce toxicity symptoms and growth inhibition in most plants (Hall, 2002). Excessive amount of heavy metal may stimulate the formation of free radicals and reactive oxygen species, often resulting in oxidative stress (Ferrat et al., 2003). The antioxidative responses of plants exposed to metals and other stress factors have been reviewed (Sharma and Dietz, 2009). In resistant forms, stress conditions may enhance protective processes such as accumulation of compatible solutes and increase detoxifying enzymes activities. Malondialdehyde (MDA) is a cytotoxic product of lipid peroxidation and an indicator of free radical production and consequent tissue damage, both in plants and animals (Ohkawa et al., 1979; Wood et al., 2006; Zhang et al., 2008).

Besides, plant antioxidative defense response can be evaluated by analyzing the activity of antioxidant enzymes such as guaiacol peroxidase (GPOX, EC 1.11.1.7) or cell antioxidant contents. Among substances able to protect plant cell from oxidative attack, a specific role has been attributed to polyamines (PAs) in preventing photo-oxidative damages (Franchin et al., 2007). The antioxidative effect of polyamines is due to a combination of their anion and cation-binding properties involving a radical scavenging function (Bors et al., 1989) and capability to inhibit both lipid peroxidation (Kitada et al., 1979) and metal-catalysed oxidative reactions (Tadolini, 1988). Moreover, polyamine catabolism produces hydrogen peroxide that can enter the stress signal transduction chain, promoting the antioxidative defense response activation (Agazio and Zacchini, 2001).

In the present study, we characterize how rapeseed plants, obtained, as described earlier (Ghnaya et al., 2007), cope with Zn exposure. We especially analyze oxidative stress markers in response to subsequent metal exposure. The influence of high zinc concentration (up to 1 mM) during regeneration of *Brassica napus* cv Jumbo and of subsequent Zn application (2 mM) will be presented on accumulation of the metal and on antioxidative defense responses such as PAs and MDA contents, GPOX activity and also chlorophyll and carotenoid contents.

MATERIALS AND METHODS

Plant material and zinc treatments

Experiments were carried out on rapeseed (*B. napus* cv. Jumbo) provided by National Institute of Agronomic Research (INRA), Rennes (France). This genotype was a pure, genetically fixed line and was obtained by autofertilization. *Brassica napus* L. was a plant with an important biomass and a deep root system, known for its capacity to accumulate metals (Ghnaya et al., 2009).

Seeds, sterilized as previously described (Ghnaya et al., 2007), were sown in test tubes on MS medium (Murashige and Skoog, 1962) supplemented with 2% (w/v) sucrose and 0.65% (w/v) agar at pH 5.8. Seeds were allowed to germinate for 2 weeks under

conditions detailed earlier (Ghnaya et al., 2008). Then, transverse thin cell layers (tTCLs of 0.3 to 0.5 mm) were excised from hypocotyls and petioles and placed in Petri dishes on MS medium containing 0.3 mg L^{-1} of 1-naphthaleneacetic acid (NAA), 3 mg L^{-1} of 6-benzylaminopurine (BAP), sucrose (3%) and agar (0.65%), supplemented with various concentrations of $ZnSO_4$ (0 to 1 mM final concentration) dissolved in MES-buffer, pH 5.8.

The plants regenerated from tTCLs, without $ZnSO_4$ (control) or in the presence of $ZnSO_4$, were acclimatized under a 16 h photoperiod, at 22 ± 1 °C. The surviving plants were transferred to 750 ml pots containing sterile compost and cultured in a greenhouse. During three weeks, each pot was watered twice a week, alternating water and Hoagland's nutrient solution (Hoagland and Arnon, 1950).

Subsequently, the three-week regenerants were treated or not every day with 2 mM $ZnSO_4$ supplied to the roots (150 ml per pot). After 7 days, the plants were harvested, washed and sampled for analysis. Every sample was repeated 6 times. The 5 plants from each pot were pooled and used for analysis of metal or pigment contents (plants from 3 pots for each type of analysis and chemical assays.

Determination of growth and zinc accumulation

Fresh weight of leaves, roots and stems of the control and treated plants were measured after harvest and dry weight was estimated by heating tissues at 70 °C for three days. Two hundred milligrams of dried plant tissues were ground up and digested by a mixture of HNO_3, HF and H_2O_2 (4/3/1, v/v/v).

Then, the vessels were closed and exposed to microwave digestion as described by Weiss et al. (1999). Zinc content was measured by electrothermal atomic absorption spectrometer (Perkin-Elmer SIMAA 6100, USA) operated in the single-element mode. All solutions were prepared with high-purity water from a MilliQ-system (Millipore, Milford, MA, USA).

Pigment content

The pigment content was determined using the method of Lichtenthaler (1987). Leaves samples (1 g fresh weight) from each plant were ground in a mortar in 100% acetone. The extracts were centrifuged at 5000 rpm for 10 min at 4 °C before reading the absorbance at 470, 663 and 645 nm.

Lipid peroxidation (MDA)

The level of lipid peroxidation was determined using malondialdehyde (MDA) as a marker. Its content was measured according to Minotti and Aust (1987) and Iturbe-Ormaetxe et al. (1998). About 1 g leaf fresh tissue was homogenized with 6 ml of metaphosphoric acid (5%, w/v) and 120 µl of 2% butyl hydroxytoluene (in ethanol). After 30 min at 4 °C, homogenate was centrifuged at 5000 rpm for 20 min.

The chromogen was formed by mixing 4 ml supernatant with 400 µl 2% butyl hydroxytoluene, 2 ml of 1% (w/v) thiobarbituric acid (in 50 mM NaOH) and 2 ml of 25% HCl. The reaction mixture was heated for 30 min at 95 °C and then rapidly cooled in ice. Chromogen was extracted by adding 1.5 ml of 1-butanol. After 30 s vortexing, organic phase was separated by centrifugation (5000 rpm, 5 min) and the thiobarbituric acid reactive-substances (TBARS) determined by measuring absorbance at 532 nm. TBARS concentration was calculated using the extinction coefficient of 155 mM^{-1}cm^{-1}.

Table 1. Effect of $ZnSO_4$ (2 mM) treatments on *B. napus* (cv. Jumbo) growth and accumulation of zinc in roots (R), stems (S) and leaves (L) of plants regenerated in the presence of (0.1-1 mM) or absence of $ZnSO_4$ and cultivated in greenhouse.

ZnSO₄ treatments (mM)		Plant growth						Zn content (µg.g⁻¹ DW)		
During *in vitro* regeneration	After regeneration	Fresh weight (g)			Dry weight (g)					
		L	S	R	L	S	R	L	S	R
0.00	0.00	7.88[a]	4.63[b]	1.25[d]	0.60[ab]	0.21[b]	0.14[b]	268.00[e]	222.33[f]	171.33[f]
0.00	2.00	6.47[b]	3.91[c]	1.70[b]	0.52[b]	0.22[b]	0.22[a]	429.67[d]	693.00[d]	1821.33[d]
0.10	2.00	7.95[a]	4.92[a]	1.49[c]	0.68[a]	0.30[a]	0.19[a]	519.67[d]	545.33[e]	751.00[e]
0.25	2.00	7.80[a]	4.73[b]	1.38[d]	0.61[ab]	0.27[a]	0.20[a]	912.67[c]	929.00[c]	2168.33[c]
0.50	2.00	5.03[c]	5.08[a]	0.90[e]	0.37[c]	0.29[a]	0.21[a]	1069.00[b]	2215.33[b]	5368.00[b]
1.00	2.00	4.67[c]	2.19[d]	1.90[a]	0.53[b]	0.20[b]	0.12[b]	1947.67[a]	4466.67[a]	9089.00[a]

The results were calculated from three replicated experiments. In each column, the values with different letters are significantly different according to ANOVA and Duncan's test at the 5% level.

Guaiacol peroxidase activity (GPOX, EC 1.11.1.7)

Fresh plant leaves, stems or roots (500 mg) were homogenized in 6 ml of cold 100 mM potassium phosphate buffer (pH 6.5) and centrifuged at 5000 g for 10 min. Guaiacol peroxidase activity was assayed in a mixture containing 2.7 ml of 0.1 M K-phosphate buffer (pH 6.0),100 µl of 0.18 M guaiacol and 100 µl of 0.03 M hydrogen peroxide (MacAdam and Sharpe, 1992). Reaction was initiated by adding 50 µl of supernatant, and increase in A $_{470}$ min⁻¹ was measured. The amount of oxidized guaiacol was calculated from the extinction coefficient of 26.6 mM⁻¹ cm⁻¹. Enzyme activity was expressed as µmol of min⁻¹ g⁻¹ DW.

Polyamine analysis

Extraction and dansylation of free polyamines were performed using the procedure described by Flores and Galston (1982) and slightly modified by Féray et al. (1992). Tissues were extracted in 5% (w/v) cold $HClO_4$. The mixture was kept at 4°C for 1 h and then pelleted at 5000 g for 15 min. Aliquots (200 µl) of the supernatant were mixed with 200 µl saturated sodium carbonate and 200 µl dansyl chloride in acetone (5 mg. ml⁻¹). The samples were vortexed (30 s) and incubated overnight at room temperature. Dansylated polyamines were extracted in 2 × 500 µl benzene, 200 µl were evaporated and re-dissolved in 500 µl methanol for analysis. Standards were treated in the same way.

The HPLC system consisted of two solvent pumps (Kontron 422) coupled to a high performance mixer, 7125 Rheodyne injection valve fitted with 20 µl loop and stainless precolumn spheri ODS 5 mm (Brownlee) (30 × 4.6 mm) in conjunction with stainless steel analytic column (Ultrasphere). The mobile phase consisted of 0.01 M NaH_2PO_4, pH 4.4 (A) and methanol/ acetonitrile (50/50, v/v) (B) mixed at ambient temperature to produce a gradient flow rate of 1 ml. min⁻¹ with 5 min B, 80%; 2 min B, 80 to 89%; 5 min B, 89 to 100%; 3 min B, 100 to 80%; 10 min B, 80%. The gradient and data analysis were controlled by a microcomputer data system (Kontron Inst.). SFM 25 spectrofluorimeter (Kontron) was used to detect with excitation at 360 nm and emission at 510 nm.

Statistical analysis

The experiments were realised according to randomized block design. The values in the tables were compared by analysis of variance (ANOVA) and the differences among means (5% level of significance) were tested by the Duncan's Multiple Range Test using StatGraphics Plus 5.1.

RESULTS

Plant growth and biomass

Control plants that were not exposed to Zinc (neither during nor after regeneration) showed the highest leaf fresh and dry weight and lowest root weight (Table 1). The 2 mM of $ZnSO_4$ greenhouse treatment of plants regenerated in the absence of $ZnSO_4$ induced a significant decrease in fresh weight of aerial parts and increase in root fresh weight (Table 1).

Dry weight was increased in roots and not affected in leaves and stems (Table 1). The aerial parts of plants regenerated in the presence of 0.1 or 0.25 mM of $ZnSO_4$ and treated with 2 mM of $ZnSO_4$ in the greenhouse revealed significantly higher fresh and dry weights than those regenerated without $ZnSO_4$ (Table 1),whereas in roots, the fresh weight was decreased and the dry weight remained unchanged.

In contrast, the leaves of plants regenerated in the presence of higher Zn concentrations(0.5 or 1 mM) showed significant decrease of leaves fresh, but for dry weight we observed the significant decrease only with 0.5 mM. The other organs were less affected (Table 1).

Zinc accumulation

The lowest content of zinc in control plants was

Table 2. Effect of $ZnSO_4$ (2 mM) treatments on leaf pigment contents (Chlorophyll and Carotenoid) of *B. napus* (cv. Jumbo) regenerated in the presence (0.1-1 mM) or absence of $ZnSO_4$.

ZnSO$_4$ treatments (mM)		Leaf pigments (µg g^{-1} FW)			
		Chlorophyll			Carotenoid
During *in vitro* regeneration	After regeneration	C_a*	C_b	C_{a+b}	C_{x+c}
0.00	0.00	196.22[a]	59.16[b]	255.39[a]	56.31[a]
0.00	2.00	163.57[b]	62.33[b]	225.90[b]	50.87[b]
0.10	2.00	207.13[a]	75.87[a]	282.99[a]	62.89[a]
0.25	2.00	195.50[a]	78.20[a]	273.67[a]	60.83[a]
0.50	2.00	114.25[c]	51.28[b]	165.53[c]	40.79[c]
1.00	2.00	93.07[d]	68.42[a]	161.49[c]	35.67[c]

*C_a= Chlorophyll a, C_b = chlorophyll b, C_{a+b} = total chlorophyll and C_{x+c} = carotenoids. Data are means of 3 independent experiments. In each column, the values with different letters are significantly different according to ANOVA and Duncan's test at the 5% level.

observed in roots (171 µg g^{-1} DW) while the highest (268 µg g^{-1} DW) was in the leaves (Table 1). Exposing the plants to 2 mM of $ZnSO_4$ in the greenhouse resulted in an increase in zinc content by ten folds in roots, three folds in stems, and only one and a half fold in leaves.

The plants regenerated *in vitro* in the presence of 0.1 mM of Zn resulted in an accumulation of zinc lower than in plants that had been regenerated in the absence of zinc. In these plants, zinc content in roots (751 µg. g^{-1} DW) was less than half the content of plants regenerated without zinc (1821 µg g^{-1} DW). When the thin layers were submitted *in vitro* to the presence of higher Zn concentrations (0.25, 0.5 and 1 mM), the zinc content of regenerated plants was significantly higher and proportional to zinc concentration in regeneration medium and was most pronounced in the roots where it reached 9089 µg. g^{-1} DW.

The concentration of zinc in stems and roots relative to the one of the leaves of plants that had been regenerated with 0.5 or 1 mM zinc is comparable to that of plants regenerated in the absence of the metal, the roots containing from 424 to 502% of the amount in leaves. In contrast, when plants were regenerated with low $ZnSO_4$ concentration, the relative distribution of zinc in the tissue was comparable to that observed in control plants.

Chlorophyll and carotenoid contents

Exposure of plants regenerated in the absence of $ZnSO_4$ to 2 mM $ZnSO_4$ resulted in a decrease in chlorophyll a and carotenoid contents (Table 2). By contrast, when plants were regenerated *in vitro* in the presence of 0.1 and 0.25 mM of $ZnSO_4$. Chlorophyll a, b and carotenoid contents were similar or higher than in plants regenerated without the metal. Increasing Zn concentration (0.5 and 1mM) in the regeneration medium caused a significant decrease in chlorophyll a and in carotenoid contents in regenerated plants leaves (Table 2).

Lipid peroxidation (MDA)

The occurrence of oxidative stress induced by zinc treatment was monitored by analyzing membrane damage through measurement of MDA levels: exposure of plants regenerated in the absence of $ZnSO_4$ to 2 mM of $ZnSO_4$ in the greenhouse resulted in a significant increase of MDA, whose content has doubled (Table 3). This increase was even more important when the plants were regenerated *in vitro* in the presence of Zn (0.1 to 1 mM), the maximum (39.92 nmol.g^{-1} fresh weight) being for plants regenerated with 1 mM Zn, approximately 4 times more than the untreated control (Table 3).

Guaiacol peroxidase activity

GPOX activity was significantly higher in all plant parts after a 2 mM $ZnSO_4$ treatment in the greenhouse than in those of untreated plants (Table 3). This activity was further increased in the leaves of plants regenerated in the presence of 0.1 or 0.25 mM Zn (reaching 202% of untreated control activity for 0.25 mM) and recovered the levels of normal untreated plants with 0.5 and 1 mM Zn. In the shoots, the GPOX activity was reduced proportionally to the Zn level and with 1 mM, the activity was similar to the one of unstressed plants. The highest activity was observed in roots and the selection *in vitro* (0.1 to 0.5 mM Zn) produced plants with normal activity (similar to unstressed plants), but it was strongly reduced with 1 mm Zn during regeneration.

This stimulation was less important in stems and roots of plants regenerated *in vitro* in the presence of zinc. In these plant organs, the higher the zinc concentration the lower the GPOX activity was for 1 mM Zn the activity was down to untreated control activity in stems and 56% less than untreated control activity in roots.

In the case of leaves, activity of GPOX in response to 2 mM Zn was further stimulated when the plants had been

Table 3. Effect of ZnSO$_4$ treatments (2 mM) on GPOX activity of leaves (L), stems (S), and roots (R) and on MDA content of *B. napus* cv Jumbo regenerated in the presence (0.1-1 mM) or in the absence of ZnSO$_4$.

ZnSO$_4$ treatments (mM)		GPOX(μmol min^{-1} g^{-1}DW)			MDA(nmol g^{-1} FW)
During *in vitro* regeneration	After regeneration	L	S	R	L
0.00	0.0	3.12^c	6.89^d	34.30^b	9.77^e
0.00	2.00	5.02^b	10.28^a	40.88^a	18.16^d
0.10	2.00	5.69^a	9.11^b	34.63^b	25.27^c
0.25	2.00	6.29^a	8.71^c	29.51^b	32.70^b
0.50	2.00	3.85^c	7.50^c	30.37^b	34.79^b
1.00	2.00	2.92^c	5.95^d	15.17^c	39.92^a

The results were calculated from three replicated experiments. In each column, the values with different letters are significantly different according to ANOVA and Duncan's test at the 5% level.

regenerated with 0.1 and 0.25 mM Zn reaching 202% of untreated control activity for 0.25 mM. Beyond that, GPOX activity in plants regenerated in the presence of 0.5 and 1 mM of ZnSO$_4$ was no longer stimulated by a treatment with 2 mM of ZnSO$_4$ in the greenhouse (Table 3).

Free polyamines

Non metal treated plants (neither during nor after regeneration) presented a gradation in the distribution of free polyamines, concentration increasing from leaves to roots. The richest organs contained 163.82 nmol. g^{-1} DW PAs (Table 4).

In all organs, spermidine (Spd) was the predominant polyamine whereas spermidine (Spm) was the less abundant. Watering the control plants in the greenhouse with ZnSO$_4$ (2 mM) induced a significant increase in the levels of the 3 PAs in leaves and roots (respectively + 115 and + 100%) but not in stem fraction (Table 4). When the plants were regenerated *in vitro* in the presence of ZnSO$_4$, the total concentration of PAs in the leaves was significantly increased (Table 4).This increase was observed for all the plants treated *in vitro*, with the maximum for those exposed to 0.1 mM during regeneration (+ 402% vs. treated control). This increase of global PAs levels resulted from higher levels of Put, Spd and Spm.The analyses of stems, showed that greenhouse exposure to the metal induced no significant change in the level of total (Table 4) and individual PAs. Nevertheless, the addition of 0.25 and 1 mM of ZnSO$_4$, in the regeneration medium induced an increase in the total PAs due to a shift in Spd content (+ 28 and 49%, respectively).

In the roots of plants regenerated in the presence of 0.1 or 0.25 mM ZnSO$_4$ a further increase in the level of total and individual Pas was observed (Table 4). This stimulating effect of the *in vitro* selection was maximum for 0.25 mM of ZnSO$_4$, but inversely the PAs content was diminished for 0.5 and 1 mM ZnSO$_4$ (+ 28 and + 48%,

respectively). Furthermore, for high concentrations of 0.5 and 1 mM of ZnSO$_4$ in the regeneration medium, the increase of PAs contents in response to the subsequent Zn treatment was lower than observed for control plants treated in the greenhouse with the metal (-36 and 26%, Table 4). This was mainly due to a reduction in Put and Spd.

DISCUSSION

In this study, some effects of Zn treatment on plants regenerated *in vitro* from tTCLs with or without the metal were investigated. The results indicate that rapeseed metabolism was strongly perturbed when the plants were exposed to zinc treatment during regeneration and the responses were dependent on the ZnSO$_4$ concentration. Toxic effect of Zn was evident from curtailed growth and reduced biomass. When ZnSO$_4$ (2 mM) was applied to plants regenerated in the absence of ZnSO$_4$ and cultivated in the greenhouse, reduction in biomass was drastic in leaves (Table 1). Growth inhibition was consistent with data reported for zinc in Brassicacae andother species (Alia and Saradhi, 1995; Gisbert et al., 2006; Ghnaya et al., 2009). Biomass distribution between roots and leaves may be a plant strategy to tolerate high soil metal contents (Audet and Charest, 2008).

However, the treatment was applied to plants regenerated *in vitro* in the presence of 0.1 or 0.25 mM of ZnSO$_4$, the growth reduction was totally alleviated. This alleviation could not be explained by reduced zinc accumulation in plant organs except for roots of plants regenerated with 0.1 mM Zn. When ZnSO$_4$ was applied to plants regenerated in the absence of ZnSO$_4$, a significant increase in the Zn content in the whole plant was observed. Accumulation of metal was maximal in roots (+963%). Below this value, stems (+212%) accumulated more than leaves (+60%). Such metal-partition in plants is generally observed and is considered a way to avoid the metal (Audet and Charest, 2008).

Although it is not possible to exclude that zinc has

Table 4. Effect of ZnSO$_4$ greenhouse treatments (2 mM) on total polyamine content (PAs) of leaves (L), stem (S), and roots (R) of *B. napus* cv Jumbo regenerated in the presence (0.1-1 mM) or in the absence of ZnSO$_4$.

ZnSO$_4$ treatments (mM)		PAs (nmoles g^{-1} DW)		
During *in vitro* regeneration	After regeneration	L	S	R
0.00	0.0	84.55[f]	101.64[c]	163.82[f]
0.00	2.00	181.57[e]	105.52[c]	328.31[c]
0.10	2.00	424.35[a]	104.29[c]	390.77[b]
0.25	2.00	339.44[c]	130.22[b]	511.79[a]
0.50	2.00	274.44[d]	104.66[c]	209.98[e]
1.00	2.00	365.82[b]	170.21[a]	243.25[d]

The data are means from three independent experiments. In each column, the values with different letters are significantly different according to ANOVA and Duncan's test at the 5% level.

mainly accumulated in root apoplasm, it should be noted that this high zinc concentration in the roots (1821 µg.g^{-1} DW) was associated with a negligible , indicating a relative tolerance of this cultivar (cv. Jumbo) to zinc treatment. When ZnSO$_4$ was applied to plants regenerated *in vitro* in the presence of 0.1 mM of ZnSO$_4$, a significant decrease of Zn accumulation was observed, mainly in roots and also in stems, but not in leaves.

Nevertheless, for plants regenerated in presence of 0.25 mM of ZnSO$_4$ and more, accumulation of zinc increased with the increase of the ZnSO$_4$ concentration in the regeneration medium. A considerable elevation of zinc content was observed in the different organs but particularly in roots. An exclusion mechanism may have been positively regulated as a consequence of the presence of low zinc concentrations during regeneration. It could also be proposed that plants habituation to low zinc during regeneration resulted in a partition of biomass and metal different from that observed without zinc habituation and modelized by Audet and Charest (2008) in case of high soil contamination. At higher concentration *in vitro*, the further accumulation of zinc is no more alleviated but clearly enhanced.The treatment of plants with 2 mM of ZnSO$_4$ in the greenhouse diminished significantly chlorophyll and carotenoid contents as reported in other plant systems for zinc (Monnet et al., 2001; Singh and Sinha, 2005) and other metals (Chatterjee and Chatterjee, 2000; Zengin and Munzuroglu, 2005; Groppa et al., 2007).

Chlorophyll content's reduction may be attributed to inhibition of their synthetic enzymes (Stobart et al., 1985). As discussed below, since zinc induced an oxidative stress in plants, it is also very likely that this stress was in turn responsible for pigment destruction (Singh and Sinha, 2005). The regeneration *in vitro* in the presence of 0.1 and 0.25 mM of ZnSO$_4$ induced a significant increase in chlorophyll and carotenoid contents.

This finding is consistent with previous studies (Ferrat et al., 2003; Ghnaya et al., 2007) showing higher carotenoid content in plants subjected to zinc and various stresses. Carotenoid, as non enzymatic antioxidant,

would play a vital role in the protection of plants from the adverse impact of reactive oxygen species (Halliwell, 1987). Similarly, stress of low intensity was already found to result in higher chlorophyll content (Teisseire et al., 1999) and formation of shade type chloroplasts (Lichtenthaler, 1984). That this increase was observed in case of substances inhibiting non-cyclic electron transport during photosynthesis would confirm an effect of zinc on this process as observed for *Lolium perene* (Monnet et al., 2001).

However, increasing Zn concentration (0.5 and 1 mM) in the regeneration medium was responsible of significant inhibition in chlorophyll and carotenoid contents in leaves from regenerated plants (Table 2). This strong reduction of pigments may be the result of partial destruction due to a strong production of ROS as suggested by GPOX activity and MDA contents. Indeed, MDA accumulated in response to Zn treatment and was further accumulated in a concentration dependent manner when Zn was present during regeneration. MDA is a major cytotoxic product of lipid peroxidation and as such it is an indicator of free radical production (Ohkawa et al., 1979; Ferrat et al., 2003). In the regenerated plants, the higher production of MDA, that increased with Zn concentration applied *in vitro* during regeneration, indicates an enhanced level of lipid peroxidation and a higher ROS production, which are also in accordance with the metal accumulation observed in the leaves.

Antioxidants and antioxidative enzymes are considered to play an important role in detoxification of oxygen species generated in the presence of metal ions. Among enzymes allowing the destruction of these ROS are peroxidases, particularly guaïacol peroxydase (GPOX). A significant increase of GPOX activity in aerial parts of plants in response to watering with Zn was observed. This increase was even higher in plants regenerated in the presence of 0.1 to 0.25 mM of ZnSO$_4$. Increase in GPOX activity may be attributed to its role in the elimination of H$_2$O$_2$ which is produced in excess, the increase in H$_2$O$_2$ leading to a higher GPOX activity. Many studies suggested the implication of GPOX in the oxidative stress could be correlated with the amount of the accumulated metal (Cuypers et al., 2002; Sharma and Dubey, 2007). However, at higher concentrations of Zn applied *in vitro* (1 mM) Zn concentration in the leaves increased (Table 1), and a decrease of GPOX activity was observed. The decrease in GPOX activity suggests a stress too severe for the defense capacity of these plants (Siedlecka and Krupa, 2002). Zn treatment did not induce any change in roots GPOX activity despite a strong Zn accumulation in this organ. It then could be that the main effects of Zn are ROS production in green tissues upon illumination leaving roots unaffected by the 9089 µg g^{-1} DW of metal accumulated.

The importance of oxidative stress in plants upon heavy metal exposure has been repeatedly shown (Cuypers et al., 1999, 2001, 2002). Zinc is one of the metals that stimulate reactive oxygen species (ROS) production and antioxidative defense system (Cakmak, 2000). Our study revealed the importance of oxidative stress effects and antioxidative defense mechanisms after application of an environmentally realistic toxic zinc concentration (2 mM), that is, concentration similar to what is measured in zinc highly polluted soils.

As far as polyamines are concerned, Zn treatment in greenhouse caused an important increase in Put, Spd and Spm contents in leaves and roots. When plants were regenerated *in vitro* in the presence of 0.1 or 0.25 mM of ZnSO$_4$, the further increase of the PAs contents observed was significant in leaves and roots but (Table 4). Accumulation of PAs in plants exposed to oxidative stress caused by heavy metals such as Cu, Cd, and Zn has been previously reported by several authors (Wettlaufer et al., 1991; Groppa et al., 2003; Franchin et al., 2007). Particularly, a protective role for polyamines in plant cells exposed to this stress has been described by many authors who showed found that these substances may act as radical scavengers (Bors et al., 1989; Sharma and Dietz, 2006) and as inhibitor of lipid peroxidation (Velikova et al., 2000; Groppa et al., 2001).

This protective role for polyamines was recently confirmed by the increase of tolerance to copper and cadmium created by an exogenous supply of polyamines (Groppa et al., 2001; Rhee et al., 2007; Wang et al., 2007). When ZnSO$_4$ treatment was applied to plants regenerated *in vitro* in the presence of 0.5 or 1 mM of ZnSO$_4$, total PAs contents in leaves was significantly higher than in untreated plants, but clearly diminished in roots (Table 4). The relationship between Zn increase of oxidative stress that would in turn lead to a PAs contents increase was observed in most cases, however, this does not explain the increase of PAs contents in the roots despite no increase of GPOX activity, further investigations are required to clarified this point.

In this study, it was clearly demonstrated that deep changes occurred in the stress response (antioxidant system and metal accumulation) of plants regenerated from cells submitted *in vitro* to ZnSO$_4$, in a dose-dependent manner. Plants regenerated in the presence of

zinc showed a higher capacity of zinc accumulation in roots and aerials parts. This accumulation increased with Zn concentration during the regeneration phase and this selective pressure applied during the regeneration process may have induced a physiological adaptation (Hall, 2002). Although it is unlikely, mutation which affected major genes responsible for zinc absorption or accumulation cannot be excluded. The present investigation showed that application of selective pressure during regeneration of rapeseed from tTCLs can improve zinc tolerance capacity and accumulation in aerial parts (×5) for more efficient phytoextraction and this strategy may complement plant genetic engineering that is being developed for enhanced phytoremediation.

ACKNOWLEDGEMENTS

The authors are grateful to the French government for providing financial support to the first author via the French Embassy in Tunisia.

REFERENCES

Agazio M, Zacchini M (2001). Dimethylthiourea, a hydrogen peroxide trap, partially prevents stress effects and ascorbate peroxidase increase in spermidine-treated maize roots. Plant Cell Environ., 24: 237-244.

Alia VSK, Saradhi PP (1995). Effect of zinc on free radicals and praline in *Brassica* and *Cajanus*. Phytochemistry, 39: 45-47.

Audet P, Charest C (2008). Allocation plasticity and plant-metal partitioning: Meta-analytical perspectives in phytoremediation. Environ. Pollut., 156: 290-296.

Ben Ghnaya A, Charles G, Hourmant A, Ben Hamida J, Branchard M (2007). Morphological and physiological characteristics of rapeseed plants regenerated *in vitro* from thin cell layers in the presence of zinc. C. R. Biol., 330: 728-734.

Ben Ghnaya A, Charles G, Branchard M (2008). Rapid shoot regeneration from thin cell layers excised from petioles and hypocotyls explants in four cultivars of *Brassica napus* L. Plant Cell, Tissue Organ Cult., 92: 25-30.

Ben Ghnaya A, Charles G, Hourmant A, Ben Hamida J, Branchard M (2009). Physiological behaviour of four rapeseed cultivars (*Brassica napus* L.) submitted to metal stress. C. R. Biol., 332: 363-370.

Bors W, Langebartels C, Michel C, Sandermann JH (1989). Polyamines as radical scavengers and protectants against ozone damage. Phytochemistry, 28: 1589-1595.

Cakmak I (2000). Possible roles of zinc in protecting plant cells from damage by reactive oxygen species. New Phytol., 146: 185-205.

Chatterjee J, Chatterjee C (2000). Phytotoxicity of cobalt, chromium and copper in cauliflower. Environ. Pollut., 109: 69-74.

Cuypers A, Vangronsveld J, Clijsters H (1999). The chemical behaviour of heavy metals plays a prominent role in the induction of oxidative stress. Free Radical Res., 31: 39-43.

Cuypers A, Vangronsveld J, Clijsters H (2001). The redox status of plant cells (AsA and GSH) is sensitive to zinc imposed oxidative stress in roots and primary leaves of *Phaseolus vulgaris*. Plant Physiol. Biochem., 39: 657-664.

Cuypers A, Vangronsveld J, Clijsters H (2002). Peroxidases in roots and primary leaves of *Phaseolus vulgaris*; copper and zinc phytotoxicity: A comparison. J. Plant Physiol., 159: 69-876.

Féray A, Hourmant A, Penot M, Moisan-Cann C, Caroff J (1992). Effects of interaction between polyamines and benzyladenine on the betacyanin synthesis in *Amaranthus* seedlings. J. Plant Physiol., 139: 680-684.

Ferrat L, Pergent-Martini C, Roméo M (2003). Assessment of the use of biomarkers in aquatic plants for the evaluation of environmental quality: application to seagrasses. Aquat. Toxicol., 65: 187-204.

Flores HE, Galston AW (1982). Analysis of polyamines in higher plants by high performance liquid chromatography. Plant Physiol., 69: 701-706.

Franchin C, Fossati T, Pasquini E, Lingua G, Castiglione S, Torrigiani P, Biondi S (2007). High concentrations of zinc and copper induce differential polyamine responses in micropropagated white poplar (*Populus alba*). Physiol. Plant, 130: 77-90.

Gisbert C, Clemente R, Navarro-Avino J, Baixauli C, Ginér A, Serrano R, Walker DJ, Bernal MP (2006). Tolerance and accumulation of heavy metals by *Brassicaceae* species grown in contaminated soils from Mediterranean regions of Spain. Environ. Exp. Bot., 205: 19-27.

Groppa MD, Benavides MP, Tomaro ML (2003). Polyamine metabolism in sunflower and wheat leaf discs under cadmium or copper stress. Plant Sci., 164: 293-299.

Groppa MD, Ianuzzo MP, Tomaro ML, Benavides MP (2007). Polyamine metabolism in sunflower plants under long-term cadmium or copper stress. Amino Acids, 32: 265-275.

Groppa MD, Tomaro ML, Benavides MP (2001). Polyamines as protectors against cadmium or copper-induced oxidative damage in sunflower leaf discs. Plant Sci., 161: 481-488.

Hall JL (2002). Cellular mechanisms for heavy metal detoxification and tolerance. J. Exp. Bot., 53: 1-11.

Halliwell B (1987). Oxidative damage, lipid peroxidation and antioxidant protection in chloroplasts. Chem. Phys. Lipids, 44: 327-340.

Hoagland DR, Arnon DI (1950). The water-culture for growing plants without soil. Cal. Agric. Exp. Sta Cir., 347: 1-32.

Iturbe-Ormaetxe I, Escuredo PR, Arrese-Igor C, Becana M (1998). Oxidative damage in pea plants exposed to Water deficit or Paraquat. Plant Physiol., 116: 173-181.

Kitada M, Igarashi K, Hirose S, Kitagawa H (1979). Inhibition by polyamines of lipid peroxide formation in rat liver microsomes. Biochem. Biophys. Res. Commun., 87: 388-394.

Lichtenthaler HK (1984). Chlroplast biogenesis, its inhibition and modification by new herbicide compounds. Z. *Naturforsch*, 39c: 492-499.

Lichtenthaler HK (1987). Chlorophylls and carotenoïds: Pigments of photosynthetic biomembrane (Colowick, S.P., Kaplan, N. O., Dones, R. et al. Eds.). Methods Enzymol., 148: 350-381.

MacAdam JW, Nelson CJ, Sharpe RE (1992). Peroxidase activity in the leaf elongation zone of tall fescue. Plant Physiol., 99: 872-878.

Minotti G, Aust SD (1987). The requirement for iron (III) in the initiation of lipid peroxidation by iron (II) and hydrogen peroxide. J. Biol. Chem., 262: 1098-1104.

Monnet F, Vaillant N, Vernay P, Coudret A, Sallanon H, Hitmi A (2001). Relationship between PSII activity, CO_2 fixation, and Zn, Mn and Mg contents of *Lolium perenne* under zinc stress. J. Plant Physiol., 158: 1137-1144.

Murashige T, Skoog F (1962). A revised medium for rapid growth and bioassay with tobacco tissue culture. Plant Physiol., 15: 473-497.

Ohkawa H, Ohishi N, Yagi K (1979). Assay for lipid peroxides in animal tissues by thiobarbituric acid reaction. Anal. Biochem., 95: 351-358.

Pilon-Smits E (2005). Phytoremediation. Annu. Rev. Plant Biol., 56: 15-39.

Rhee HJ, Kim EJ, Lee JK (2007). Physiological polyamines: simple primordial stress molecules. J. Cell. Mol. Med., 11: 685-703.

Rugh CL, Wilde HD, Stack NM; Thompson DM, Summers AO, Meagher RB (1996). Proceedings of the National Academy Sciences, 93: 3182-3187.

Sharma P, Dubey RS (2007). Involvement of oxidative stress and role of antioxidative defense system in growing rice seedlings exposed to toxic concentrations of aluminum. Plant Cell Rep., 26: 2027-2038.

Sharma SS, Dietz KJ (2006). The significance of amino acids and amino acid-derived molecules in plant responses and adaptation to heavy metal stress. J. Exp. Bot., 57: 711-726.

Sharma SS, Dietz KJ (2009). The relationship between metal toxicity and cellular redox imbalance. Trends Plant Sci., 14: 43-50.

Siedlecka A, Krupa Z (2002). Functions of enzymes in heavy metal treated plants. In: Physiology and biochemistry of metal toxicity and tolerance in plants (Prasad, M. N. V., and Strzalka, K., Eds). Kluwer. Academic Publishers, pp. 303-324.

Singh S, Sinha S (2005). Accumulation of metals and its effects in *Brassica juncea* (L.) Czern. (cv Rohini) grown on various amendments of tannery waste. Ecotoxicol. Environ. Saf., 62: 118-127.

Stobart AK, Griffiths WT, Bukhari IA, Sherwood RP (1985). The effect of cadmium on the biosynthesis of chlorophyll in leaves of barley. Physiol. Plant, 63: 293-298.

Tadolini B (1988). Polyamine inhibition of lipid peroxidation. Biochem. J., 249: 33-36.

Teisseire H, Couderchet M, Vernet G (1999). Phytotoxicity of diuron alone and in combination with copper or folpet on duckweed (*Lemna minor*). Environ. Pollut., 106: 39-45.

Velikova V, Yordanov I, Edreva A (2000). Oxidative stress and some antioxidant systems in acid rain-treated bean plants: Protective role of exogenous polyamines. Plant Sci., 151: 59-66.

Wang X, Shi G, Xu Q, Hu J (2007). Exogenous polyamines enhance copper tolerance of *Nymphoides peltatum*. J. Plant Physiol., 164: 1062-1070.

Weiss DJ, Shotyk W, Schafer J, Loyall U, Grollimund E, Gloor M (1999). Microwave digestion of ancient peat and lead determination by voltammetry. Fresenius. J. Anal. Chem., 363: 300-305.

Wettlaufer SH, Osmeloski J, Weinstein LH (1991). Response of polyamines to heavy metal stress in oat seedlings. Environ. Toxicol. Chem., 10: 1083-1088.

Wood LG, Gibson PG, Garg ML (2006). A review of the methodology for assessing *in vivo* antioxidant capacity. J. Sci. Food Agr., 86: 2057-2066.

Zengin FK, Munzuroglu O (2005). Effects of some heavy metals on content of chlorophyll, proline and some antioxidant Chemicals in Bean (*Phaseolus vulgaris* L.) Seedlings. Acta Biol. Cracov. Bot., 47: 157-164.

Zhang GW, Liu ZL, Zhou JG, Zhu YL (2008). Effects of Ca $(NO_3)_2$ stress on oxidative damage, antioxidant enzymes activities and polyamine contents in roots of grafted and non-grafted tomato plants. Plant Growth Regul., 56:7-19.

Acute toxicity and behavioral response of freshwater fish, *Mystus vittatus* exposed to pulp mill effluent

A. Mishra[1], C. P. M. Tripathi[2], A. K. Dwivedi[3] and V. K. Dubey[3]*

[1]Aquatic Toxicology Division, IITR Lucknow, Uttar Pradesh, India.
[2]D. D. U. Gorakhpur University, Gorakhpur, India.
[3]National Bureau of Fish Genetic Resources, Canal Ring Road, PO, Dilkusha, Lucknow- 26002, Uttar Pradesh, India.

The present study deals with the acute toxicity (LC_{50} evaluation) of paper mill effluent to a freshwater fish, *Mystus vittatus*. A monthly record covering all the spawning season (pre-spawning phase, spawning phase and post-spawning phase) was evaluated at different exposure periods (24-, 48-, 72- and 96-h) using the whole paper mill effluent. Annual variation in LC_{50} values in relation to annual breeding cycle and water temperature were also taken into consideration. A well marked variation in the LC_{50} values in different exposure periods as well as in various months of the spawning phases of the test fish were observed.

Key words: Acute toxicity, paper mill effluent, annual variation, spawning phases.

INTRODUCTION

The Bagridae family of fish is the richest and most important of the teleostei class and its members are distributed throughout the world (Day, 1878). In the Bagridae family, the fish *Mystus vittatus* (Smith, 1945) is economically important and distributed in the semitemporal freshwater system of South India (Perennou and Santharam, 1990).

Under laboratory conditions, toxicity testing procedures (mortality studies LC_{50} estimates) may provide information regarding the harmfulness of industrial stress for aquatic animals, including fishes (Marier, 1973). In the acute toxicity of contaminants in static bioassays, the use of 96-h, LC_{50} has been widely recommended as a preliminary step in toxicological studies on fishes (McLeay, 1976; Whittle and Flood, 1977; USEPA, 1973, 2005; APHA, 1998, 2005; Chapman, 2000; Ali and Sree-Krishnan, 2001; ASTM, 2002; Parrott et al., 2006; Moreira et al., 2008). The static bioassay procedures to study the toxicity of pulp and papermill effluent to variety

of fishes have been reported like the Cohosalmon, *Onchorhynchus kisutch* (McLeay, 1973, 1975, 1976, 1977; Gordon and McLeay, 1977; McLeay and Howard, 1977; McLeay and Gordon, 1977; McLeay and Brown, 1979) the rainbow trout, *Salmo gairdneri* (McLeay, 1976; Gordon and McLeay, 1977; Whittle and Flood, 1977; McLeay and Gordon, 1978; Couillard et al., 1988) and other fishes (Gordon and Servizi, 1974; Davis, 1976; Walden, 1976; Hewitt et al., 2006).

The paper mill, when it is playing directly influx to the streams and rivers, without previous treatment of effluents it can change the native fish or biota. Therefore, in the present study, the acute toxicity of paper mill effluent has been studied with regards to the mortality of *M. vittatus*. The 24-, 48-, 72- and 96-h, LC_{50} values were determined during the different phases of the reproductive cycle of the fish, and the correlation of LC_{50} values with environmental temperature has also been established.

The main objective of this study is to evaluate the acute toxicity of paper mill effluent on the behavioral responses of freshwater fish *M. vittatus* in the context of annual breeding cycle and water temperature.

*Corresponding author. E-mail: vineet_dubey26@yahoo.co.in.

Table 1. Characteristics of WPME (whole paper mill effluent), Magahar, Santkabeer Nagar (U.P.) India.

Characteristics	Variable constituents through the year (mean values)			Yearly average ± S. E.
	January-April	May-August	September-December	
Color	Dark brownish	Dark brownish	Dark brownish	-
Sodium, Na^+ (mg/L)	350	320	351	340±12.2
Cl^- (mg/L)	420	450	425.7	431.9±11.3
SO_4 (mg/L)	1.12	5.8	3.5	3.5±1.7
Nitrate (mg/L)	7.6	7.3	7.2	7.4±0.1
Total nitrogen (mg/L)	1.7	6.3	3.8	3.9±1.6
PO_4 (mg/L)	0.77	0.52	0.69	0.7±0.1
pH	7.3	7.3	7.4	7.3±0.04
Temp (°C)	23.5	28	26.5	26±1.6
Suspended solid (mg/L)	5021	4643	4256	4640±270
Dissolved solid (mg/L)	1111	1222	1013	1115±74
Total solid (mg/L)	6132	5865	5269	5755±312
BOD (mg/L)	552	538	498	526±19
COD (mg/L)	2379	2551	2326	2418±83
Fe (mg/L)	9.6	13.5	10.4	11.2±1.5
Mg (mg/L)	1.9	1.65	1.28	1.6±0.2
K (mg/L)	8.6	6.4	4.4	6.5±1.5
Cu (mg/l)	0.11	0.14	ND	0.1±0.1
Total Cr (mg/L)	ND	0.077	0.072	0.07±0.03
Mn (mg/L)	0.34	0.071	0.53	0.31±0.12
Co (mg/L)	-	0.001	0.0012	0.001±0.002
Cd (mg/L)	0.025	0.016	0.018	0.02±0.004
Zn (mg/L)	0.06	0.08	0.106	0.08±0.021

Data based on samples taken during the morning shift of the normal course of mil operation at 8 am.

MATERIALS AND METHODS

The freshwater teleost fish, *M. vittatus* were collected from a local freshwater river Ami of Sant Kabeer nagar district (U. P.). They were transported to laboratory and washed with $KMNO_4$ solution (1 *mg*/L) for 5 min and then transferred to the acclimation tank. The fishes having an average body length of 7.6 ± 0.18 cm, and body weight of 73 ± 0.23 g. were selected for the present study.

The fishes were acclimatized in laboratory conditions for 3 to 4 weeks at room temperature (22 to 30°C) using dechlorinated tap water. The medium of the acclimation tanks was changed daily. The dissolved oxygen content of the medium was maintained at 60% to 100% by regular aeration. The whole paper mill effluent was collected in polyethylene container of 10 to 20 L capacity during the morning hours. The effluent samples were taken from two places, that is, the point where the effluent comes out of the mill and the point where it enters the collecting tank and the mixed unfiltered samples were used within 24 h of collection. The effluent samples were chemically analyzed per month and the characteristics are summarized in Table 1.

Acute toxicity of effluent on test fish was measured in terms of LC_0, LC_{50} and LC_{100}. Static bioassay procedures, as outlined by the USEPA (2005) were followed. A minimum of 8 concentrations of effluent and 20 animals were used for each concentration. The control set had the same number of fish kept in normal, dechlorinated tapwater. No food was provided to either the control or the test fishes during the period of the toxicity experiments. Experiments were carried out up to 96-h, and the observations were recorded after every 24-h. Experiments were conducted every month, and LC_{50} values for 24-, 48-, 72- and 96-h, were determined. The tested concentrations of effluent and the observed percentage of the fish mortality were subjected to linear regression for determining the LC_{50} values for the different exposure periods. The yearly average LC_{50} values were calculated, and the 95% confidence interval determined for this mean.

RESULTS

The mortality of test fish exposed to paper mill effluent is summarized in Table 2.

Yearly average LC_{50}

The yearly average LC_{50} value was found to be 60.04 ± 2.13; for 24 h, 53.6 ± 2.16, for 48 h; 48.6 ± 2.1, for 72 h, and 44.4 ± 2.2, for 96 h exposure periods (Table 2).

Annual variation in LC_0 and LC_{100}

The lowest LC_0 and LC_{100} values were recorded in the spawning phase of the fish, whereas the values were the highest in the post spawning phase (October to January)

Table 2. Seasonal variation of WPME toxicity (LC$_{50}$) to *M. vittatus* in different spawning phases.

Spawning phases	Month	Water temperature (°C)	LC$_{50}$ % (v/v)			
			24 h	48 h	72 h	96 h
Pre spawning phase	February	23.5	64.5	54.0	49.5	45.6
	March	24.0	62.5	54.5	50.5	45.7
	April	25.0	60.4	53.5	46.05	41.5
	May	27.0	54.0	48.5	43.5	42.5
	Mean		60.35	52.6	47.8	43.8
Spawning phase	June	29.0	50.0	42.55	38.0	35.5
	July	28.0	50.5	42.55	38.0	37.0
	August	28.5	53.0	46.0	41.5	38.0
	September	27.0	57.25	53.75	49.0	44.25
	Mean		52.7	46.2	41.6	38.7
Post-spawning phase	October	26.5	64.5	59.5	50.5	46.0
	November	25.0	68.5	62.5	58.0	51.5
	December	23.0	71.75	64.0	59.0	53.0
	January	22.0	68.25	61.75	58.0	53.0
	Mean		68.2	62.0	56.4	50.8
Yearly average ±S. E.		25.7±0.7	60.4±2.13	53.6±2.16	48.6±2.1	44.4±2.2
95% Confidence interval		-	65.1-55.7	58.3-48.8	53.2-43.9	49.2-39.6

of the annual reproductive cycle. Values observed in three phases are recorded in Table 3.

Annual variation in LC$_{50}$

A well marked variation was recorded in the LC$_{50}$ values for the different months of the year. The values for the spawning phase were generally lower than those of other phases (Table 2).

Variation in relation to water temperature

The annual variation in LC$_{50}$ values exhibited an inverse relationship with water temperature. The LC$_{50}$ value was generally recorded to be higher in the colder months of the year and lower in the hotter months. Maximum LC$_{50}$ values were observed in the month of December (water temperature 23°C), while lowest in the month of July (water temperature 28°C.).

The Cr and Cd contents of test effluent were found to be much higher; the total content is 0.497 mg/L for Cr and 0.0073 mg/L for Cd. Regarding Cu, the maximum permissible concentration is 0.006 ppm or alternatively 0.1 times the LC$_{50}$ 96 h value. In case of the presently used papermill effluent, the Cu content was found to be 0.083 mg/L, which is about 2.5 times its estimated LC50 96 h value (that is, estimated concentration present in the LC50 96 h test effluent).

The matter suspended within the papermill effluent used in the present study could also be contributing to the mortality of the *M. vittatus* individuals maintained in this effluent. The concentration of suspended matter in WRPBILE ranged from 4256 to 5021 mg/L.

Behavioral responses

The surfacing behavior as well as the rate of opercular movement of the fishes was observed to be increased within an hour of the commencement of the toxicity experiments. With the continuation of the exposure, the fishes progressively became sluggish and lethargic. The majority of them became completely inactive after 72 h of exposure to 60 to 70% of test effluent, and all died after 96 h of exposure to 70% effluent. Prior to their death in the contaminated medium, they showed abnormal swimming and the loss of equilibrium.

DISCUSSION

The stressed fishes were found to be the most sensitive to effluent during the spawning phases of the reproductive cycle. Increased sensitivity under pollution stress during the spawning season has also been reported in pacific *herrings* exposed to benzene toxicity (Korn et al., 1976). Higher temperature nature has been shown to increase the toxicity of pulp and papermill

Table 3. Seasonal variation of WRPBILE toxicity (LC_0 and LC_{100}) to *M. vittatus* in different spawning phases.

Spawning phase	Months	Water temperature (°C)	LC_{50} % (v/v) 24 h		48 h		72 h		96 h	
			LC_0	LC_{100}	LC_0	LC_{100}	LC_0	LC_{100}	LC_0	LC_{100}
Pre- spawning phase	February	23.5	30.0	85.0	30.0	80.0	25.0	75.0	25.0	70.0
	March	24.0	30.0	80.0	30.0	75.0	25.0	75.0	25.0	70.0
	April	25.0	25.0	80.0	25.0	75.0	20.0	70.0	20.0	65.0
	May	27.0	25.0	75.0	20.0	70.0	20.0	65.0	20.0	60.0
	Mean		27.5	80.0	26.3	75.0	22.5	71.3	22.5	66.3
Spawning phase	June	29.0	20.0	75.0	20.0	70.0	20.0	65.0	20.0	60.0
	July	28.0	20.0	70.0	20.0	65.0	15.0	60.0	15.0	55.0
	August	28.5	25.0	75.0	20.0	70.0	20.0	65.0	15.0	60.0
	September	27.0	30.0	80.0	25.0	75.0	25.0	75.0	20.0	70.0
	Mean		23.7	75.0	21.3	70.0	20.0	66.3	17.5	61.3
Post-spawning phase	October	26.5	30.0	80.0	30.0	75.0	25.0	75.0	25.0	70.0
	November	25.0	40.0	85.0	35.0	80.0	35.0	80.0	30.0	75.0
	December	23.0	40.0	90.0	40.0	85.0	35.0	80.0	30.0	75.0
	January	22.0	40.0	85.0	40.0	80.0	35.0	75.0	30.0	70.0
	Mean		37.5	85.0	36.3	80.0	32.5	77.5	28.8	72.5
Yearly average ±S.E.		25.7±0.7	29.6±2.2	80.0±2.1	28.0±1.8	75.0±2.1	25.0±1.8	71.7±1.9	22.9±1.5	66.7±1.9
95% confidence Interval		-	34.4 -24.7	84.6 -75.3	31.9-24.0	79.6-70.3	28.9-21.0	75.8-67.5	26.2-19.6	70.8-62.5

effluent for fishes like *O. kisutch* and *S. gairdneri* (McLeay, 1976; Gordon and McLeay, 1977). Increase in temperature reduces the solubility of oxygen in water and hence, could raise the metabolic rate (oxygen demand) of the fish, thus limiting the oxygen carrying capacity of the blood (Gordon and McLeay, 1977). But in the present study, the LC_{50} played the higher toxicity in the colder months, when the temperatures were very low. The result indicates that toxicity can occur independent of the high or low temperatures. Simultaneously, the higher BOD (Biological oxygen demand) and COD (Chemical oxygen demand) could also contribute to the mortality (Polak and Palmer, 1977; Whittle and Flood, 1977; McLeay et al., 1979a, b; McLeay and Brown, 1979).

It is known that many pollutants, such as heavy metals, like Mg, Cu and Zn, and industrial wastes become more harmful at higher water temperatures. In view of this, it may be presumed that the toxic metals, especially Cu and Zn, present in the tested effluent would acquire greater toxicity during the warmer months of the year and thus contribute to the overall increased mortality during these months in *M. vittatus*. As regard the contribution of the different constituents of effluent to *M. vittatus* mortality, the content of many of these constituents is noticeably higher than the upper limits considered as safe for fish health (Wedemeyer, 1976). For example, among metals, the upper limit for Cr is 0.03 ppm, and that for Cd 0.0004 ppm. The total suspended solids, absorbable organic halogens and also certain other toxic compounds like wood extractives, present in pulp and papermill waste water are also known to affect fishes (Sprague and McLeese, 1968; Kelso,1977; Hewitt et al., 2006; Parrott et al., 2006). The concentration of the suspended matter in WRPBILE is much above the TSS concentration considered (Wedemeyer, 1976) optimum for fishes (that is, 80 to 100 ppm), and falls within that range of suspended matter of papermill effluents (that is, 530 to 17900 mg/L approximately) which has earlier been considered injurious to fishes.

The opercular movement of the dying fishes became extremely slow and they were seen to secrete thick coat of mucus profusely around their opercular region. Similar types of increased opercular movement, erratic and rapid movements have been observed by Kumar and Gopal (2001) in *Channa punctatus* exposed to distillery effluent. Clotfelter et al., (2006) have observed aggressive behaviour in fighting fish, increased surface breathing and opercular movement in the stressed *M. vittatus* would point a sustained respiratory discomfort in the toxic WPME (whole paper mill effluent) (Lioyd, 1961). The stressed fishes were observed to secrete mucus around their opercular region which has been considered to be a symptom of the inflammatory reaction of gill towards the pollutants (Durve and Jain, 1980 ; Pandey and Pandey, 1988), while the abnormal swimming and disturbed orientation has been considered as a generalized sign of diseases in fishes (Sindermann, 1970).

The kinds of behavioral faults in orientation and locomotion, as observed in the present study, can be related to the impairment of sensory organ systems particularly the mechano and chemo-receptor systems. Sensory organs like lateral line, olfactory organs and membranous babyrinth helps the fishes in maintaining harmony with their environments and also control their vital behaviors (Gardner, 1975). Hence, any impairment of these organs would produce behavioral faults in the fishes. Therefore, the behavioral changes, particularly those concerned with respiratory insufficiency, observed in *M. vittatus* exposed to mill effluent, might be contributing to the mortality in these stressed fishes.

ACKNOWLEDGEMENTS

The authors are grateful to Prof. C. P. M. Tripathi, HOD Department of Zoology, D. D. U., Gorakhpur University and Dr. Krishna Gopal, for encouragement and valuable support.

REFERENCES

Ali M, Sreekrishnan TR (2001). Aquatic toxicity from pulp and paper mill effluents: A review. Adv. Environ. Res., 5: 175-196. doi: 10.1016/S1093-0191/(00)00055-1.

American Public Health Association (APHA). American Water Work Association (AWWA) and Water Pollution Control Federation (WPCF) (1998). Standard method for the examination of water and waste water. 20th ed. Washington, DC.

APHA (2005). American Public Health Association. Standard methods for examination of water including bottom sediments and sludges. Standard Methods, (19th ed.), p. 874.

ASTM (American Society for Testing and Materials) (2002). Standard guide for conducting acute toxicity tests on test materials with fishes, macro vertebrates and amphibians. E729/96. In Annual Book of ASTM standards, ASTM, Philadelphia, PA., 11.05: 179–200.

Chapman PM (2000). Whole effluent toxicity test-usefulness, level of protection and risk assessment. Environ. Toxicol. Chem., 19: 3-13. doi:10.1897/1551-5028(2000). 019<0003: WETTUL>2.3 Co:2.

Couillard CM, Berman RA, Panisset JC (1988). Histopathology of rainbow trout exposed to a bleached kraft pulp paper mill effluent. Environ. Contam, Toxicol., 17: 319-323.

Clotfelter D, Ethane C, Alison R (2006). Behavioural changes in fish exposed to phytoestrogens. Environ. Pollut., 144: 833-839.

Davis JC (1976). Progress in sublethal effect studies with kraft pulp papermill effluent and salmonids. J. Fish. Res. Board Canada 33: 2031-2035.

Day F (1878). The Fishes of India. William Dawson, London. 1.2: 210-215.

Durve VS, Jain SM (1980). Toxicity of distillery effluent to the cyprinid weed fish *Rasbora daniconius* (Ham). Acta. Hydrochim., 8(4): 329-336.

Gardner GR (1975). Chemically induced lesions in estuarine or marine teleosts. In: The Pathology of fishes (Eds:Ribelin, W.E. and Migaki, G.M.). Wisconsin Press, Madison, pp. 657-693.

Gordon MR, McLeay DJ (1977). Sealed-Jar bioassays for pulp mill effluent toxicity: Effects of fish species and temperature. J. Fish Res. Board, Canada, 30: 1389-1396.

Gordon RW, Servizi JA (1974). Acute toxicity and detoxification of kraft pulp mill effluent. Int. Pac. Salmon Fish. Comm. Rep., p. 31.

Hewitt LM, Parrott JL, McMaster ME (2006). A decade of research on the environmental impacts of pulp and papermill effluents in Canada: Sources and characteristics of bioactive substances. J. Toxicol. Environ. Health, 9: 341-356. doi:10.1080/15287390500195976.

Kelso JRM (1977). Density, distribution and movement of Nipigon Bay fishes in relation to a pulp and paper mill effluent. J. Fish. Res. Broard Canada, 34: 879-885.

Korn S, Hirsch N, Struhsaaker JN (1976). Uptake, distribution and depuration of 14C-benzene in Northern anchovy, *Encraulis mordax*, and striped bass, *Monero saxatilis*, Fish. Bull., 74: 545.

Kumar S, Gopal K (2001). Impact of distillery effluent on physiological consequences in the fresh water teleost *Channa punctatus*. Bull. Environ. Contam. Toxicol., 66: 617-622.

Lioyd R (1961). Effect of dissolved oxygen concentrations on the toxicity of several poisons to rainbow trout (*Salmo gairdneri. Rich.*). J. Exp. Biol., 38: 447.

Marier JR (1973). The effects of pulp and paper-wastes on aquatic life with particular attention to fish and bioassay procedures for assessment of harmful effects. NRC (*Nat. Res. Counc. Com.*) Publ. No. 73-3(ES). p. 33.

McLeay DJ (1973). Effects of a 12-h and 24-day exposure to Kraft pulp mill effluent on the blood and tissue of Juvenile cohosalmon (*Onchorhynches kisutch*). J. Fish. Res. Board Canada, 30: 395-400.

McLeay DJ (1975). Sensitivity of Blood cell counts in Juvenile cohosalmon (*Onchorhynches kisutch*) to stressors including sublethal concentrations of pulpmill effluent and Zinc. J. Fish. Res. Board Canada, 32: 2357-2364.

McLeay DJ (1976). A rapid method for measuring the acute toxicity of pulpmill effluents and other toxicants to salmoid fish at ambient room temperature. J. Fish. Res. Board Canada, 33: 1303-1311.

McLeay DJ (1977). Development of a blood sugar bioassay for rapid measuring stressful levels of pulp effluent to salmoid fish. J. Fish. Res. Board Canada, 34: 477-485.

McLeay DJ, Brown DA (1979). Stress and chronic effects of untreated and treated bleached kraft pulpmill effluent on the biochemistry and stamina of Juvenile cohosalmon (*Onchorhynchus kisutch*). J. Fish. Res. Board Canada, 36: 1049-1059.

McLeay DJ, Gordon MR (1977). Leucocrit: A simple hematological technique for measuring acute stress in salmonid fish including stressful concentration of pulpmill effluent. J. Fish. Res. Board Canada, 34: 2164-2175.

McLeay DJ, Gordon MR (1978). Effect of seasonal photoperiod on acute toxic responses of Juvenile rainbowtrout (*Salmo gairdneri*) to pulpmill effluent. J. Fish. Res. Board Canada, 32: 1388-1392.

McLeay DJ, Howard TE (1977). Comparison of rapid bioassay

procedures for measuring toxic effects of bleached kraft mill effluent to fish. Proc. 3rd Aquatic Toxicity workshop, Halifax, N.S., Nov. 2-3, 1976. Env. Prot. Serv. Tech. Report No. EPS-5_AR-77-1, Halifax. Canada, pp. 141-155.

McLeay DJ, Walden CC, Munro JR (1979a). Influence of dilution water on the toxicity of kraft pulp and papermill effluent including mechanisms of effect. Water Res., 13: 151-158.

McLeay DJ, Walden CC, Munro JR (1979b). Effect of pH on toxicity of kraft pulp and papermill effluent to salmonid fish in fresh and sea water. Water Res., 13: 249-254.

Moreira-Santos M, Donato C, Lopes I, Ribeiro R (2008). Avoidance tests with small fish: determination of the median avoidance concentration and of the lowest-observed-effect gradient. Environ. Toxicol. Chem., 27: 1576-1582. doi:10.1897/07-094.1.

Pandey RK, Pandey SK (1988). Tolerance measurement and histopathological observations on gills of Mystus. M. vittatus under the toxic stress of the fertilizer NPK. Acta hydrochim. Hydrobiol. 17(5): 597-601.

Parrott JL, McMaster ME, Hewitt LM (2006). A decades of research on the environmental impacts of pulp and papermill effluents in canada: Development and application of fish bioassays. J. Toxicol. Environ. Health, 9: 297-317 doi:10.1080/15287390500195752.

Perennou C, Santharam V (1990). Anthropological survey of some wetlands in south-east India. J. Bombay Nat. Hist. Soc., 87: 354–363.

Polak J, Palmer MD (1977). Concentration pattern of chemical constituents in papermill effluent plume: Dynamics and model. J. Fish. Res. Board Canada, 34: 805-816.

Sindermann CJ (1970). Principal diseases of fish and shellfish. Academic Press, London/New York, p. 369

Sprague JB, McLeese DW (1968). Different toxic mechanisms in kraft pulpmill effluent for two aquatic animals. Water Res., 2: 761-765.

Walden CC (1976). The toxicity of pulp and papermill effluents and corresponding measurement procedures. Water Res., 10: 639-664.

Whittle DM, Flood KW (1977). Assessment of the acute toxicity, growth impairment, and flesh taining potential of a bleached Kraft mill effluent on rainbow trout Salvo gairdneri. J. Fish Res. Board Canada, 34: 869-878.

U.S. Environmental Protection Agency, (USEPA) (1973). Proposed Criteria for Water Quality. US Environmental Protection Agency, Washington DC, USA, Vol. 1.

United States Environmental Protection Agency (USEPA) (2005). Aquatic Toxicity Information Retrieve AQUIRE aquatic toxicology database. Available:/www.epa.gov/ecotox/accessed:August 2003.

Wedemeyer GA (1976). Physiological response of juvenile coho salmon (Oncorhynchus kisutch) and rainbow trout (Salmo gairdneri) to handling and crowding stress in intensive fish culture. J. Fish. Res. Board Can., 33: 2699-2702.

Evaluation of the levels of organochlorine pesticide residues in water samples of Lagos Lagoon using solid phase extraction method

David Adeyemi[1], Chimezie Anyakora[1*], Grace Ukpo[1], Adeleye Adedayo[2] and Godfred Darko[3]

[1]Department of Pharmaceutical Chemistry. Faculty of Pharmacy, University of Lagos, Nigeria.
[2]Nigerian Institutes of Oceanography and Marine Research, Victoria Island, Lagos, Nigeria.
[3]Department of Chemistry, Rhodes University, Grahamstown 6140, South Africa.

Water samples were collected from 6 different locations of the Lagos lagoon at 0.5 and 2.5 m depths, respectively and from 2 other river bodies in Lagos and analyzed for the presence of 9 OCPs (organochlorine pesticides) residues. Samples were extracted using solid phase extraction method and were analyzed with gas chromatography. The analytes quantified included chlordane (0.006 to 0.950 µg/L), heptachlor (N.D: 0.067 µg/L), methoxychlor (N.D: 0.123 µg/L), hexachlorobenzene (0.015 to 0.774 µg/L), endosulfan (0.015 to 0.996 µg/L), dichlorodiphenyltrichloroethane (0.012 to 0.910 µg/L), Dichlorodiphenyldichloroethylene (0.005 to 0.477 µg/L), dieldrin (0.015 to 0.996 µg/L) and aldrin (0.080 to 0.790 µg/L). The ratio of DDE/∑DDT obtained was 0.812 and this indicates that the degraded metabolite formed a significant proportion of the total DDTs. The overall mean total concentration of aldrin was the highest while heptachlor was the lowest. Concentration of methoxychlor and heptachlor were relatively low and below the detection limit in 50 and 57% of samples analysed respectively. The mean concentrations of organochlorine pesticides were higher than European Community allowable residual limit (0.1 µg/L) for individual OCPs in drinking water in 37.3% of samples analyzed.

Key words: Organochlorine, pesticides, Lagos, gas-chromatography, solid phase extraction.

INTRODUCTION

Pesticides are synthetic organic compounds that despite their benefits also pose considerable hazards to the environment. They are broadly divided into many classes of which the most important are organochlorine pesticides (OCPs). The organochlorine pesticides are broad spectrum insecticides, and are the most widely used in many countries including Nigeria for agricultural purposes and control of mosquitoes (Bouman, 2004; Blaso et al., 2005). Organochlorine pesticides are very stable compounds and it has been cited that the degradation of dichlorodiphenyltrichloroethane (DDT) in soil ranges from 4 to 30 years, while other chlorinated

stable for many years after application, due to a high resistance to chemical and biological degradations (Afful et al., 2010). The chemicals are very persistence lipo-soluble compounds and are capable of bioaccumulating in the fatty parts of biological beings such as breast milk, blood and fatty tissues (William et al., 2008) via the food chains. As a result of its position in the food chain (end of the food chain), man is greatly exposed to the effect of the micropollutants by eating foods either coming from contaminated earth or water (Belta et al., 2006; Raposo et al., 2007).

Organochlorine pesticides have become ubiquitous contaminant and have been implicated in a broad range of deleterious health effects in laboratory animals and man. The toxic effect include reproductive failures (Bouman, 2004), immune system malfunction (Kolpin et al., 1998), endocrine disruption (Ize et al., 2007) and

*Corresponding author. E-mail: canyakora@gmail.com.

breast cancers (Garabrant et al., 1992). Previous studies have shown that DDT has the ability to block potassium influx across membranes of nerve fibres, thereby causing increased negative after-potentials. It also induces mixed function oxidize system thereby, alters the metabolism of xenobiotics and steroid hormones (Colborn and Smolen, 1996).

The ability of the most prevalent metabolite of DDT, dichlorodiphenyldichloroethylene (DDE) to bind to androgen receptor in animals has been reported (Saxena et al., 1981), while a long exposure to dieldrin and aldrin could results in chronic convulsions (Belta et al., 2006). Epidemiological reports have also suggested an etiological link between exposure to organochlorine pesticides and Parkinsons diseases (Fleming et al., 1994). The aquatic environment often serves as a sink for many potentially harmful organic pollutants emitted from industrial and domestic sources and contamination of water bodies by pesticide residues has been an issue of serious concerns due to the health risks associated with them (Golfinopoulos et al., 2003).

The US-EPA has set allowable residual limit of individual and total concentration in drinking water set at 0.1 and 0.5 µg/L, respectively. Despite the ban on the production and use of OCPs in accordance with Stockholm convention in 2001 (Ennacer et al., 2008) and replacement with less persistent organophosphates and carbamates, some developing countries have resumed the use of OCPs such as DDT, because of its early spectacular success in malaria eradication and fighting of life-threatening typhus. Although, so far there is no local production in Nigeria, over 100 different brands of pesticides are being imported in to the country. It is worth noting that dieldrin, aldrin and other organochlorine pesticides are still in use for the control of pest of cotton, cocoa, fruits, cereals and vegetables (Ize et al., 2007).

Water bodies are a key recipient of xenobiotic conta-minants such as pesticides in the ecosystem (Barcelo, 1991). These contaminants can enter water either through point sources like sewage plants, sewer over-flows and poor management practices of farmers or through diffuse sources such as atmospheric deposition, run offs and leaching from agricultural applications (Guruge et al., 2001; Mayon et al., 2006; Shukla et al., 2006). The environmental persistence and toxicity of OCPs have led to its ban in many countries and are in the priority list of European Community.

Pesticides are often found in trace quantities in complex matrices and for efficient study of fate of pesticides in natural waters, very low detection limit must be reached. As a result, both efficient sample preparation and a high performance analytical instrument are required for accurate determination at trace levels in which they occur in complex matrices. The determination of pesticide residues in water is necessary for solving various environmental and biological problems. The pollution of water bodies in Nigeria (Lagos lagoon especially) through agricultural activities and direct

deposition is on the rise. Meanwhile, a large proportion of the populace depends on it for potable and recreational water, as well as a source of cheap and affordable protein in form of fish. There are general reports on ground water contami-nation from pesticides in the country (Nwankwoala and Osibanjo, 1992; Ize et al., 2007) but relatively few studies have focused on samples from Lagos lagoon.

The objective of this study was to evaluate the spectrum of organochlorine pesticides residues in water samples of Lagos Lagoon by solid phase extraction method and analysis with GC-ECD. The OCPs investigated include 1,1,1-trichloro-2,2-bis(4-chlorophenyl)ethane (DDT), 1,1-dichloro-2,2-bis (p-chlorophenyl)ethylene (DDE), 1,2,3,4,10,10-hexachloro-6,7-epoxy-1,4,4a,5,6,7,8,8a,octahydro-1,4,5,8-dimethanonaphthalene (Dieldrin), Hexachlorobenzene (HCB), 1,4,5,6,7,8,8-Heptachloro-3a,4,7,7a-tetrahydro-4,7-methano-1H-indene (heptachlor), 1,2,3,4,10,10-hexachloro-1,4,4a,5,8,8a-hexahydro-1,4:5,8-dimethanonaphthalene, (aldrin), 6,7,8,9,10,10-hexachloro-1,5,5a,6,9,9a-hexahydro-6,9-methano-2,4,3-benzodioxathiepine-3-oxide (endosulfan), octachloro-4,7-methanohydroindane (chlordane), and 1,1,1-trichloro-2,2-bis(4-methoxyphenyl)ethane (methoxychlor).

MATERIALS AND METHODS

Sampling area

Lagos is a port and populous conurbation city in Nigeria with more than 12 million people and estimated to be the fastest growing city in Africa. Lagos lagoon (Figure 1) is about 50 km long and 3 to 13 km wide, separated from Atlantic Ocean by long sand and spit 2 to 5 km wide with swampy margins on the Lagoon side. Lagos Lagoon empties into the Atlantic via the Lagos harbor, a main channel through the heart of the city, 0.25 to 1 km wide and 10 km long. The principal ocean port of Lagos is located at Apapa in a broad western branch off the main channel of the harbor. Lagos Lagoon is fairly shallow and the city spread along more than 30 km of the Lagoons South western and western shoreline. The area west of the lagoon is not well provided with road, and many communities there traditionally relied on water transport.

Chemicals and reagents

The OCPs standards; aldrin (98.1%), chordane (98.43%), DDE (99.5%), DDT (99.6%), dieldrin (97.9%), β-endosulfan (99.9%), heptachlor (99.7%), methoxychlor (99.5) and hexachlorobenzene (99.6%) were obtained from Supelco (Belle—Fonte, USA). The stock solutions of each analyte (1.0 mg/L) were prepared in acetone and a fresh working solution containing a mixture of each analyte was prepared by stepwise dilution of the stock solution. Solvents used were of pesticide grade standard. HPLC/UV grade acetone, dichloromethane and n-hexane were purchased from Merck Chemical (Bonn, Germany). Ultra high purity water was generated from a Millipore alpha-Q-system supplied by Millipore (Molsheism, France). Granular anhydrous sodium sulfate (AR grade) was heated to 400°C for 4 h prior to use. C-18 SPE catridges(62ESEB) containing ultra pure silica gel were obtained from Gueboc OC, Canada. The standard solution of OCPs, each containing the target

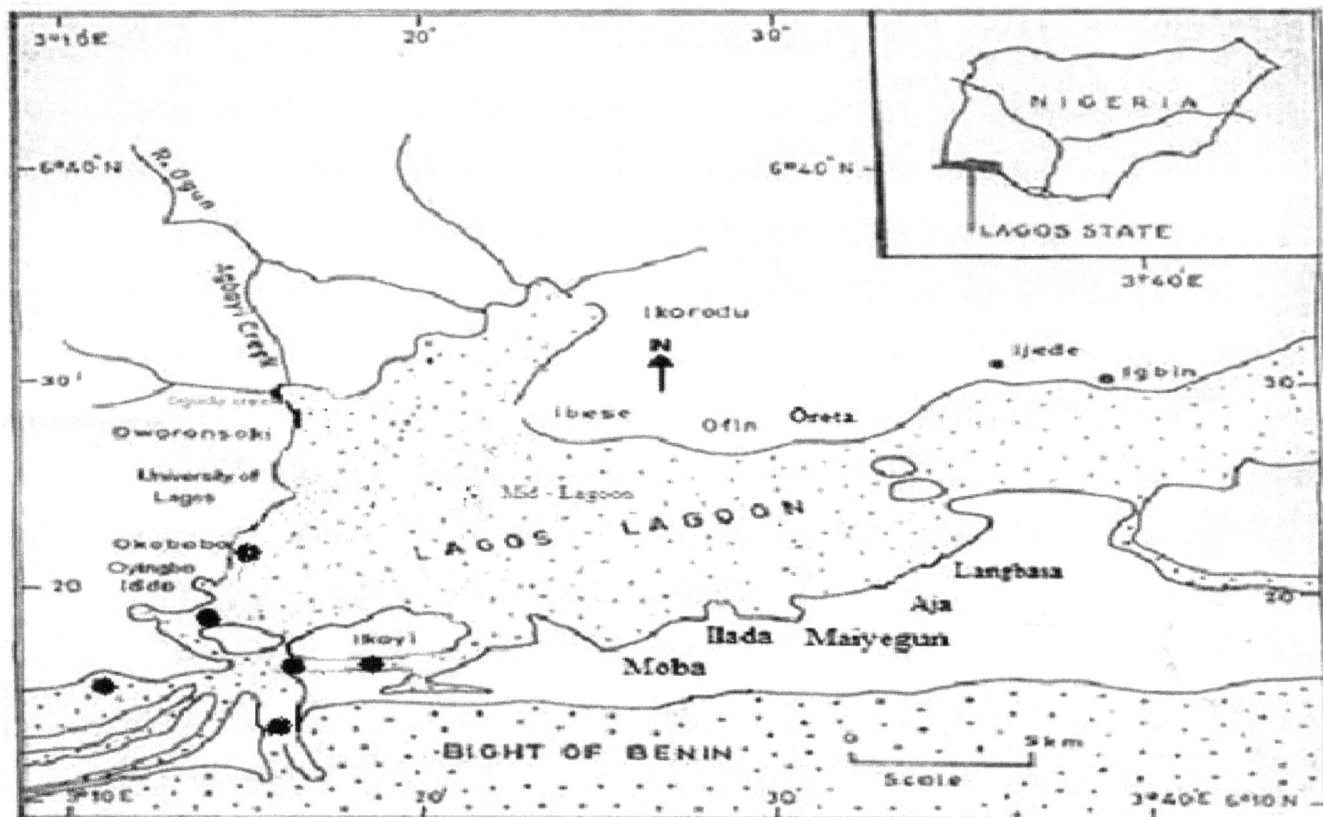

Figure 1. Sketch of the Map of Lagos lagoon; Represents the sampling points.

analyte (Aldrin, chlordane, DDE, DDT, dieldrin, endosulfan, heptachlor, methoxychlor and hexachlorobenzene) were prepared by diluting the standard mixtures to a desired concentration (0.1 to 2.0 µg/L) with n-hexane.

Extraction procedure

Water samples were collected from 6 different locations during a high tidal in Lagos lagoon at 0.5 and 2.5 m depth and from 2 other river bodies within Lagos metropolis which are quite distant from the Lagoon with low instances of intense Agricultural practices and industrialization (thereafter referred to as OKB, AP, OB, VI, EE, CMS, Ag and Aj, between March and April, 2010) in a pre-cleaned glass jars. Global positioning system (Global tech Ltd, Hong Kong) was used to identify the location where samples were collected were collected, while the physicochemical parameters of water samples including total suspended solids, pH, conductivity, turbidity, salinity, alkalinity, temperature and dissolved oxygen (Table 1) were measured with Horiba U-10 (Horiba Ltd, Japan) multiparameter water quality metres. On arrival in the laboratory, samples were filtered through 0.45 µm glass filter (Whatman GF/F) to remove particulate matter, acidified with dilute hydrochloric acid to pH 2.5 and stored in the dark at temperature between 0 to 4°C prior to extraction. SPE using pre-packed reversed-phase octadecyl (C-18 bonded silica) contained in cartridges was used for sample extraction. Methanol (5 ml) was added to the water samples to aid extraction. Prior to the extraction, the C-18 bonded phase, which contains 500 mg of bonded phase was conditioned with 5 ml methanol and equilibrated with 2 ml ultra high purity water. 10 ml of the sample was passed at a rate of 0.3 ml/min through the

conditioned cartridge. The analyte was allowed to percolate through the cartridge under vacuum for 15 min after which time they were eluted with 5 ml n-hexane. The eluate was evaporated to complete dryness under a gentle stream of nitrogen gas and analyte was reconstituted in 0.5 ml n-hexane for GC analysis. For recovery studies, 10 ml aliquots of ultra high purity Milli-Q water samples (pH adjusted to 6.5) and spiked with standard solutions (0.05, 0.5 and 1.5 µg/L) of the analytes were similarly extracted in triplicates as described (Spyros et al., 2003).

Gas chromatography analytical conditions

Analyses were performed on an Agilent GC-6820 (Santa Clara, C. A, U.S.A) equipped with [63]Ni electron capture detector (ECD). Zebron ZB-multiresidue-1 column (30 × 0.25 mm i. d × 0.25 µm film thickness) manufactured by phenomenex (Torrence, USA) was employed in the separation of the analytes. Helium was used as the carrier gas while nitrogen was used as the make-up gas. The analytical conditions are as shown in Table 2. A 1 µl volume was injected in a splitless mode. The residues of organochlorine pesticides were determined by comparing the peak areas of the samples and the calibration curves of the standards.

Quality assurance

For every set of 10 samples, a procedural blank and spiked sample consisting of all reagents was run to check for interference and cross-contamination. The range of linearity of the detector was evaluated from the curves generated by plotting the detector signal

Table 1. Phsicochemical parameters of water samples. Each sampling location was marked with coordinate. ^{X}; Water samples collected from other water bodies in Lagos. OKB; Oko baba, AP (Apapa), OB (Obalende), VI (Victoria Island), EE (Ebute- Ero),CMS (CMS), Ag (Agbelekale), Aj (Ajasa). Value in bracket represents the depth at which the water is collected in metres.

Sample code	Coordinate	Total suspended solid (Mg/L)	pH	Conductivity (S/M)	Turbidity (Mg/L)	Salinity (Mg/L)	Alkalinity	Temperature (°C)	Dissolved oxygen (Mg/L)
OKB (0.5)	$06^0 28^1 56^{11N}$	10	7	31.5	12	20.1	20	29	1.5
OKB(2.5)	$03^0 23^1 34^{11E}$	45	8	31.8	77	20.1	22	29	1.8
AP(0.5)	$06^0 26^1 11^{11N}$	7.5	7	21.8	45	17.5	18	30	1.7
AP(2.5)	$03^0 21^1 39^{11E}$	21	8	28.8	59	18.0	24	29	1.6
OB (0.5)	$06^0 26^1 02^{11N}$	5.0	8	28.6	4	17.8	20	30	1.2
OB(2.5)	$03^0 21^1 52^{11E}$	8.0	8	32.4	3	2.4	20	29	1.6
VI(0.5)	$06^0 21^1 57^{11N}$	1.0	8	28.4	10	17.5	22	31	1.6
VI(2.5)	$08^0 22^1 05^{11E}$	8.0	8	37.1	6	23.8	22	30	2.3
EE(0.5)	$06^0 27^1 16^{11N}$	6.0	8	34.7	3	20.1	20	31	6.8
EE(2.5)	$03^0 23^1 03^{11E}$	8.0	8	34.2	1	21.8	16	30	3.0
CMS(0.5)	$06^0 28^1 56^{11N}$	8.0	8	29.6	13	18.4	20	29	3.2
CMS(2.5)	$03^0 23^1 34^{11E}$	2.9	8	32.9	1	21.8	18	28	1.7
Ag(0.5)^x	$05^0 22^1 24^{11N}$	13	9	27.6	13	16.7	16	32	1.5
Aj (0.5)^x	$04^0 23^1 34^{11N}$	21	8	25.4	47	15.8	28	31	1.6

Table 2. Gas chromatograph conditions.

Conditions	Values
Career gas flow	2 mL/min
Make up gas flow	20 mL/min
Oven temperature	50 °C (1 min).
	- 120 °C (2 min) at 30 °C/min
	- 250 °C (5 min) at 15 °C/min
	- 280 °C (5 min) at 30 °C/min
Injector temperature	280 °C
Detector temperature	300 °C

versus the amount injected.

RESULTS AND DISCUSSION

Analytical quality assurance

The GC chromatogram of 2.0 µg/L OCPs standards mixture is as shown in Figure 2. The method detection limits of OCPs in samples were determined as the concentration of analyte in a sample that gives rise to a peak with a signal-to-noise ratio (S/N) of 3. The detection limit was lowest for methoxychlor (0.004 µg/L) and highest for dieldrin (0.012 µg/L), while the detection limit of other analytes lies within the range. The average recoveries (n = 3) for OCPs through the analytical procedures (laboratory treatment) was 82.5 to 100.2% as determined by spiking OCPs mixtures in ultrapure milli-Q water samples. The relative standard deviations (RSD) were below 6.0%. They all met the requirement of US-EPA

(Recovery: 70 to 130%, RSD is < 30%) showing that the analytical protocols used in this study can effectively determine OCP residue in water. The quantitative data were however not corrected for surrogate recoveries.

Concentrations of OCPs

The sample location and the relative values of physico chemical parameters of water samples are as shown in Table 1, while the mean concentration of OCPs residues in water samples are as shown in Table 3. Turbidity is a measure of cloudiness, while total suspended solids is a measure of suspended particles and salinity is a measure of dissolved salt content of the water samples. Total suspended solids are relatively higher at OKB and CMS while turbididy were relatively higher at OKB, AP and AJ. The temperature of the water bodies lies between 28 to 32°C, while the pH of the samples was either neutral or slightly alkaline. It appears that there is no correlation

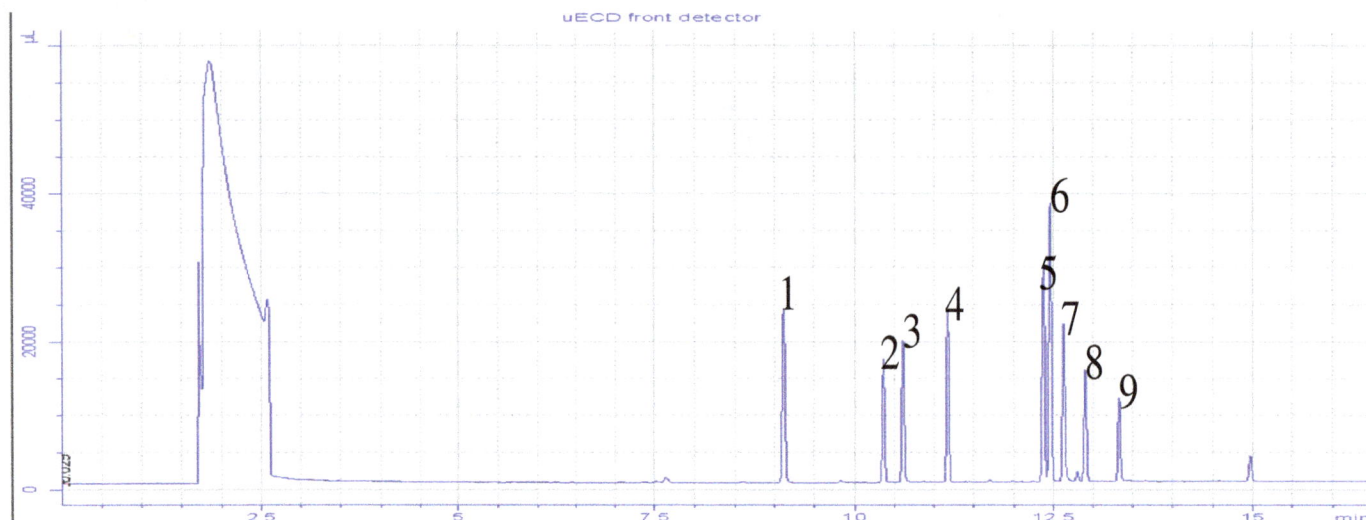

Figure 2. Gas-Chromatogram of OCPs standard mixture (2.0 µg/L): 1, DDT, 2, Dieldrin, 3, DDE, 4, Chlordane, 5, heptachlor, 6, HCB, 7, Endosulfan, 8, Methoxychlor, 9, Aldrin.

Table 3. Mean concentrations of OCPs residues (µg/L) at various locations sampled.

Sample code	DDT	DDE	Dieldrin	Aldrin	Chlordane	Heptachlor	Methoxychlor	HCB	Endosulfan
OKB(0.5)	0.049	0.030	0.037	0.417	0.006	0.032	ND	0.015	0.084
OKB(2.5)	0.056	0.005	0.038	0.287	0.007	ND	ND	0.015	0.067
AP(0.5)	0.910	0.063	0.954	0.128	0.818	0.060	0.032	0.608	0.936
AP (2.5)	0.245	0.073	0.015	0.270	0.904	0.040	0.005	0.034	0.616
OB(0.5)	0.368	0.324	0.150	0.359	0.342	0.020	ND	0.488	0.660
OB(2.5)	0.032	0.087	0.996	0.295	0.950	0.064	0.038	0.774	0.996
VI(0.5)	0.032	0.078	0.017	0.402	0.078	ND	0.005	0.048	0.068
VI(2.5)	0.095	0.061	0.071	0.637	0.312	0.052	0.052	0.077	0.298
EE(0.5)	0.026	0.123	0.235	0.125	0.010	ND	ND	0.034	0.050
EE(2.5)	0.012	0.477	0.137	0.736	0.008	ND	ND	0.156	0.236
CMS(0.5)	0.083	0.270	0.780	0.790	0.034	ND	ND	0.320	0.200
CMS(2.5)	0.071	0.064	0.296	0.700	0.009	ND	ND	0.034	0.050
Ag(0.5)[x]	0.051	0.030	0.076	0.080	0.012	0.067	0.016	0.087	0.015
Aj(0.5)[x]	0.052	0.032	0.140	0.100	0.009	0.061	0.123	0.532	0.321

N.D; Non detectable; *: Water samples collected from other water bodies in Lagos. OKB: Oko baba, AP: Apapa; OB: Obalende; VI: Victoria Island; EE: Ebute- Ero; CMS: CMS; Ag: Agbelekale; Aj: Ajasa; Value in bracket represent the depth at which the water is collected in metres.

between the organochlorine pesticides residues distribution and the measured physicochemical parameters of the samples.

The results indicate that 88% of samples analysed recorded positive for the presence of OCPs. Substances such as chlordane, endosulfan, hexachlorobenzene, dieldrin and aldrin were the pesticides detected in larger concentration (0.001 to 0.996 µg/L), while heptachlor, methoxychlor, DDT and DDE were detected in lower concentrations (< 0.368 µg/L). On the whole, the overall mean total concentration of aldrin was the highest while overall mean total concentration of heptachlor was the lowest. The concentration of heptachlor and methoxychlor were below the detection limit in 57 and 50% of samples analysed respectively. 37.3% of samples analyzed recorded mean concentrations higher than the European Community allowable residual limits in drinking water set at 0.1 µg/L.

In this study, the concentration of OCPs detected in water samples collected from other sources (freshwater) in Lagos were generally lower compared to samples from Lagos lagoon. This can be attributed to the fact that these other water bodies are located near residential areas where agricultural and industrial activities are very low. The mean concentration of OCPs detected in water samples were much lower than was detected in our

Table 4. Comparison of the mean concentration of OCPs (μg/L) residues in water samples from Lagos Lagoon with other water bodies in some other cities elsewhere.

OCPs	Lagos lagoon (Present study)	Ibadan (Nwakwoala et al., 1991)	Culturama Brazil (Raposo Junior et al., 2007)	Hyderabad city, India. (Shukla et al., 2005)	Okavango Delta, Botswana (Mmualefe et al., 2009)
HCB	0.015-0.217	N.D-0.092	N.D-0.0325	-	61.4
Heptachlor	ND- 0.022	0.004-0.202	N.D-0.0046	-	-
Aldrin	0.08-0.375	N.D-0.04	N.D-0.00116	-	-
Endosulfan	0.015-0.355	N.D-0.43	N.D-0.003	0.21-0.87	-
Dieldrin	0.017-0.311	0.018-0.657	N.D-0.0016	-	-
DDT	0.012-0.169	N.D-1.3	N.D-0.0023	0.15-0.19	-
Chlordane	0.006-0.300	-	N.D-0.0037	-	3.2
DDE	0.005-0.169	-	N.D-0.0149	-	5.3

previous studies (Adeyemi et al., 2008) on fish samples from the same lagoon. This outcome is expected because of the high lipophilic and hydrophobic nature of the compound, and the possibility of being retained on the organic phase of sediment and organisms (Sarka et al., 1997).

The concentrations of OCPs in water samples from Lagos lagoon were also compared to water bodies in other parts of the world as shown in Table 4. The mean concentrations of OCPs in water samples in this study were lower compared to levels detected in water samples from Hyderabad City in India (Shukla et al., 2006) and Okavango Delta in Botswana (Mmualefe et al., 2009) but were higher than those detected in previous studies on water bodies from Ibadan, Nigeria (Nwankwoala et al., 1992) and Culturama in Brazil (Raposo et al., 2007). Although the use of the OCPs has either been banned or restricted, same cannot be said about developing nations like Nigeria. The OCPs are still being used in developing countries such as Indian, Botswana and Nigeria for disease control and increased food production for the growing population.

Relatively low levels of DDT detected could be as a result of metabolism of DDT to DDE. The relative concentration of the parent DDT compound and its biological metabolite DDD and DDE are often used as indices for assessing the possible sources and time of application (Hites and Day, 1992). This is because DDT can be biodegraded to DDE and DDD under aerobic and anaerobic condition respectively. In this study, the ratio DDE/$\sum$DDT was greater than 0.5, which is an indication of less tendency of recent exposure to new sources of DDT and accumulation was probably through indirect sources such as long range transport or historical application (Hong et al., 1999; Zhang et al., 1999). Furthermore, the result reveals a wide-range distribution of the OCPs in the sampled location and the possibility of contamination arising from point sources by way of direct discharge and recent applications of most of the OCPs through agricultural run offs. Since the improvement in crop yield by pesticides application is always concomitant

with the occurrence of pesticides residues in soil and water samples, there is need for regulatory control on application and point sources of pesticides in order to forestall serious health hazards on the environment.

ACKNOWLEDGEMENTS

The authors like to thank the technical staff of Chemical and Physical Department, Nigerian Institute of Oceanography and Marine Research (NIOMR), Victoria Island, Lagos for the support during the sampling campaign. The authors are also grateful to Dr E. A. Ajao, J. P. Unyimandu for assistance and providing facilities in their laboratory.

REFERENCES

Adeyemi D, Ukpo G, Anyakora C, Unyimandu J (2008). Organochlorine pesticides residues in fish samples from Lagos Lagoon, Nigeria. Am. J. Environ. Sci., 4: 649-653.

Barcelo D (1991). Occurrence, handling and chromatographic determination of pesticides in aquatic environment. Analyst., 116 : 681-689.

Belta GD, Likata P, Bruzzese A, Naccarri C, Trombetta D, Turco VL, Dugo C, Richetti A, Naccari F (2006). Level and congener pattern of PCBs and OCPs residues in blue-fin tuna (*Thunnus thynnus*) from the straits of Messina (Sicily, Italy). Environ. Int., 32: 705-710.

Blaso C, Font G, Pico Y (2005). Analysis of pesticides in fruits by pressurized liquid extraction and liquid chromatography-ion trap-triple stage mass spectrometry. J. Chromat., A 1098: 37-43.

Bouman H (2004). South Africa and the Stockholm on Persistent organic pollutants. Afr. J. Sci., 100: 323-328.

Colborn T, Smolen MJ (1996). Epidemiological analysis of persistent organochlorine contaminants in cetaceans. Rev. Environ. Contam. Toxicol., 146: 91-172.

Ennacer S, Gandaoura N, Driss R (2008). Distribution of Polychlorinated biphenyls and Organochlorine pesticides in human breast milk from various locations in Tunisia: Levels of contamination, influencing factor and infant risk assessment. Environ. Res., 108: 86-93.

Fleming L, Mann JB, Briggle T, Sanchez-Ramos JR (1994). Parkinson disease and brain levels of organochlorine pesticides. Ann. Neurol., 36: 100-103.

Garabrant DH, Held J, Langholz B, Peter JM, Mark TM (1992). DDT and related compounds and risk of pancreatic cancer, J. Natl. Cancer Inst., 84: 764-771.

Golfinopoulos SK, Nikolaou AD, Kostopoulou MN, Xilourgidis NK, Vagi MC, Lekkas DT (2003). Organochlorine pesticides in the surface waters of Northern Greece. Chemosphere, 50: 507-516.

Guruge KS, Tanabe S (2001). Contamination by persistent organochlorine and butylin compounds in West coast of Sri Lanka. Mar. Pollut. Bull., 42: 179-186.

Hites RK, Day HR (1992). Unusual persistence of DDT in some Western USA soils. Bull. Environ. Contam. Toxicol., 48: 259-264.

Hong H, Chen W, Xu L, Wang X, Zhang L (1999). Distribution and fate of organochlorine pollutants in the Pearl River Estuary. Mar. Pollut Bull., 39: 376-382.

Ize Iyamu OK, Abia IO, Egwakhide PA (2007). Concentration of residues from Organochlorine pesticides in water and fish from some rivers in Edo state, Nigeria. Int. J. Phys. Sci., 2: 9.

Kolpin DW, Thurman EW, Lingart SM (1998). The environmental occurrence of herbicides: The importance of degradates in ground water. Bull. Environ. Contam. Toxicol., 35: 385-90.

Mayon N, Berrand A, Lerroy D, Malbrouck C, Mandiki SNM, Silvestere F, Golffart A, Thorme J, Kestemont P (2006). Multiscale approach of fish responses to different types of environmental contamination: A case study. Sci. Total Environ., 367: 715-731.

Raposo Jr LJ, Nilva Re-Poppi (2007). Determination of organochlorine pesticides in ground water samples using solid-phase microextraction by gas chromatography-electron capture detection. Talanta, 72: 1833-1841.

Mmualefe LC, Torto N, Huntsman-Mapila P, Mbongwe B (2009). Headspace solid phase microextraction in the determination of pesticides in water samples from the Okavango Delta with GC-ECD and time-of-flight mass spectrometry. Micro Chemical J., 91: 239-244.

Nwankwoala AU, Osibanjo O (1992). Baseline levels of selected organochlorine pesticide residues in surface waters in Ibadan (Nigeria) by electron capture gas chromatography. Sci. Total Environ., 119: 179-190.

Sarka A, Nagarajan R, Chaphadkar S, Pal. S, Singabal SYS (1997). Contamination of organochlorine pesticides in sediments from the Arabian-sea along the west coast of India. Water Res., 31: 195-200.

Saxena MC, Siddique MKJ, Bhargava AK, Murtti CRK, Kutty D (1981). Placenta transfer of pesticides in humans. Arch Toxicol., 62: 199-206.

Shukla G, Kumar A, Bhanti M, Joseph PE, Taneja A (2006). Organochlorine pesticide contamination of ground water in the city of Hyderabad. Environ. Int., 32: 244-247.

Afful S, Anim A, Serfor-Armah Y (2010). Spectrum of Organochlorine Pesticide Residues in Fish Samples from the Densu Basin. Research J. Environ. Earth Sci., 2(3): 133-138

Spyros K, Golfinopoulos A, Anastasia D N, Maria N, Kostopoulou A, Nikos K, Xilourgidis A, Maria C, Vagi A, Dimitris T, Lekkas A (2003). Organochlorine pesticides in the surface waters of Northern Greece. Chemosphere, 50:507–516.

William J, Tagoe L, Drechsel P, Kelderman P, Gijzen H, Nyarko E (2008). Accumulation of Persistence Organochlorine contaminants in milk and serum of farmers from Ghana. Environ. Res., 106: 17-26.

Zhang G, MinYS, Mai BX, Sheng GY, Fu JM, Wang ZS (1999). Time trend of BHCs and DDTs in a sedimentary core in Macao estuary, Southern China, Mar. Pollut. Bull., 39: 326-330.

Ruditapes decussatus embryo-larval toxicity bioassay for assessment of Tunisian coastal water contamination

Salem Fathallah*, Mohamed Néjib Medhioub, Amel Medhioub and Mohamed Mejdeddine Kraiem

Laboratoire d'Aquaculture – Institut National des Sciences et Technologies de la Mer BP59, route de Khniss 5000 Monastir, Tunisie.

This paper aims to assess Tunisian coastal water quality using the clam *Ruditapes decussatus* embryos and larvae bioassays tests. Water samples collected from four stations (Monastir lagoon, Chebba, Mahres and Zarat), was used for chemical analysis and clam bioassay tests (embryogenesis, larval growth and metamorphosis). The results, based on chemical analysis, showed the highest metal contamination in Mahres and Zarat while the lowest ones were recorded in Chebba station. The Monastir lagoon is characterized by the highest level of mercury ($18.1 \pm 2.16 \ \mu gL^{-1}$). Compared to control, reduction of clam embryogenesis is up to 40, 60 and 67% respectively in Monastir, Zarat and Mahres while no reduction were observed in Chebba ($p > 0.05$). Larval growth rate was significantly ($p < 0.05$) reduced in all stations except Chebba this also shown when daily growth rate (DGR) is calculated for larvae reared in water collected from each station. A significant ($p < 0.05$) reduction in larval survival is also shown in Monastir (74%), Zarat (70%) and Mahres (52.8%) compared to control (94.5%). Compared to control (82%), metamorphosis success is significantly reduced only on two stations (Zarat: 68% and Mahres: 64%) conversely survival in this stage was affected in three stations (Monastir: $69 \pm 6.7\%$; Zarat: $52 \pm 6.1\%$ and Mahres: $44 \pm 5.2\%$) compared to the control ($83 \pm 4.7\%$). This work showed that both clam embryos and larvae are sensitive to contaminants and can be used to evaluate seawater contaminations and monitoring pollution.

Key words: Bioassay, coastal water, embryogenesis, larvae, metamorphosis, *Ruditapes decussatus*.

INTRODUCTION

Estuaries are often highly productive ecosystems affected by urban development and industrial activities that strongly increase background levels of potentially harmful chemical and physical agents. The ever increasing number of xenobiotics and the effects of physicochemical parameters on their availability to marine organisms greatly complicate monitoring based on chemical analyses. Even if we had *a priori* knowledge of the kind of pollutants present, analytical chemistry allows determination of the degree and nature of pollution, but it does not provide evidence for biological

consequences. Bioassays allow the detection of these effects by measuring biological responses on marine organisms, and particularly in their highly sensitive early life stages (His et al., 1999).

Criteria for the choice of target organisms for bioassays have been evaluated. The embryos and larvae of marine organisms are generally more sensitive to toxic substances than adults, and gametes and embryos of bivalve have been recognized as valuable tools in toxicological studies since Prytherch (1924) tested *Crassostrea virginica*. Toxicity bioassays are now used worldwide to help assess sea water and sediment quality because they can integrate the various complex effects of contaminants. The bivalve embryo bioassay, one of these procedures, has been shown to be reliable, sensitive, and ecologically relevant (Pamela and O'Halloran, 2001;

*Corresponding author. E-mail:salem.fathallah@yahoo.fr.

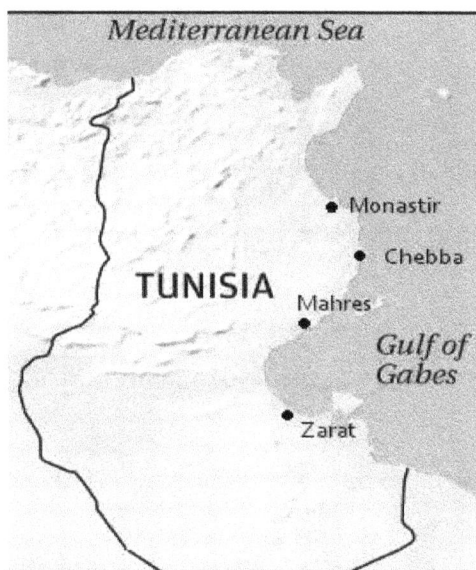

Figure 1. Tunisian map, showing the location of seawater sampling stations for bioassay and chemical analysis.

Beiras, 2002; Nendza, 2002; Volpi et al., 2005; Novelli et al., 2006). During the past decades, numerous studies have been published on the use of bivalve embryos, either concerning the effects of individual contaminants, industrial effluents, and sediments or the assessment of sea and brackish water quality (Losso et al., 2004; Dalmazzone et al., 2004; His et al., 1999; Quiniou et al., 2005, 2007; Stronkhorst et al., 2004). Because of its sensitivity, we considered this test as the most suitable for toxicity testing to better understand environmental disturbances affecting the Tunisian coasts.

Although these methods are used globally, the potential of such toxicity tests has not been adequately explored for Tunisian coastal waters. This study describes bioassay utilizing *R. decussatus* embryos and larvae to monitor coastal water quality and is based on the sensitivity these life stages to different concentrations of water samples. Four stations were selected and water samples were chosen according to their potential effects on sensitive biological components (embryos and larvae) that could be exposed to pollutants. The aims were: a) to evaluate the replicability of all procedures; b) to rank sites (make a classification among sites) for toxicity. This work seeks to form the methodological basis for providing reliable toxicological data to contribute to seawater quality assessment and monitoring in transitional environments.

MATERIALS AND METHODS

Sites description and sampling procedure

Four sampling sites located along Tunisian coasts were chosen (Figure 1). Sampling sites were chosen on the basis of population

abundance of the most important local bivalve, the clam (*R. decussatus*). Zarat is an important site used to grow bivalve juveniles (Tunisian-japaneese project for growing clams), Mahres is also one of the most important site. Our laboratory and the nursery were located at the site of Monastir. Chebba is considered as the most distant site, all sites (Zarat, Mahres, Monastir and Chebba) have potential for clam juveniles' culture.

Triplicate of surface water sample, serving for bioassay, were taken in polypropylene flasks from each site and stored at room temperature (22°C). For chemical analysis, water was sampled in 250 ml polypropylene flasks and stored at 4°C. All flasks used were previously already washed with ultrapure acids. The hydrographic conditions (temperature, salinity, pH and dissolved oxygen) were analyzed in situ by means of Orion and Hanna electrodes.

Metal analysis

The background metal concentrations in seawater were determined for the entire samples collected from the four stations are given in Table 1.

The metal concentrations were determined in seawater collected from the four stations. After filtration, the metal samples were extracted by using ammonium pyrrolidine dithiocarbamate (APDC) and methyl isobutyl ketone (MIBK) and the heavy metals in seawater were estimated according to the method of Tewari et al. (2001). The samples were examined for Cu, Pb, Cd, Zn, and Ni by using an air-acetylene flame atomic absorption spectrophotometer (AAS; Spectra AA-10 Varian). Mercury was analyzed by automated cold vapor AAS, according to Weltz and Schubert-Jacobs (1991). To avoid contamination, all glassware and equipment used were acid-washed. To check for contamination, procedural blanks were analyzed once for every five samples.

R. decussatus embryotoxicity bioassay

Clams (*R. decussatus*) used in this work were conditioned in our laboratory for at least 5 month and fed on the microalgae *Isochrysis galbana* and *Chaetoceros calcitrans*. Handling conditions of adult stock were 20- 22°C temperature, 34.5–36.0 ppt salinity, 6.2– 6.6 mg/L O2, and 7.5– 8.4 pH.. The embryo toxicity test was performed according to the method proposed by His et al. (1997). Adults were induced to spawn by thermal stimulation (temperature cycles at 20 and 28°C). Gametes of good quality derived from the best males and females were selected (sperm with high motility and eggs with homogeneous dimensions and regular shape) and filtered at 30 μm (sperm) and 100 μm (eggs) to remove impurities. Eggs (1000 ml) were fertilized by injecting 10 ml of sperm; fecundation was verified by microscope, controlling the presence of the fertilization membrane and the number of sperm cells (10–20) around each egg (His et al., 1997). Egg density was determined by counting four subsamples of known volume. Fertilized eggs were delivered at a density of 60 eggs/ml into experimental vials containing seawater collected from studied sites and artificial seawater (control), reconstituted according to ASTM (1998) at 34 ppt salinity, and were incubated for 24 h at 23°C. The assay was terminated after 24 h using 10% formalin solution when D-shape larvae were observed in control. One hundred larvae were counted, distinguishing between normal larvae (D-shape) and abnormalities (malformed larvae and prelarval stages). The acceptability of test results was based on negative control for a percentage of normal D-shape larvae ≥ 80% (His et al., 1999).

Growth and survival experiment

After 24 h, swimming veliger larvae (D-larvae stage) were re-

Table 1. P values registered after analysis of variances using one-way ANOVA comparison test between Control and studied sites.

	Chebba	Monatir	Zarat	Mahres
Embryogenesis	1	0.035	0.0026	0.0021
Larval growth	0.92	0.03	0.036	0.006
Larval mortality	0.95	0.039	0.031	0.007
metamorphosis	1	0.97	0.048	0.045

Table 2. Heavy metals contents (μg/L) in water collected from studied stations (Mean ± SD).

Metals stations	Cd	Pb	Cu	Ni	Zn	Hg
Mahres	15.48±1.73	31.7±2.67	78.9±5.4	4.4±0.21	124±6.97	8.5±0.14
Monastir	7.24±1.54	19.5±2.60	15.7±3.56	3.8±0.37	48.7±3.95	18.1±2.16
Zarat	11.40±1.45	25.8±1.49	108.5±6.67	1.8±0.25	94±4.79	4.6±0.15
Chebba	3.05±0.21	4.3±0.44	4.3±2.17	0.6±0.1	18.4±1.50	2.8±0.07

suspended in a 2 L glass beakers, approximately $2×10^4$, containing seawater collected from all studied sites and artificial seawater for the control preparation (3 replicates per treatment) which was replaced daily. Larvae were fed daily with phytoplankton mixture (*Isochrysis Tahiti* and *Chaetoceros calcitrans*). The mean length of the larvae (30 individuals per treatment) was recorded after 2, 4, 6, 8, 10, 12 and 14 d of rearing. Larval survival was assessed at the end of the experiment by counting under microscope dead and live larvae in the first 100 larvae encountered.

Metamorphosis experiments

This experiment was carried out when pediveligers were competent to metamorphose, as indicated by a larval size (180 μm) and appearance of feat. The number of pediveligers was calculated first, and placed in 10 ml (4-6 ind. ml^{-1}) Petri dishes (3 replicates) filled with different water samples and ASW (control). The number of postlarvae (metamorphosed larvae) was recorded 72 h later using microscopy. This experiment was conducted at 23°C and salinity was adjusted at 37 ppt in all preparations.

Statistical analysis

Responses to each treatment (percentage of abnormality in embryo development) were corrected for effects in control tests by applying Abbott's formula (ASTM, 1998). Comparisons among sites and the control were conducted for embryogenesis, larval survival, growth and metamorphosis using one-way analysis of variance and Dunnett's multiple comparison tests (Table 1). Differences were considered significant at p ≤ 0.05. All statistical analysis were done using SPSS 14.0 for Windows software.

RESULTS

Metal analysis

Heavy metal concentrations of surface water sampled from studied stations are shown in Table 2. Mahres showed maximum values for all pollutants in water except for mercury (maximum in Monastir) and copper (maximum in Zarat). The concentrations of metals found in water collected from Mahres, Monastir and Zarat are considered relatively high compared to concentrations found in Chebba water. We note that concentrations of copper and zinc in Mahres and Zarat were higher than EC50 recorded for *R. decussatus* (Fathallah et al., 2010) and As Chebba site showed the least heavy metals amounts, it can be considered as reference site.

Embryotoxicity bioassay

Percentages of embryos successfully developing to normal 'D' stage after 24 h incubation in water collected from the four selected stations and control water are shown in Figure 2. Analysis of variance revealed a significant reduction of embryogenesis in three stations among the four selected ones. Mahres and Zarat waters showed the highest toxicity effect on *R. decussatus* embryos, they reduced significantly ($p < 0.01$) embryogenesis success to up to 50% comparing to control. Reduction of embryogenesis is also significant ($p < 0.05$) in Monastir station but lower than reduction observed in the two previous stations (Mahres and Zarat); we estimate that this reduction is due to high level of mercury found in this station. Results showed that number of normal 'D' stage larvae is not significantly (p > 0.05) different in Chebba water comparing to control. These founding are in concordance with water metal concentrations in studied stations (Table 2).

Growth and survival experiment

Growth

Larval growth (shell height increase), for all studied

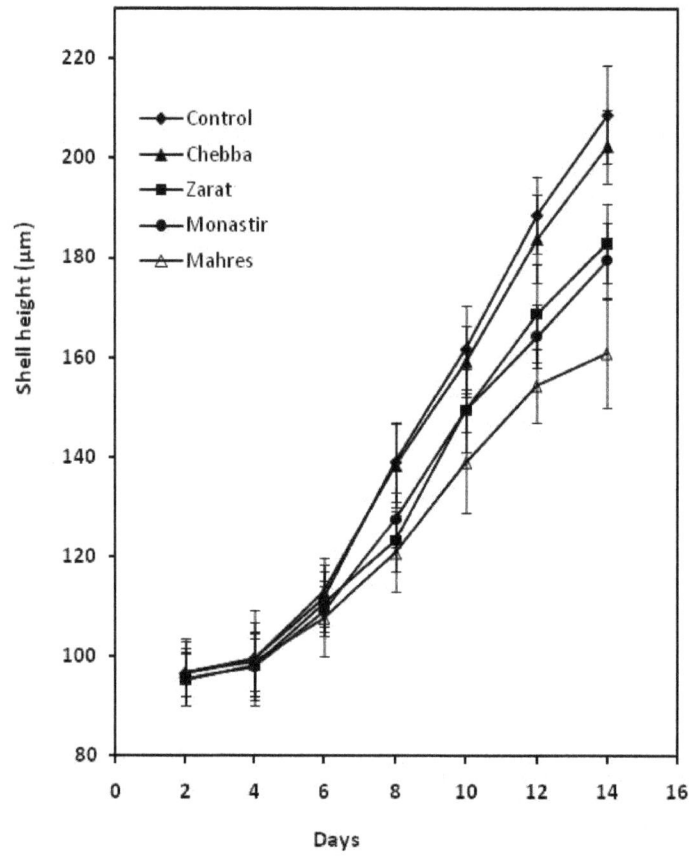

Figure 3. *Ruditapes decussatus* height increase in larvae reared from D-shaped stage to pediveliger stage in sediment elutriates of different studied sites.

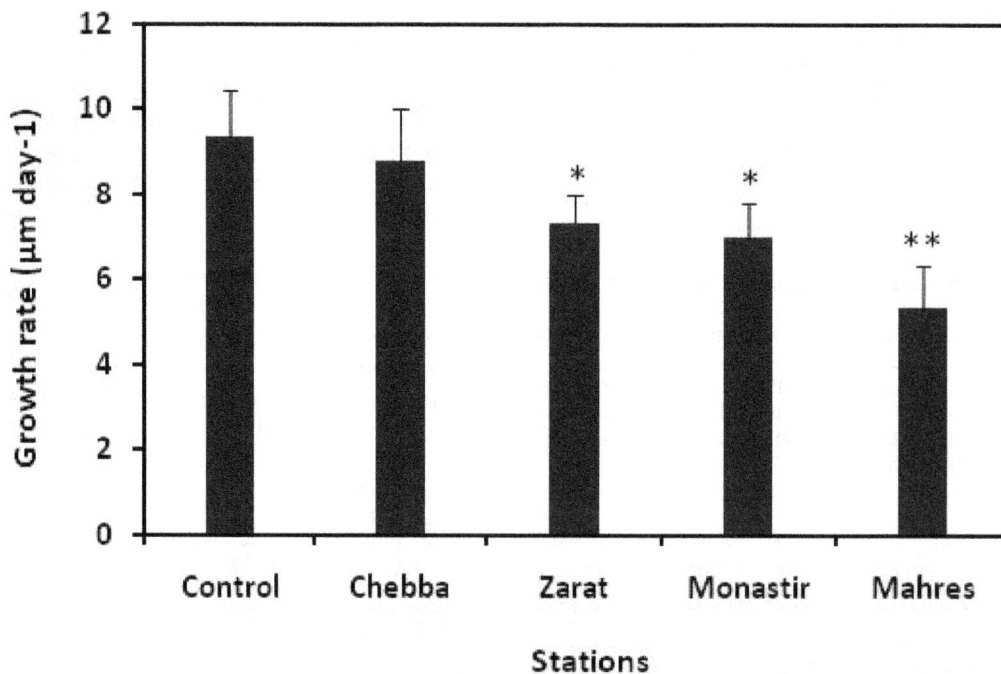

Figure 4. Mean growth rates (μm day^{-1}) of *R. decussatus* larvae exposed to sediment elutriate of different studied sites for 14 days. * indicates treatments significantly less than the control ($p<0.05$).

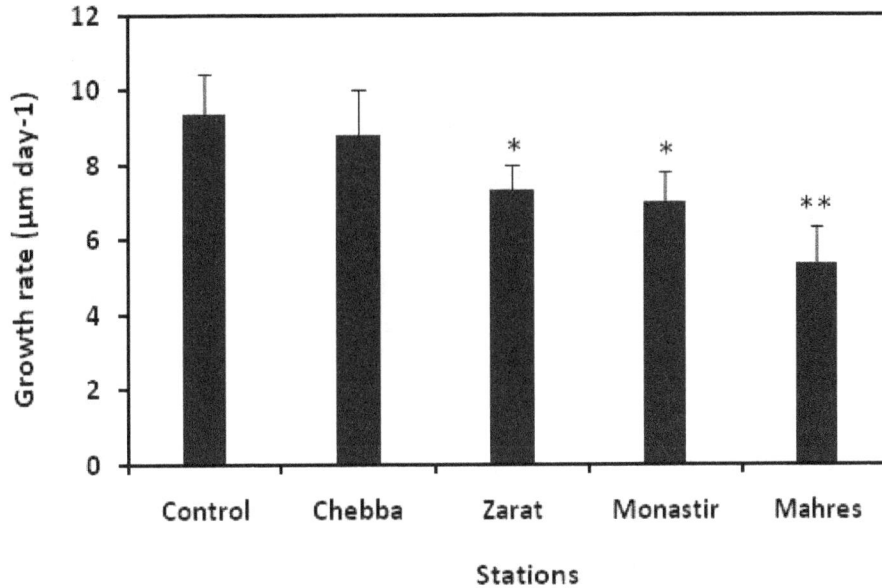

Figure 4. Mean growth rates (µm day⁻¹) of *R. decussatus* larvae exposed to sediment elutriate of different studied sites for 14 days. * indicates treatments significantly less than the control (p<0.05).

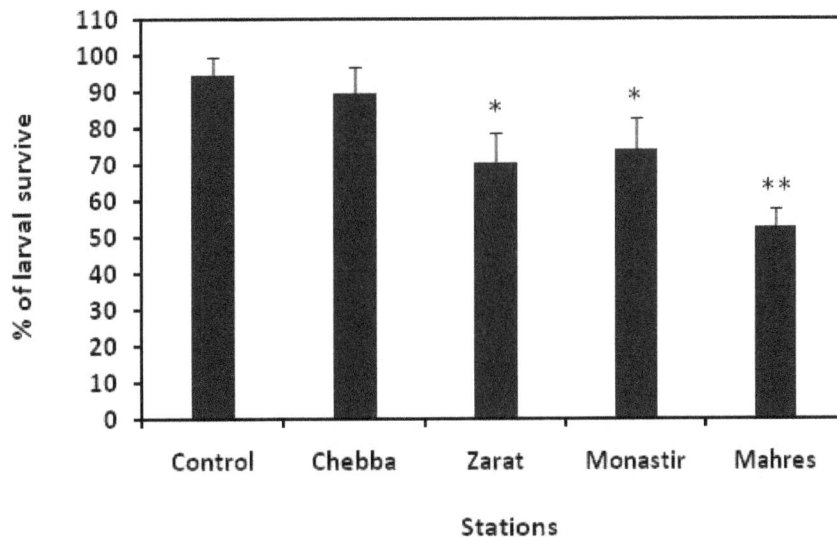

Figure 5. Survival at the end of larval stage (14 days) of *R. decussatus* larvae exposed to sediments elutriates of differents studied sites. Asterisks indicate the samples which significantly differ from the controls, *p<0.05 while ** at p<0.01.

stations, is illustrated in Figure 3. Growth was significantly (*p*<0.05) reduced, compared to control, in water collected from Mahres, Monastir and Zarat from Day 6 to Day 14. Conversely, larvae reared in water collected from Chebba didn't show a significant height reduction (p>0.05). The calculation of the daily growth rates (DGR) of larvae confirmed those results (Figure 4). In effect, compared to the control, larval DGR is significantly (*p*<0.05) lower in Mahres, Monastir and

Zarat waters.

Survival

Larval survival recorded after 14 days of rearing (pediveliger stage reached) is shown in Figure 5. Mean *R. decussatus* larval survival ranged from 52.9 to 89.8%. Significant reduction in survival, relative to the control,

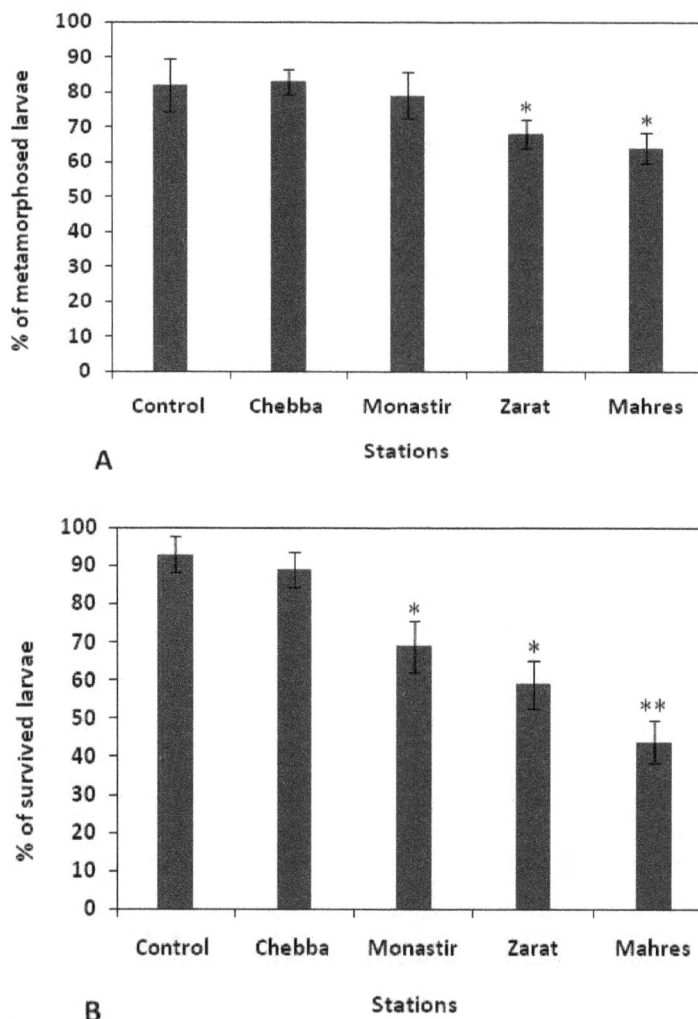

Figure 6. Percentage of metamorphosed individuals (A) and survival (B) at the end of metamorphosis stage of *R. decussatus* pediveligers exposed to surface water collected from different studied stations. Asterisks indicate the samples which significantly differ from the controls, *$p<0.05$ while ** at $p<0.01$.

was exhibited in water collected from three stations Monastir (74.1%), Zarat (70.7%) and Mahres (52.8%) (Dunnett's tests, p < 0.05). Conversely no significant statistical differences in survival were shown between the control and the fourth site (Chebba: 89.8%).

Metamorphosis experiments

The percentages of metamorphosed individuals, obtained in competent pediveligers exposed to water from studied stations, and survival rates were presented respectively in Figures 6A and 6B. This experiment showed a significant ($p<0.05$) reduction of the number of metamorphosed larvae exposed to waters collected from Zarat and Mahres compared to control. In contrast, no significant differences ($p>0.05$) was recorded between

metamorphosis success in control water and the rest of studied stations (Monastir and Chebba). Larval survival, recorded at the end of settlement stage, is also affected in all stations except Chebba. In effect, after 72 h of exposure to different stations water, the number of larvae is reduced significantly ($p<0.05$) compared to control (survival = 92%) and survival percentages obtained are 69, 58 and 44% respectively in Monastir, Zarat and Mahres. However, in Chebba's water, no significant ($p>0.05$) effect was shown (survival = 89%).

Discussion

Pollution in estuarine and marine environments is considered a critical environmental issue due to anthropogenic pollution and to the high variation in several abiotic factors that impose severe restrictions to organisms living in these areas (Matthiessen and Law, 2002; Amado et al., 2006). The present work is aimed to assess the pollution statue of four sites belonging to the Tunisian coasts by exposing R. decussatus embryos and larvae to water samples, collected from these sites.

In this work we selected 4 sites along the Tunisian coastal areas characterized by different levels of heavy metal contamination. Sampling sites were chosen because of their geographical distribution near urban, industrial and agricultural areas and their importance for clam culture. Monastir lagoon, where located our nursery, is located in mid Tunisia and seems to be relatively polluted since the presence of textile industry as contamination source. Chebba site is also located in mid Tunisia, seems to be relatively unpolluted since no contamination source is present. The other two sites belong to South Tunisia and are represented by Zarat and Mahres. The latter is located south to Sfax that is the most important industrial site of the country, and close to a phosphogypsum plant, that can be a source of heavy metal release (Figure 1).

Heavy metals analysis clearly showed different degrees of heavy metal loads in water sampled from selected sites. Mercury content was higher at Monastir station. Chebba station showed the lowest heavy metals levels, this can be explained by the fact that is the most distant site from industry. The highest levels of heavy metals, except mercury and copper where found in water collected from Mahres station. In fact, Mahres were characterized by relative high heavy metal loads (Cd, Zn and Cu), as previously reported by several authors. Others studies performed at Mahres site demonstrated the presence of high concentrations of such heavy metals with a corresponding increase of Metallothioneins protein levels in tissues of R. decussatus (Hamza-Cheffai et al., 2003 and Banni et al., 2005) and the Mediterranean mussel Mytilus galloprovincialis (Banni et al., 2007) and Nereis diversicolor (Bouraoui et al., 2009). The results confirm the pollution statue of Sfax city coasted area, due essentially to the presence of continuous discharge of heavy metals and also of organic compounds from local industrial activities as described by Banni et al. (2007); Jebali et al. (2007); Smaoui-damak et al. (2004) and Banni et al. (2003).

Bivalve (clam, oyster and mussel) embryo bioassay was one of the techniques selected for monitoring and surveillance of sediment and coastal water (Chapman and Morgan, 1983; Thain, 1992; Geffard et al., 2002; Phillips et al., 2004; King et al., 2004; Volpi Ghirardini et al., 2005). The results of the present study clearly document the sublethal and lethal responses of the development of clam embryos exposed to water samples collected from different stations. Our data provide a very good demonstration of a strong relation between the level of water pollution and R. decussatus embryogenesis. The clams embryos that were allowed to develop in contact water collected from different stations have shown marked developmental effects.

R. decussatus embryos development was significantly reduced in water collected for all studied stations except Chebba. This confirms chemical analysis that demonstrated Mahres as the most polluted site and that Chebba is the less contaminated one. Compared to control, reduction of embryogenesis is up to 40, 60 and 67% respectively in Monastir, Zarat and Mahres. Reduction of embryogenesis observed in Monastir is due especially to the high level of mercury recorded in the water (18.1 ± 2.16 µg L^{-1}); this level is slightly lower than the EC50 recorded for mercury on R. decussatus embryos (Fathallah et al., 2010) for that reason inhibition did not reached 50 %.

The pediveliger larva of the European clam, R. decussatus, was explored as a bioassay organism because of nearly year round commercial availability, and common use in previous work with culture and toxicity testing. In this study clam larvae showed a normal growth in Chebba and a reduction in growth rate in Mahres, Monastir and Zarat waters comparing to the control treatment. Highest mortality is occurred in Mahres station. The present study confirm sensitivity of clam larvae to heavy metals like shown by Manahan and Crisp (1982) and can be easily used for field monitoring of water and sediment quality. Beiras and His (1994) showed a reduction in growth rate in Crassosstrea gigas larvae exposed to 8 and 10 µg L^{-1} of Hg from Day 6 of rearing. Wang et al. (2009) showed that Metrix metrix larvae exposed to 187 µg µg L^{-1} Hg grew little or died after 24 h and only grew about 20 µm when exposed to 18.5 µgL^{-1} of Hg and that larval growth was limited at a Cd concentration of 1,046 µg L^{-1} and was significantly retarded at 104 µg µg L^{-1}, they showed also that metal affect significantly larval survival. Many other works showed that heavy metals caused failure in metamorphosis (Beiras et His 1994; Hoare et al., 1995). In fact, Wang et al. (2009) showed that 104 µg l^{-1} Cd reduced larval attachment by 44.5%, while 187 µg l^{-1} Hg inhibited metamorphosis by 41.8%. 11.0 µg l^{-1} Cd and 18.5 µg l^{-1} Hg had no adverse effect on larval settlement; on the other hand, enhanced average metamorphosis was also observed at 18.5 µg L^{-1} Hg. This was in concordance with our study where metamorphosis success is affected in the most polluted stations (Mahres and Zarat), survival is also reduced at the end of this step at Mahres, Zarat and Monastir stations. But the use of larvae for bioassay is problematic because sensitivity is age dependant (Beiras and His, 1994). In our knowledge this work is this first studying contamination in Monastir lagoon, but the complex nature of the environment studied with regard to the wide range of possible sources of contamination (waste outlets, inputs from rivers, local industries, tourism and harbor activities) makes interpretation difficult. Concerning Zarat and Mahres stations, inhibition of embryos development and larval growth and survival is due to the combined effect of the high levels of three metals (Zn, Cd and Cu) known to be toxic to bivalve embryos and larvae. Mahres is located south to Sfax that is the most important industrial site of the country, and close to a phosphogypsum plant, that can be a source of heavy metal release. This station is characterized by relative high heavy metal loads (Cd and Cu), as previously reported by several authors. In fact, many studies performed at Mahres site demonstrated the presence of high concentrations of such heavy metals and their toxic effect on adult *R. decussatus* (Hamza-Cheffai et al., 2003; Banni et al., 2005) and the Mediterranean mussel *Mytilus galloprovincialis* (Banni et al., 2007). Results obtained for all realized test and chemical analysis confirms that Mahres station is the most contaminated site and contamination found in Monastir and Zarat is lethal for embryogenesis, larval development and survival, contrarily, Chebba can be considered as a reference site because no significant differences were observed compared to control treatment for embryos and larvae bioassays.

Conclusion

It is apparent from this and other studies that embryos are preferable for acute bioassay tests due to (1) higher sensitivity to metal pollutants, (2) more rapid and simple evaluation of the lethal effects of the pollutant, and (3) more simple standardization of the bioassay, avoiding interference of variables such as larval age, larval condition and presence of algal food.

ACKNOWLEDGEMENTS

This work was partially funded by the Tunisian Ministry of Higher Education, Scientific Research and Technology (Laboratoire d'Aquaculture de l'Institut National des Sciences et Technologies de la Mer).

REFERENCES

Amado LL, Da Rosa CE, Meirelles L, Moraes L, Vaz Pires W, Pinho GL, Martins CM, Robaldo RB, Nery LE, Monserrat, JM, Bianchini Martínez PE, Geracitano LA (2006). Biomarkers in croakers Micropogonias furnieri (Teleostei, Scianidae) from polluted and non polluted areas from the Patos Lagoon estuary (Southern Brazil): evidences of genotoxic and immunological effects. Mar. Pollut. Bull., 52: 199–206.

ASTM (1998). Standard guide for conducting static acute toxicity tests starting with embryos of four species of saltwater Q5 bivalve molluscs. E7 24–98, 21.

Banni M, Ben DR, El Abed A, Boussetta H (2003). Genotoxicity, catalase and acetylcholinesterase in the biomonitoring of the Tunisian coasted areas. Bull. Environ. Contam. Toxicol., 70:167-174.

Banni M, Dondero F, Jebali j, Guerbej H, Boussetta H, Viarengo A (2007). Assessment of heavy metal contamination using real-time PCR analysis of mussel metallothionein mt10 and mt20 expression: a validation along the Tunisian coast. Biomark., 12(4): 369-383.

Banni M, Jebali J, Daubeze M, Clerendeau C, Guerbej H, Narbonne JF, Boussetta H (2005). Monitoring Pollution in Tunisian coasts: application of a classification scale based on biochemical markers. Biomark., 10: 105-116.

Beiras R (2002). Comparison of methods to obtain a liquid phase in marine sediment toxicity bioassays with Paracentrotus lividus sea urchin embryos. Arch. Environ. Contam. Toxicol., 42: 23–28.

Beiras R, His E (1994). Effects of dissolved mercury on embryogenesis, survival, growth and metamorphosis of Crassostrea gigas oyster larvae. Mar. Ecol. Prog. Ser., 113: 95-103.

Bouraoui Z, Banni M, Chouba L, Ghedira J, Clerandeau C, Jebali J, Narbonne JF, Boussetta H (2009). Monitoring pollution in Tunisan coasts using a scale of classification based on biochemical markers in worms Nereis diversicolor (Hediste). Environ. Monit. Assess., 164, 691-700.

Chapman PM, Morgan JD (1983). Sediment bioassays with oyster larvae. Bull. Environ. Contam. Toxicol., 34: 438-444

Dalmazzone S, Blanchet D, Lamoureux S (2004) Impact of drilling activities in warm sea: recolonization capacities of seabed. Oil Gas Sci. Technol., 59(6):625–647.

Fathallah S, Medhioub MN, medhioub A, Kraiem MM (2010). Toxicity of Hg, Cu and Zn on early developmental stages of the European clam (Ruditapes decussatus) with potential application in marine water quality assessment. Environ. Monit. Assess. doi: 10.1007/s10661-010-1311-0.

Geffard A, Geffard O, His E, Amiard JC (2002). Relationships between metal bioaccumulation and metallothionein levels in larvae of Mytilus galloprovincialis exposed to contaminated estuarine sediment elutriate. Mar. Ecol. Prog. Ser., 233: 131–142.

Hamza-Chaffai A, Pellerin JC (2003). Health assessment of a marine bivalves Ruditapes decussatus from the Gulf of Gabés (Tunisia). Environ. Int., 28: 609-617.

His E, Beiras R, Seaman M (1999). The assessment of aquatic contamination: bioassays with bivalve larvae. Adv. Mar. Biol., 37:1-178

His E, Seaman MNL, Beiras R (1997). A simplification: the bivalve embryogenesis and larval development bioassay method for water quality assessment. Water Res., 31 (2): 351–355.

Hoare K, Davenport J, Beaumont AR (1995). Effects of of exposure and previous exposure to copper on growth of veliger larvae and survivorship of Mytilus edulis juveniles. Mar. Ecol. Prog. Ser., 120: 163 – 168.

Jebali J, Banni M, Alves de Almeida E, Boussetta H (2007). Oxidative DNA damage levels clams Ruditapes decussatus as pollution biomarkers of Tunisian marine environment. Environ. Monit. Assess., 124:195-200.

King CK, Dowse MC, Simpson SL, Jolley DF (2004). An assessment of five Australian polychaetes and bivalves for use in whole-sediment toxicity tests: toxicity and accumulation of copper and zinc from water and sediment. Arch. Environ. Cont. Toxicol., 47: 314–323.

Losso C, Arizzi Novelli A, Picone M (2004). Evaluation of superficial sediment toxicity and sediment physico-chemical characteristics of representative sites in the Lagoon of Venice (Italy). J. Mar. Sys., 51:281–292.

Manahan DT, Crisp DJ (1982). The role of dissolved organic material in the nutrition of pelagic larvae: amino acid uptake by bivalve veligers. Amer. Zool., 22: 635-638.

Matthiessen P, Law RJ (2002). Contaminants and their effects on estuarine and coastal organisms in the United Kingdom in the late twentieth century. Environ. Pollut., 120:739–757.

Nendza M (2002). Inventory of marine biotest methods for the evaluation of dredged material and sediments. Chemosphere, 48: 865–88.

Novelli AA, Losso C, Libralato G, Tagliapietra D, Pantani C, Volpi Ghirardini A (2006). Is the 1:4 elutriation ratio reliable? Ecotoxicological comparison of four different sediment:water proportions. Ecotoxicol. Environ. Saf., 65: 306–313.

Pamela AB, John O'H (2001). The role of bivalve molluscs as tools in estuarine sediment toxicity testing: a review. Hydrobiologia, 465: 209–217.

Phillips IR, Lamb DT, Hawker DW, Burton ED (2004). Effects of pH and salinity on copper, lead, and zinc sorption rates in sediments from

Moreton Bay, Australia. Bull. Environ. Cont. Toxicol., 73: 1041–1048.

Prytherch HF (1924) Experiments in the artificial propagation of oysters. *Report U.S. com fisheries, 1923, XI, 14 p.*

Quiniou F, Damiens G, Gnassia-Barellib M (2007). Marine water quality assessment using transplanted oyster larvae. Environ. Int., 33(1):27–33.

Quiniou F, His E, Delesmont R, Caisey X, Thebaud MJ (2005) Bioindicator of potential toxicity in aqueous media: a bivalve embryo-larval development bioassay. Methodes d'analyse en milieu marin. *Ifremer [Methodes Analytiques du Milieu Marin Ifremer], Ifremer Editor, France, 22 pp.*

Smaoui-Damak W, Hamza-Chaffaia A, Bebianno MJ, Amiard JC (2004). Variation of metallothioneins in gills of the clam *Ruditapes decussatus* from the Gulf of Gabes (Tunisia). Comp. Biochem. Physiol. Part C., 139:181–188.

Stronkhorst J, Ciarelli S, Schipper C (2004). Inter-laboratory comparison of five marine bioassays for evaluating the toxicity of dredged material. Aquatic Ecosystem Health Manag., 7(1):147–159.

Tewari A, Joshi HV, Trivedi RH (2001). The effect of ship scrapping industry and its associated wastes on the biomass production and biodiversity of biota in situ condition at Alang. *Mar.* Pollut. Bull., 42:461–468.

Thain J (1992). Use of the oyster *Crassostrea gigas* embryo bioassay on water and sediment elutriate samples from the German Bight. Mar. Ecol. Prog. Ser., 91: 211 – 213.

Volpi GA, Losso C, Arizzi NA, Baugrave A, His E, Pier Francesco G (2005). *Mytilus galloprovincialis* as bioindicator in embryotoxicity testing to evaluate sediment quality in the lagoon of Venice (Italy). Chem. Ecol., 21 (6), 455 – 463.

Wang Q, Liu B, Yang H, Wang X, Lin Z (2009). Toxicity of lead, cadmium and mercury on embryogenesis, survival, growth and metamorphosis of *Meretrix meretrix* larvae. Ecotoxicol., 18:829–837.

Weltz B, Schubert-Jacobs M (1991). Evaluation of a flow injection system and optimization of parameters for hydride generation atomic absorption spectrometry. Atomic Spectroscopy 12:91–103.

Metal concentrations in sediments and water from Rivers Doma, Farinruwa and Mada in Nasarawa State, Nigeria

Aremu, M. O. [1]*, Atolaiye, B. O.[1], Gav, B. L.[1], Opaluwa, O. D.[1], Sangari, D. U.[2] and Madu, P. C.[1]

[1]Department of Chemistry, Nasarawa State University, P. M. B. 1022, Keffi, Nigeria.
[2]Department of Geography, Nasarawa State University, P. M. B. 1022, Keffi, Nigeria.

The Rivers Doma, Farinruwa and Mada are important rivers in Nasarawa State, Nigeria especially with regard to domestic use, irrigation and aquatic food. Levels of sodium, potassium, nickel, copper, magnesium, iron, calcium, zinc, lead, cadmium, arsenic, selenium, chromium and manganese were determined in samples of sediment and water collected from different points of the three rivers at two distinct seasons (dry and wet) using atomic absorption spectrophotometer. The results showed that calcium had the highest concentration in sediment and water samples for both seasons. It was found that the cadmium, arsenic and selenium were completely not within the detection limit of AAS for all the water samples for dry and wet seasons. The results further revealed that Ni, Fe, Pb, and Mn concentrations in all the water samples are above the deleterious level based on the standard limits set by World Health Organization (WHO) for drinking water. However, source protection is proposed for these bodies of water for the benefit of mankind, because they were not fit for human consumption.

Key words: River-water, sediments, metal, AAS.

INTRODUCTION

It is a well known fact that adequate supply of fresh and clean drinking water is a basic need for all human beings on earth; yet it has been observed that millions of people are deprived of this, particularly in the developing countries, including Nigeria. This is due not only to over exploitation and poor management of fresh water but also due to their ecological degradation by man's activities in this part of the country (Okuo et al., 2007).

Industrial growth, urbanization and the increasing use of synthetic organic substances have serious and adverse impacts on freshwater bodies due to the intro-duction of various pollutants such as organic compounds, heavy metals, agricultural waste, etc. (Kakulu and Osibanjo, 1988). Metals are introduced into aquatic systems as result of the weathering of rocks and soils, for example, volcanic eruptions and also from several human and industrial materials that contain metals contaminants

(Marr and Creaser, 1983; Adeyeye, 1993; Gutenmann et al., 1988). The increased use of metal containing fertilizers due to the agricultural revolution could lead to pollutants in fresh water reservoirs due to water run-off (Aremu and Inajoh, 2007). Vehicle emissions, and tire and engine wear contribute sizeable concentration of all metals, particularly zinc and copper. Thus, significant correlations are found between traffic volumes and metal concentrations (Ademoroti, 1996; Aremu et al., 2006). Heavy metals are commonly found in natural waters. Though some are essential to living organisms, yet they may become highly toxic when present in high con-centration. Some of these heavy metals, for examples, lead, zinc, manganese, iron, nickel and copper occur in nature in ore deposits. They are released through leaching and weathering into the rivers. Therefore areas characterized by the presence of metal bearing formations and mining operations are expected to have elevated levels of metals in water and sediments (Foster and Wiltman, 1983; Preston and Chester, 1996). Sediments in the aquatic environments form the major repository of heavy metals, holding > 90% of the total amount (Preston and Chester, 1996).

Potable water supply to communities in Nasarawa State is the responsibilities of the government which in most cases has been characterized by low productivity, small coverage and inefficient service delivery. Doma, Wamba and part of Akwanga local government areas are one of the areas that do not enjoy potable water supply. Most rural dwellers therefore depend on various available water sources. The qualities of these sources are generally not guaranteed and cases abound where health problems have risen as a result of consumers drinking from such sources. Thus for people in Doma, Wamba and part of Akwanga local government areas of Nasarawa State, Nigeria to meet their daily water needs and households' requirement, they source water from a few privately and government owned boreholes and wells while majority depend largely on Rivers Doma, Farinruwa and Mada located in Doma, Wamba and Akwanga local government areas, respectively.

This work aims at investigating the pollution levels of Na, K, Ni, Cu, Mg, Fe, Ca, Zn, Pb, Cd, As Se, Cr and Mn in the sediments and water from the three rivers which are the main sources of potable water for the inhabitants in these areas.

MATERIALS AND METHODS

Study area

Nasarawa State is one of 36 States in the Federal Republic of Nigeria. It is located in north-central geopolitical zone of Nigeria otherwise known as the middle belt region. The State is made up of thirteen (13) local government areas. Rivers Doma, Farinruwa and Mada are found in Doma, Wamba and Akwanga local government areas, respectively. Their latitudes are 08° 66" – 08° 72", 08° 52 – 08° 58" and 08° 49 – 08° 52" while their longitudes are 07° 64" – 07° 69", 07° 53" – 07° 57" and 07° 51" – 07° 56", respectively. These local government areas share boundaries with Benue, Plateau and Kaduna States in Nigeria (Figure 1).

The physical features of the area are largely mountainous, most of which are rocky and of undulating highlands of average height. It has a typical climate of the tropical zone because of its location. Its climate is quite pleasant with a maximum temperature of 95°F and a minimum of 50°F. Rainfall varies 131.73 cm in some places to 145 cm in others (Obaje et al., 2005). Mineral resources such as marble deposits, granite rocks, baryte and mica are found in some areas. The climate is characterized by two distinct seasons, dry and wet. The dry season spans from October to March while the raining season is from April to September. The months of December, January and February are cold due to harmattern wind blowing across the local government areas from the north-east of Nigeria. The sediments are generally comprised of sandstones, silt stones and forest soils which are rich in humus and very good for crop production. More than 80% of the inhabitants are predominantly farmers while few engage in fishing business.

Samples collection

Water samples were collected between May to July, 2009 and November, 2009 to February, 2010 for wet and dry seasons,

respectively. Representative water samples were taken just below the water surface at three different locations of each river using one litre acid leached polythene bottle. The water samples were stored in a deep freezer at −18°C prior to analysis. A diver was used to take soil sediment samples from the surface down to a depth of about 15 cm at locations where water samples were taken, stored in a polythene bag which had been washed and leached accordingly (Aremu et al., 2008) and kept in deep freezer prior to analysis.

Samples treatment

A known volume (5 cm^3) of concentrated hydrochloric acid was added to 250 cm^3 of water sample and evaporated to 25 cm^3. The concentrate was transferred to 50 cm^3 standard flask and diluted to the mark with distilled deionized water (Aremu et al., 2006). The soil sample was air-dried, and then sieved using 200 mm mesh. Five gramme of the soil sample was weighed into 150 cm^3 conical flask, digested using 150 cm^3 nitric acid, 2 cm^3 perchloric acid and placed on a hot plate for 3 h (Adeyeye, 1993). On cooling, the digest was filtered into 100 cm^3 volumetric flask and make up to mark with distilled water.

Mineral analysis

The elemental analysis (except Na and K) was done in the water samples using Perkins Elmer and Oak Brown (UK) atomic absorption spectrophotometer. The instrument settings and operational conditions were done in accordance with the manufacturer's specifications. Na and K were determined by using a flame photometer (Model 405, Corning, UK).

All the chemicals used were of analytical grade and obtained from British Drug Honks (BDH, London).

Statistical analysis

All the data generated were analyzed statistically (Steel and Torrie, 1960). Parameters evaluated were grand means, standard deviation and coefficient of variation. All the determinations were in triplicate.

RESULTS AND DISCUSSION

The result of trace metals in sediments and ambient water of both dry and wet seasons in Doma River is displayed in Tables 1A and 1B. The highest concentrated mineral in sediment and water samples for both seasons was Ca while the least was Cd (dry season) and Se (wet season). The following metals were not within the detection limit Cd, As, Se and Cr in water (dry season); Ni, Zn, Cd, As, Se and Mn in water and Cd in sediments (wet season). When compared, the level of trace metals in the water and sediment samples, the highest variability was found in Mn (90.48%) and Pb (88.64%) for dry and wet seasons, respectively while the least were Mg (33.05%) and Ca (40.06%) (Tables 1A and 1B). The orders of variation were found to be Mn > Ni > Zn = Pb > Fe > Cu > K > Na > Ca > Mg (dry season) and Pb > Fe > Cr > Na > Cr > K > Mg > Ca (wet season). The results of metal analysis in Farinruwa River water sample is

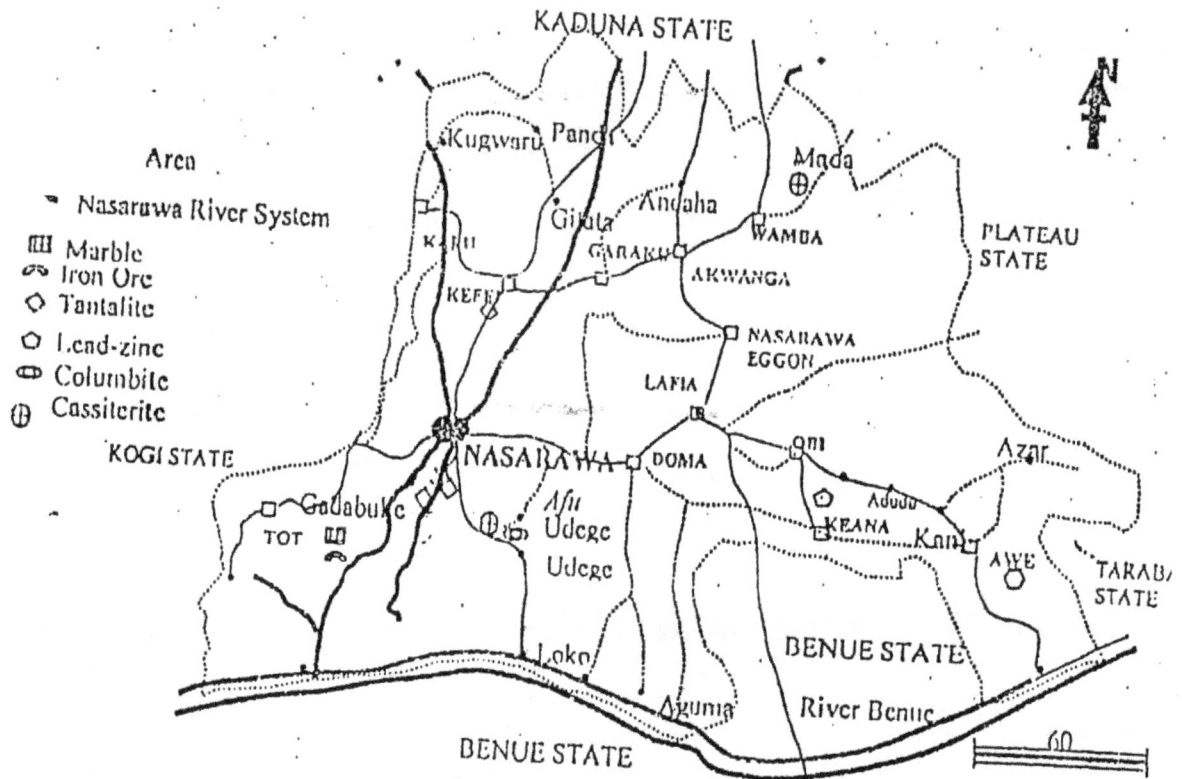

Figure 1. Map of Nasarawa State showing the study area.

Table 1a. Metals concentration (ppm)[a] of sediments and ambient water in River Doma dry season.

Mineral	Sediments	Water	Mean	SD	CV%
Na	5.64 ± 2.50	2.82 ± 1.20	4.23	1.41	33.33
K	3.15 ± 0.50	2.01 ± 1.05	3.46	1.45	41.91
Ni	3.15 ± 1.00	0.18 ± 1.20	1.67	1.48	88.62
Cu	2.36 ± 1.50	0.49 ± 1.15	1.43	0.94	65.75
Mg	17.19 ± 5.05	8.45 ± 2.10	12.92	4.27	33.05
Fe	10.88 ± 1.50	1.19 ± 1.50	6.04	4.85	80.3
Ca	22.66 ± 2.50	11.34 ± 1.00	17	5.66	33.29
Zn	2.88 ± 1.01	0.31 ± 0.01	1.6	1.29	80.63
Pb	1.36 ± 0.10	0.15 ± 0.15	0.76	0.61	80.63
Cd	0.21 ± 1.02	ND	nd	nd	nd
As	2.28 ± 1.50	ND	nd	nd	nd
Se	1.14 ± 1.30	ND	nd	nd	nd
Cr	2.10 ± 0.00	ND	nd	nd	nd
Mn	2.40 ± 0.10	0.12 ± 2.10	1.26	1.14	90.48

ND = not detected; 3D – standard deviation; CV% = coefficient of variation percent; nd = not determined; [a]Values are mean ± standard deviation of triplicate determinations

presented in Table 2. Cd, As and Se in water sample for dry season; As, Se and Mn in water sample and Cd (both samples) for wet season were not within detection limit of AAS. Ca was still found to be the highest concentration in both samples for the two seasons. The orders of variability are: Cr > Mn > Pb > Zn > Ni > Cu > Fe > K > Na > Mg > Ca and Ni > Zn > Cr > Cu > Na > Mg > K > Ca > Fe for dry and wet seasons, respectively (Tables 2a and 2b).

The result of metal analysis of sediment and water

Table 1b. Metals concentration (ppm)[a] of sediments and ambient water in River Doma in wet season.

Mineral	Sediments	Water	Mean	SD	CV%
Na	4.73 ± 1.20	0.88 ± 0.20	2.81	1.93	68.68
K	2.48 ± 1.50	0.72 ± 1.50	1.6	0.88	55
Ni	1.12 ± 1.01	ND	nd	nd	nd
Cu	0.89 ± 1.50	0.19 ± 1.01	0.54	0.35	64.81
Mg	16.89 ± 2.50	6.81 ± 1.05	11.55	5.04	42.53
Fe	10.77 ± 0.60	0.72 ± 1.20	5.75	5.03	87.48
Ca	19.99 ± 150	8.56 ± 1.01	14.28	5.72	40.06
Zn	1.28 ± 0.50	ND	nd	nd	nd
Pb	0.82 ± 1.20	0.05 ±0.50	0.44	0.39	88.64
Cd	ND	ND	nd	nd	nd
As	0.87 ± 1.50	ND	nd	nd	nd
Se	0.59 ± 1.50	ND	nd	nd	nd
Cr	1.20 ± 0.20	0.10 ± 0.60	0.65	0.55	84.62
Mn	1.20 ± 1.05	ND	nd	nd	nd

ND = not detected; SD = standard deviation; CV% = coefficient of variation percent;   nd = not determined; [a]Values are mean ± standard deviation of triplicate determinations

Table 2a. Metals concentration (ppm)[a] of sediments and ambient water in Farinruwa River in dry season.

Mineral	Sediments	Water	Mean	SD	CV%
Na	5.20 ± 2.50	2.89 ± 1.50	4.06	1.16	28.57
K	3.67 ± 1.50	1.86 ± 1.50	2.77	0.91	32.85
Ni	1.20 ± 1.60	0.20 ± 1.00	0.7	0.5	71.43
Cu	3.01 ± 1.00	0.65 ± 1.50	1.83	1.18	64.48
Mg	15.56 ± 2.50	9.95 ± 1.50	12.76	2.81	22.02
Fe	5.99 ± 1.50	1.94 ± 1.20	3.97	2.03	51.13
Ca	21.53 ± 2.50	14.05 ± 2.50	17.79	3.74	21.02
Zn	2.47 ± 0.50	0.32 ± 0.05	1.4	1.08	77.14
Pb	1.75 ± 0.00	0.20 ± 1.50	0.98	0.78	79.59
Cd	0.28 ± 1.00	ND	nd	nd	nd
As	1.94 ± 050	ND	nd	nd	nd
Se	1.10 ± 1.50	ND	nd	nd	nd
Cr	1.75 ± 2.10	0.11 ± 1.20	0.93	0.82	88.17
Mn	2.31 ± 1.10	0.19 ± 0.05	1.25	1.06	84.8

ND = not detected; SD = standard deviation; CV% = coefficient of variation percent;   nd = not determined; [a]Values are mean ± standard deviation of triplicate determinations

samples from Mada River (Tables 3a and 3b) showed that Ca was also the highest concentrated metal for both seasons. The least concentrated metal was Cd (0.11 ppm) and Se (0.23 ppm) for dry and wet seasons, respectively. Metals that were not within the detection limit of AAS for both samples were: Cd, As and Se in water sample (dry season) and Zn, Pb and As (water sample) and Cd (both samples) (wet season). The orders of variability are: Mn > Ni > Pb > Fe > Cr > Zn > Cu > Na > Ca > Mg > K and As > Mn > Cr > Cu > Fe > Ca > Na > Mg > K > for dry and wet seasons, respectively.

Comparison of the mean levels of trace metals in sediment and ambient water samples from rivers Doma, Farinruwa and Mada is displayed in Table 4. Ca had the highest concentration in all the three rivers (sediment and water samples) ranging from 17.0 ppm in Doma River to 19.06 ppm in Mada River and 14.28 ppm in Doma to 15.84 ppm in Mada for dry and wet seasons, respectively followed by Mg (11.51 – 12.92 ppm) (dry season) and Mg (8.43 – 11.85 ppm) (wet season). Similar observations made by Aremu et al. (2008) and Aremu and Inajoh (2007) described Ca and Mg as the predominant minerals in surface and ground water. Mg functions as an essential constituent for bone structure, reproduction for

Table 2b. Metals concentration (ppm)[a] of sediments and ambient water in Farinruwa River in wet season.

Mineral	Sediments	Water	Mean	SD	CV%
Na	3.88 ± 2.50	1.03 ± 2.10	2.46	1.43	58.13
K	2.20 ± 2.30	0.90 ± 4.50	1.55	0.65	41.94
Ni	0.83 ± 2.50	0.90 ± 0.010	0.42	0.41	97.62
Cu	1.82 ± 1.50	0.17 ± 0.50	0.99	0.83	83.84
Mg	12.76 ± 2.50	4.10 ± 2.10	8.43	4.33	51.36
Fe	3.91 ± 1.50	0.91 ± 1.50	2.41	1.5	6.22
Ca	19.01 ± 5.50	11.00 ± 3.50	15.01	4.01	26.72
Zn	1.29 ± 1.00	0.07 ± 1.10	0.68	0.61	89.71
Pb	0.78 ± 1.50	ND	nd	nd	nd
Cd	ND	ND	nd	nd	nd
As	1.03 ± 2.30	ND	nd	nd	nd
Se	0.81 ± 1.50	ND	nd	nd	nd
Cr	0.90 ± 0.60	0.07 ± 0.50	0.49	0.42	85.71
Mn	1.75 ± 0.30	ND	nd	nd	nd

ND = not detected; SD = standard deviation; CV% = coefficient of variation percent;   nd = not determined; [a]Values are mean ± standard deviation of triplicate determinations.

Table 3a. Metals concentration (ppm)[a] of sediments and ambient water in Mada River in dry season.

Mineral	Sediments	Water	Mean	SD	CV%
Na	6.11 ± 0.50	2.31 ± 1.10	4.21	1.9	45.13
K	3.25 ± 2.50	2.09 ± 1.20	2.67	0.58	21.72
Ni	1.65 ± 3.01	0.15 ± 2.10	0.9	0.75	83.33
Cu	3.12 ± 2.01	0.78 ± 2.50	1.95	1.17	60
Mg	14.36 ± 3.45	8.65 ± 2.50	11.51	2.86	24.85
Fe	13.45 ± 2.01	2.01 ± 1.50	7.73	5.72	74
Ca	24.97 ± 0.10	13.15 ± 2.50	19.06	5.91	31.01
Zn	1.02 ± 1.01	0.24 ± 1.50	0.63	0.39	61.9
Pb	0.76 ± 1.03	0.09 ± 1.01	0.43	0.34	79.07
Cd	0.11 ± 1.01	ND	nd	nd	nd
As	1.57 ± 2.00	ND	nd	nd	nd
Se	0.95 ± 1.30	ND	nd	nd	nd
Cr	2.61 ± 1.40	0.55 ± 2.10	1.58	1.03	65.19
Mn	1.95 ± 2.01	0.19 ± 0.40	1.57	1.38	87.9

ND = not detected; SD = standard deviation; CV% = coefficient of variation percent; nd = not determined; [a]Values are mean ± standard deviation of triplicate determinations.

normal functioning of various other systems. It also forms part of the enzyme system (Shills and Young, 1988). Ca plays an important role in blood clotting, in muscular contractions and in some enzymes assisting in metabolic processes. Ca tends to be a coordinator among inorganic elements, such that when K, Mg and Na are present in quantities beyond a particular limit in the body, Ca assumes a corrective role (Fleck, 1976). The Ca and Mg levels in all the sediment/water samples fall within the WHO (1993) recommended range of values. These are desirable for drinking without adverse effect. WHO recommended a maximum Ca level of 200 ppm above which values deposition of $CaCO_3$ in water systems can lead to major problems. On the other hand, the permissible level of Mg is fixed at 150 ppm provided the sulphate concentration is less than 250 ppm.

Potassium concentrated values ranged from 2.67 ppm in Mada River to 3.46 ppm in Doma while the values for Na varied from 4.06 – 4.23 ppm. K is primarily an intracellular cation found mostly bound to protein in the body along with sodium where they influence osmotic pressure and contribute to normal pH equilibrium (Fleck,

Table 3b. Metals concentration (ppm)[a] of sediments and ambient water in Mada River in wet season.

Mineral	Sediments	Water	Mean	SD	CV%
Na	3.87 ± 1.50	1.76 ± 1.50	2.82	1.06	37.59
K	1.88 ± 1.50	0.98 ± 1.10	1.43	0.45	31.47
Ni	0.72 ± 1.10	ND	nd	nd	nd
Cu	2.30 ± 1.50	0.13 ± 0.10	1.22	1.09	89.34
Mg	13.24 ± 2.30	5.99 ± 0.15	9.62	3.63	37.34
Fe	12.62 ± 1.50	0.83 ± 1.03	6.73	5.9	87.67
Ca	23.20 ± 0.40	8.47 ± 1.05	15.84	7.37	46.53
Zn	0.90 ± 0.10	ND	nd	nd	nd
Pb	0.26 ± 1.50	ND	nd	nd	nd
Cd	ND	ND	nd	nd	nd
As	0.73 ± 1.03	ND	0.37	0.36	97.3
Se	0.23 ± 1.50	ND	nd	nd	nd
Cr	2.12 ± 1.50	0.11 ± 1.05	1.12	0.11	91.67
Mn	1.88 ± 1.20	0.08 ± 2.05	0.98	0.9	91.84

ND = not detected; SD = standard deviation; CV% = coefficient of variation percent; nd = not determined; [a]Values are mean ± standard deviation of triplicate determinations.

Table 4a. Mean levels of metals concentration (ppm) in Rivers Doma, Farinruwa and Mada compared in dry season.

Mineral	Doma River	Farinruwa River	Mada River	WHO in ppm	FEPA in ppm
Na	4.23	4.06	4.21	na	na
K	3.46	2.77	2.67	na	na
Ni	1.69	0.7	0.9	0.05	< 1.0
Cu	1.43	1.83	1.95	1	2.0 – 4.0
Mg	12.92	12.76	11.51	30	< 30
Fe	6.04	3.97	7.73	0.3	1
Ca	17	17.79	19.06	45	< 45
Zn	1.6	1.4	0.63	5	20
Pb	0.76	0.98	4.5	0.05	< 1.0
Cd	nd	nd	nd	0.005	1.8
As	nd	nd	nd	0.05	0.5
Se	nd	nd	nd	0.01	na
Cr	nd	0.93	1.58	0.05	< 1.0
Mn	1.26	1.25	1.57	0.01	0.05

na = not available; nd = not determined

1976). The Na content in water is important for healthy reasons, except when combined with excessively high concentrations of sulphate. Such combinations can lead to gastrointestinal initiation for persons placed on low Na diet as a result of heart, kidney or circulatory ailment or complications of pregnancy. The usual low Na diet allowed in drinking water is 20 ppm (Ademoroti, 1996). The values recorded for both K and Na in the present study fall within the WHO limits. Thus, the samples of water from the different sources are good for consumption as far as these cations are concerned. The high iron content and its wide distribution throughout the sampling points (Tables 1A, 1B, 3A and 3B) reflect its presence at high concentration in Nigerian soils (Aiyesanmi, 2006; Aremu et al., 2010; Kakulu and Osibanjo, 1988). Iron is one of the essential components of haemoglobin which is responsible for the transport of oxygen in the body. It also occurs in the prosthetic group of the cytochromes which function in election transport and in some enzymes like the dehydrogenases (Wheby, 1974). Iron also facilitates the oxidation of carbohydrates, proteins and fats. It therefore contributes significantly to the prevention of anaemia, which is widespread in developing countries like Nigeria (Bender, 1992). The

Table 4b. Mean levels of metals concentration (ppm) in Rivers Doma, Farinruwa and Mada compared in wet season.

Mineral	Doma River	Farinruwa River	Mada River	WHO in ppm	FEPA in ppm
Na	2.81	2.46	2.82	na	na
K	1.6	1.55	1.43	na	na
Ni	nd	0.42	nd	0.05	< 1.0
Cu	0.54	0.99	1.22	1	2.0 – 4.0
Mg	11.85	8.43	9.62	30	< 30
Fe	5.75	2.41	6.73	0.3	1
Ca	14.28	15.01	15.84	45	< 45
Zn	nd	0.68	nd	5	20
Pb	0.44	nd	nd	0.05	< 1.0
Cd	nd	nd	nd	0.005	0.2 – 1.80
As	nd	nd	0.37	0.05	0.5
Se	nd	nd	nd	0.01	na
Cr	0.65	0.49	1.12	0.05	< 1.0
Mn	nd	nd	0.98	0.01	0.05

na = not available; **nd** = not determined.

iron contents in all the water samples (Tables 1A, 1B, 3A and 3B) are higher than WHO/USEPA value of 0.30 ppm for drinking water (USEPA, 2002). This is not unacceptable to the consumers but could give rise to iron-dependent bacteria which in-turn can cause further deterioration in the quality of water through the development of slimes and or objectionable odour. The result obtained may be due to run-offs and geological formations of the sample locations (Aremu et al., 2008). Lead concentration levels ranged from 0.45 ppm in Mada River to 0.98 ppm in Farinruwa (dry season) and 0.15 ppm in Mada River to 0.44 ppm in Doma (wet season). Lead even at low concentration is known to be toxic and has no known function in biochemical process. It can impair the nervous system and affect foetus, infants and children resulting in lowering of intelligent quotient (IQ) even at its lowest dose (UN, 1998).

The onset of lead pollution of surface waters in Nigeria has been reported (Mombershora et al., 1983; Okoye, 1991). Lead sticks to soil particles and enters drinking water only if the water is acidic or soft. However, lead content values in the present study (Tables 1A, 1B, 3A and 3B) are higher than WHO/USEPA recommended value of 0.05 ppm. Cu and Zn are essential metals and play an important role in enzyme activity (NAS, 1971). The Cu and Zn contents in the present study for all the water samples (Tables 1A, 1B, 3A and 3B) (both seasons) were found to be within the permissible limits of WHO/USEPA standards. Similar observation was made by the Federal Ministry of Environment on water standards for aquatic life to which most metals conform (FME, 2001). However, because a metal concentration in the aquatic environment is low and considered to be naturally occurring or background, does not mean that the concentration could not cause adverse ecological

effects (USEPA, 2002). The presence of one metal can significantly affect the impact that another metal may have on an organism. The effect can be synergistic, additive or antagonistic (Eisler, 1993). Cd, As and Se were below the detection limit of AAS for all the water samples (Tables 1A, 1B, 3A and 3B) (both seasons). They were found to be present in sediments. Some of the diseases caused by Se to mammals include accumulation of fluid throughout the body and destructive damage to the liver.

The early symptoms of acute toxic effects of Se are sore throat, fever, vomiting, irritation to eyes and nose, headache, drop in blood pressure, dermatitis and garlic odour of the breath (Luckey and Venugopal, 1977). As has been implicated in lung cancer, especially when the arsenic compound inhaled is of low solubility. It has also been found to have an effect on the liver by causing a disease termed cirrhosis and a rare form of liver cancer called haemongioendothelioma (Hutton, 1987). Cr levels ranged between 0.93 – 1.58 ppm (dry season) and 0.49 – 1.12 ppm (wet season). The metal is essential for life; its deficiency results in diabetic mellitus and increases the toxicity of lead (Aremu et al., 2010). Mn is an essential element and one of moderate toxicities. Its levels of concentration in the rivers varied between 1.25 – 1.57 ppm (dry season) and (nd – 0.98 ppm) (wet season) (Tables 4A and 4B). Mn has been implicated in neurological problems, especially when inhaled.

Conclusion

This study has presented metal concentrations (Na, K, Ni, Cu, Mg, Fe, Ca, Zn, Pb, Cd, As, Se, Cr and Mn) in sediment and water samples collected from Rivers Doma, Farinruwa and Mada located in Nasarawa State,

Nigeria. The results revealed that there was an indication of some heavy metals (Ni, Fe, Pb and Mn) pollution in all the water samples because their contents were found to be higher than WHO/USEPA recommended values. This work therefore will serve as baseline information for future work.

ACKNOWLEDGEMENT

Authors express their appreciation to Education Trust Fund (ETF), Nigeria for funding this research work.

REFERENCES

Ademoroti CMA (1996). Environmental Chemistry and Toxicology. Foludex Press Ltd., Ibadan, pp. 180 – 184.

Adeyeye EI (1993). Trace heavy metal distribution in *Illisha africana* (Bloch) fish organs and tissue In: Lead and Cadmium. Ghana J. Chem., 1: 377 – 384.

Aiyesanmi AF (2006). Baseline concentration of heavy metals in water samples from rivers within Okitipupa, southwest belt of the Nigerian bitumen field, Nigeria. J. Chem. Soc., 31(1-2): 30 – 37.

Aremu MO, Atolaiye BO, Labaran L (2010). Environmental implication of metal concentration in soil, plant foods and pond in area around the derelict Udege mines of Nasarawa State, Nigeria. Bull. Chem. Soc. Ethiopia, 24(3): 351 – 360.

Aremu MO, Inajoh A (2007). Assessment of elemental contaminants in water and selected sea-foods from River Benue, Nigeria. Curr. World Environ., 2(2): 167 – 173.

Aremu MO, Olonisakin A, Ahmed SA (2006). Assessment of heavy metal content in some selected agricultural products planted along some roads in Nasarawa State, Nigeria. J. Eng. Appl. Sci., 1(3): 199 – 204.

Aremu MO, Sangari DU, Musa BZ, Chaanda MS (2008). Assessment of groundwater and stream quality for trace metals and physicochemical contaminants in Toto Local Government Area of Nasarawa State, Nigeria. Int. J. Chem. Sci., 1(1): 8 – 19.

Bender A (1992). Meat and Meat Products in Human Nutrition in Developing Countries. FAO Food and Nutrition Paper 53, FAO, Rome, Italy, pp. 46 – 47.

Eisler R (1993). Zinc hazard to fish, wildlife and invertebrates: A synoptic review. US fish and wildlife service, biological report 10. Publication Unit, USFWS. Washington, DC, 20240.

Fleck H (1976). Introduction to Nutrition, 3rd edn. Macmillan, New York, USA, pp. 207 – 219.

FME, Federal Ministry of Environment (2001). National Guidelines and Standards for Water quality in Nigeria, FME, Nigeria.

Foster U, Wiltman GTW (1983). Metal pollution in aquatic environment. Springer-verlag, Berlin, p. 486.

Gutenmann WH, Bache, CA, McCahan JB, List DI (1988). Heavy metals and chlorinated hydrocarbons in marine fish products. Nutr. Rep. Int., 38: 1157 – 1161.

Hutton M (1987). In: Lead, Mercury, Cadmium and Arsenic in the Environment, Hutchinson TC, Mecma KM (Eds), Wiley, UK, pp. 85 – 94.

Kakulu SE, Osibanjo O (1988). Trace heavy metal pollution status in sediment of the Niger Delta Area of Nigeria. J. Chem. Soc. Nig., 13: 9 – 15.

Luckey TD, Venugopal B (1977). Metal Toxicity in Mammals, Vols. I & II, Plenum Press, UK, p. 78.

Marr II, Creaser MS (1983). Environmental Chemical Analysis. Pub. Blackie and Sons Ltd. London, p. 104.

Mombershora CO, Osibanjo O, Ajayi SO (1983). Pollution studies on Nigerian Rivers; the onset of lead pollution of surface waters in Ibadan. Environ. Int., 9: 81 – 84.

NAS, National Academy of Sciences (1971). In: Introduction to Nutrition. Fleck H (Ed.), 3rd edn. Macmillan Publishing Co. Inc. New York, p. 235.

UN, United Nations (1998). Global opportunities for reducing use of lead gasoline, IOMC/UNEP/CHEMICALS/98/9: Switzerland.

Obaje N, Nzegbuna AI, Moumouni A, Ukaonu CE (2005). Geology and mining resources of Nasarawa State. Bulletin of Department of Geology and Mining, Nasarawa State University, Keffi, Nigeria, p. 11.

Okoye BCO (1991). Heavy metals and organisms in the Lagos Lagoon. Int. J. Environ. Stud., 37: 285 – 292.

Preston MR, Chester R (1996). Chemistry and pollution of the marine environment. In: Pollution, Causes, Effects and Control, Harison RM (ed.), 3rd edn, Royal Society of Chemistry, UK, pp. 26 – 51.

Shills MEG, Young VR (1988). Modern Nutrition in Health and Disease. In: Nutrition Nieman DC, Butterworth DE, Nieman CN (Eds), W McBrown publishers, Dubuque, USA, pp. 276 – 282.

Steel RGD, Torrie JH (1960). Principles of Procedures of Statistics, McGraw Hill; London, pp. 1 – 360.

USEPA, United States Environmental Protection Agency (2002). Current Drinking Water Standards, office of Groundwater and Drinking Water; Government printing Office, Washington, DC, p. 63.

Wheby MS (1974). Synthetic effects of iron deficiency. In: Iron, Crosby WH (Ed.), Midicom Inc., New York, p. 39.

WHO, World Health Organization (1993). Guidelines for Drinking Water Quality, WHO, Geneva. pp. 65-78

11

Effect of cadmium on germination, growth, redox and oxidative properties in *Pisum sativum* seeds

Moêz Smiri

Bio-physiologie cellulaires, faculté des sciences de Bizerte, 7021 zarzouna, Tunisia. E-mail: smirimoez@yahoo.fr.

Pea seeds were treated with 5 mM $CdCl_2$ for 5 days. Experiments carried out in cotyledons and embryonic axes were performed to evaluate the redox and oxidative properties. Germination rate and embryonic axis growth were determined. After five days, Cd treatment caused 60% decrease in germination success, and 50% inhibition in embryo length. The reduction level [NADPH/ ($NADP^+$ + NADPH)] was used to define the redox status. The reduction level in Cd-treated mitochondrial and peroxisomal fractions was ~20 to 70% lower than that of the control (water-treated) from 120 h of exposure. Under normal conditions, the intracellular milieu is predominately reducing, but stress conditions can shift the redox balance toward an oxidizing milieu. NADPH oxidase is considered to be oxidative stress-related enzymes. NAD(P)H oxidase activities were strongly stimulated after Cd exposure. We suggest that alteration of redox and oxidative properties in both tissues of pea seeds due to treatment with $CdCl_2$ is highly responsible for decrease of germination success and inhibition of embryonic axes growth.

Key words: Cadmium, germination, oxidative stress, *Pisum sativum*, redox.

INTRODUCTION

Environmental stresses often lead to great yield losses under various agricultural production systems. Of diverse abiotic stresses, heavy metal is a pernicious problem affecting the productivity and quality of economically valuable crops (Wagner, 1993). Certain heavy metals, such as copper (Cu) or iron (Fe), can be toxic through their participation in Fenton-type reactions producing reactive oxygen species (ROS), which are known to be extremely harmful for living cells (Stochs and Bagchi, 1995). Cadmium (Cd^{2+}) is a non-redox metal unable to take part in this type of reaction. Nevertheless, it has been clearly demonstrated that Cd^{2+} induces changes in the antioxidant status in plants (Grataö et al., 2005). Cd^{2+} is regarded as a non-essential metal without any known physiological function. It is extremely toxic to plants and animals, has a long half-life and is extremely persistent in the environment (Wagner, 1993).

Moreover, in pea plants, long-term exposure to Cd^{2+} produces oxidative stress in roots, as a result of disturbances in enzymatic and no enzymatic antioxidant defenses, bringing about an increase in ROS accumulation (Rodríguez-Serrano et al., 2006). Garnier et al. (2006) demonstrated that Cd^{2+} induces a transient increase in cytosolic Ca^{2+} concentration that appears to regulate the extracellular NADPH-oxidase dependent generation of hydrogen peroxide (H_2O_2). In this way, transcriptome analysis of the antioxidative enzymes in leaves of pea plants grown with Cd^{2+} and treated with some modulators of the signal transduction cascade suggested that at least Ca^{2+} channels and H_2O_2 were involved in some steps between the Cd^{2+} signal and transcript expression of some antioxidant enzymes. This indicated the existence of cross-talk between these elements and ROS metabolism during Cd^{2+} stress (Romero-Puertas et al., 2007). Moreover, in roots and leaves of pea plants Cd^{2+} produced a significant inhibition of growth, as well as a reduction in the transpiration and photosynthesis rate, chlorophyll content of leaves, and an alteration in the nutrient status (Sandalio et al., 2001; Romero-Puertas et al., 2004).

Usually, experimental studies of heavy metals impact on adult plants have applied low concentrations (Woolhouse, 1983; Ernst, 1998). Nevertheless, high pollutant doses have been used in seed germination assays (Chugh and Sawhney, 1996; Rahoui et al., 2008, 2010), although the germination is considered as a sensitive process as compared to other stages of plant development (Ernst, 1998). This is explained, at least in part, by the fact that seed coats may be impermeable to heavy metals. This can avoid an over-accumulation of contaminant during the vital heterotrophic regime of germinating seed. Consequently, the behaviour of seed germination is not regarded according to heavy metal doses in the seed surrounding medium, but considered with respect to the real accumulation and the compartmentation of pollutant at cellular and subcellular levels (Woolhouse, 1983; Ernst, 1998; Rahoui et al., 2010). Delay in germination can be associated with disorders in the event chain of germinative metabolism which is a highly complex multistage process, but one of underlying metabolic activities following imbibition of the seed is the resumption of respiration.

Studies on the mitochondrial biogenesis during germination have focused on seed storage tissues in the post-germination period. In pea cotyledons, this has been suggested as a result of the activation of preformed mitochondria through the import of existing polypeptides in the cytoplasm (Nawa and Asahi, 1971; Sato and Asahi, 1975). Functional mitochondria capable of adenosine triphosphate (ATP) synthesis have been isolated from dry sunflower seeds (Attucci et al., 1991), suggesting that the energy for mitochondrial metabolism may be supplied by oxidative phosphorylation. As well as a general increase in mitochondrial protein in the post-germination phase, there is also a differential regulation of mitochondrial metabolism in lipid storing seeds, to support the conversion of storage lipid to sugars for export to the growing embryo (Ehrenshaft and Brambl, 1990; Hill et al., 1992). All these processes require energy and thus depend on the competence and stability of mitochondria (Prasad et al., 1995). Abiotic stresses have pronounced and complex effects on mitochondrial properties, in terms of both their function and their biogenesis. The response of mitochondrial respiration is likely to be controlled by the adenylates and the substrate supplies (Atkin and Tjoelker, 2003).

Nicotinamide adenine dinucleotide phosphate (NADP) play vital roles in signalling via the generation and scavenging of reactive oxygen species (Berger et al., 2004; Noctor et al., 2006; Hunt et al., 2007) and in systems controlling adaptation to environmental stresses such as UV irradiation, salinity, heat shock and drought (Mittler et al., 2004; Chai et al., 2006). The plant mitochondrial inner membrane contains a branched electron transport pathway that includes multiple enzymes for the oxidation of both cytosolic and matrix NAD(P)H

(Agius et al., 1998; Moore et al., 2003). When plants are grown under stressful environments, the reduced form of cytosolic NADP is produced in the pentose phosphate pathway by glucose 6-phosphate dehydrogenase (G6PDH) and 6-phosphogluconate dehydrogenase (6PGDH) and plays critical roles in the generation of ROS and in the production of anti-oxidants as ROS scavengers (Murata et al., 2001; Tamoi et al., 2005). In addition to the classical complexes, all plant mitochondria possess alternative enzymes that confer on them the capacity to oxidize external NAD(P)H or internal NAD(P)H independently of complex I and to transfer electrons directly from reduced ubiquinone to oxygen via an alternative oxidase (Douce and Neuburger, 1989; Rasmusson et al., 2004).

This work aimed to examine the impact of imbibition with Cd solution on germination rate and the behaviour of some enzyme capacities involved in the redox regulation in the mitochondrial and peroxisomal fractions of cotyledons and embryos of germinating pea seeds.

MATERIALS AND METHODS

Seeds of pea (Pisum sativum L. cv. douce province) were disinfected with 2% of sodium hypochlorite for 10 min and then rinsed thoroughly and soaked in distilled water at 4 °C for 30 min to obtain an initial stage. Seeds were germinated at 25 °C in the dark for 5 days over two sheets of filter paper moistened with distilled water or aqueous solution of 5 mM $CdCl_2$. Germinated seeds were recorded until the maximum germination of control (H_2O) was obtained. Germination rate was calculated as a percentage of the control and germinating seeds were sampled for the assays. At harvest, the coat was removed and the embryonic axes and cotyledons were weighed and stored in liquid nitrogen until analysis.

Mitochondria, from germinating pea seeds were isolated by the procedure of Smiri et al. (2009). Cotyledons and embryonic axes were ground in a mortar and pestle with sand and the following medium (w/v = 1/5): 50 mM Tris- HCl (pH 8.0), 0.4 M saccharose, 1 mM EDTA, 5 mM ascorbic acid and 1 mM $MgCl_2$. The homogenate was squeezed through double cheesecloth, centrifuged at 3,000 g for 20 min. Mitochondria from supernatant were sedimented by centrifuging it at 20,000 g for 30 min. The supernatant obtained was carefully decanted and designated as "peroxisomal fraction". The pellet was re-suspended in the homogenising media (w/v = 1/0.5) 50 mM Tris- HCl (pH 8.0), 0.4 M saccharose and referred to as "mitochondrial fraction" (Smiri et al., 2010a). Coenzyme extraction and concentration determination was processed as previously described (Smiri et al., 2010a). For NAD^+ and $NADP^+$, 0.2 N HCl was used to homogenize the burst mitochondria and the peroxisomal fractions of cotyledons and embryonic axes (w/v = 1/10); for NADH and NADPH, 0.2 N NaOH were used. Each homogenate was heated in a boiling water bath for 5 min, cooled in an ice bath, then centrifuged at 10,000 g at 4 °C for 10 min. The supernatant solutions were transferred to separate tubes and kept on ice for coenzyme assay (Zhao et al., 1987). Enzyme cycling assays were employed with 3-(4,5-dimethyl-2-thiazolyl)-2,5-diphenyl-2H tetrazoliumbromide (MTT) as the terminal electron acceptor (Matsumura and Miyachi, 1980). The rate of reduction of MTT (measured at 570 nm) was directly proportional to the concentration of coenzyme. Standard curves were prepared for each coenzyme in each experiment.

NADPH oxidase activity was determined in an assay mixture

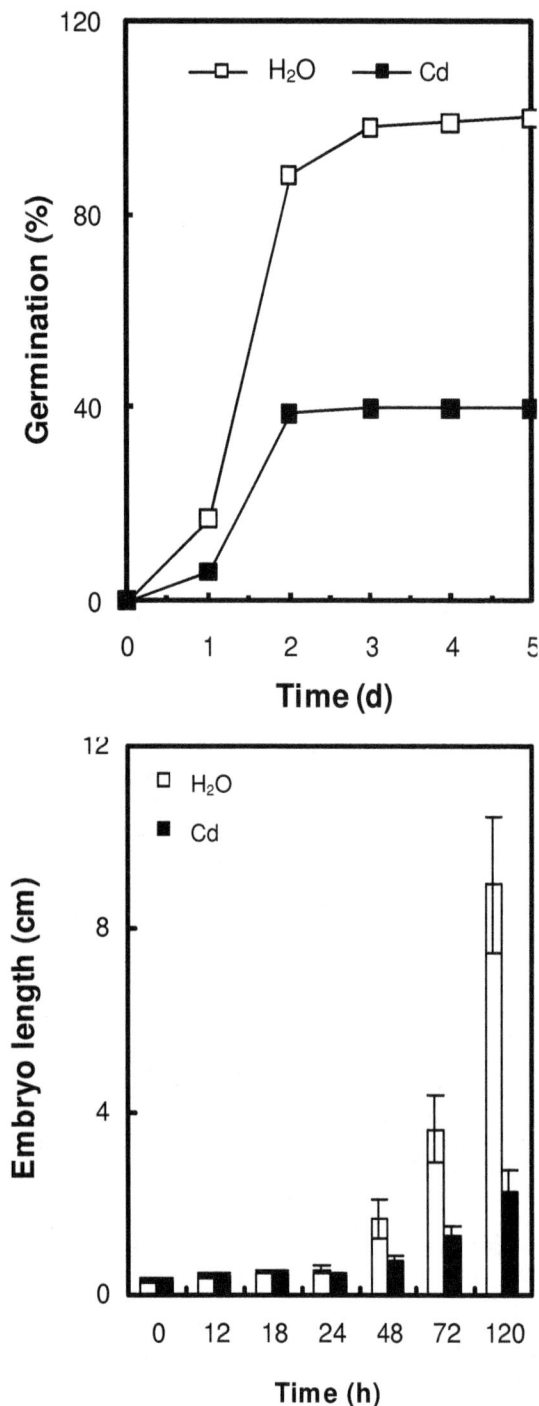

Figure 1. Germination rate and embryonic axis length of pea seeds after imbibition with H_2O or 5 mM $CdCl_2$. Each experiment was carried out with 80 germinating seeds.

containing 100 mM sodium acetate (pH 6.5), 1 mM $MnCl_2$, 0.5 mM p-coumaric acid, 0.2 mM NADPH and enzyme extract. The reaction was monitored by following the decrease in absorbance at 340 nm, with extinction coefficient of 6.22 $mM^{-1}cm^{-1}$ (Ishida et al., 1987). Each treatment consisted of six replicates and each experiment

was carried out at least twice at different times. All data were statistically analyzed using one-way ANOVA. Differences were considered significant at $P < 0.05$.

RESULTS AND DISCUSSION

The most critical stages in the life cycle of higher plants are seed germination and seedling establishment (Bewley, 1997). Germination starts with the uptake of water by the quiescent dry seed and terminates with the elongation of the embryonic axis (Bewley and Black, 1994). The decreased percentage of germination was observed after 24 h and after 5 days, it reached a 60% decrease from the control. Analysis of embryo length showed significant differences between seeds germinating on water and Cd. The inhibitory effect of the metal was increased in a time-dependent manner (Figure 1). When the medium surrounding the seed was contaminated with Cd, delays in germination were often observed (Rahoui et al., 2008, 2010). This can be associated with several disorders in the event chain of germinative metabolism.

Especially, seed germination and subsequent embryo growth are important stages of the plant life and highly sensitive to surrounding medium fluctuations, because the germinating seed is the first interface of material exchange between plant development cycle and environment (Ernst, 1998). The interactions of Cd with crucial physiological functions in adult plants have been widely investigated (Wagner, 1993), and drastic reduction of biomass production and nutritional quality have been observed in crops grown on soils contaminated with this non-essential element (Ernst, 1998). Cd^{2+}, with no reported biological function except one occasion as a cofactor for carbonic anhydrase in marine diatom (Xu et al., 2008), disturbs the cellular metabolic process by producing excessive ROS leading to oxidative stress. ROS is known to react with proteins, nucleic acids and lipids causing deleterious effects on various cellular processes (Møller et al., 2007). Its high affinity for sulfhydryl and oxygen containing groups results in blocking the essential functional groups of biomolecules (Noctor et al., 2006).

Consequently, it inhibits the uptake and transport of many macro/micronutrients and thus, induces the nutrient deficiencies. Considerable effort has been devoted to understanding the molecular basis of resistance to heavy metal, and it has been revealed that plants react to abiotic stresses by the alterations in metabolic rates, protein turnover, osmolytes, membrane function, and gene expression (Møller, 2001). All these processes require energy and thus depend on the competence and the stability of mitochondria (Prasad et al., 1995). Abiotic stresses have pronounced and complex effects on mitochondrial properties in terms of both their function and their biogenesis. Germinating seeds must adapt their metabolic and developmental programmes to the prevailing

Table 1. Effect of Cd on NADP balance in germinating pea seeds.

Time(h)			0	12	18	24	48	72	120
Plant part	Fractions	Treatments							
			NADP$^+$ Balance:[NADP$^+$/(NADP$^+$ + NAD$^+$)]						
C	P	H$_2$O	48±5	33±3	24±2	14±1	15±1	10±1	9±1
		Cd	48±5	41±1	22±2	24±1*	24±1*	24±1*	26±2*
	M	H$_2$O	20±1	15±1	15±3	16±4	14±1	14±1	14±2
		Cd	20±1	11±1	11±1	12±1*	12±1*	13±1*	21±1*
E	P	H$_2$O	8±1	4±1	5±1	3±1	1±1	1±1	1±1
		Cd	8±1	4±1	4±1	3±1	1±1	1±1	2±1
	M	H$_2$O	64±6	59±1	43±3	40±1	35±3	35±2	32±3
		Cd	64±6	59±1	47±4	39±1	33±1	35±1	33±1
			NADPH balance:[NADPH/(NADPH + NADH)]						
C	P	H$_2$O	28±1	17±1	19±1	22±1	17±1	14±1	14±1
		Cd	28±1	18±1	18±1	17±1	15±1	15±1	15±1
	M	H$_2$O	64±3	50±3	46±1	45±3	51±4	50±1	44±1
		Cd	64±3	46±1	40±5	47±2	46±3*	38±1*	36±1*
E	P	H$_2$O	21±1	18±1	16±1	16±1	14±1	21±1	21±1
		Cd	21±1	19±1	14±1	10±1*	9±1*	11±1*	12±1*
	M	H$_2$O	36±1	39±1	41±1	46±1	42±1	47±2	44±1
		Cd	36±1*	53±1*	61±1*	71±1*	77±2*	62±2*	54±1*

NADP$^+$ balance was evaluated as the percentage of NADP$^+$ from total coenzyme (NADP$^+$ + NAD$^+$). NADPH balance was evaluated as the percentage of NADPH from total coenzyme (NADPH + NADH). Data are the means ± S.E. of 6 individual measurements. Each measurement was performed in an extract obtained from several germinating seeds. Significant differences between treated and control were determined using one-way ANOVA. *P<0.05. C, cotyledons; E, embryonic axis; M, mitochondria; P, peroxisomal fraction.

environmental conditions to become photoautotrophic before their nutrient reserves become exhausted. A germinating seed relies almost exclusively on its reserves to supply metabolites for respiration. To operate in stressful conditions, plant mitochondria might have evolved distinctive features designed to increase their metabolic flexibility and stress tolerance, which is clearly illustrated by both the complexity of the mitochondrial proteome and by its dynamic response to environmental stress, and the response of mitochondrial respiration is likely to be controlled by the substrate supplies (Atkin and Tjoelker, 2003). In the present study, the NADP reduction levels and NADPH oxidase activity was determined to evaluate their possible role in combating the cadmium toxicity.

The NADP$^+$ balance [NADP$^+$/(NADP$^+$+NAD$^+$)] in Cd-treated mitochondrial and peroxisomal fractions of cotyledons was ~50 to 190% higher than that of the control (water-treated) from 120 h of exposure, and did not change in the embryonic axis. The NADPH balance [NADPH/(NADPH+NADH)] in Cd-treated mitochondria of cotyledons was ~20% lower than that of the control (water-treated) from 120 h of exposure and ~20% higher than that of the control in embryonic axis. It was ~20% lower than that of the control (water-treated) peroxisomal fractions in the embryonic axis (Table 1). The reduction

level [NADPH/(NADP$^+$ + NADPH)] in Cd-treated mitochondrial and peroxisomal fractions of both tissues was ~20 to 70% lower than that of the control (water-treated) from 120 h of exposure (Figure 2).

In a previous work, we have demonstrated that Cd interferes with the enzyme activities of the Krebs cycle and electron transport chain in germinating pea seeds (Smiri et al., 2009, 2010c). Thus, the decline in NAD(P)-dependent respiratory rates in pea mitochondria may be due to the disorder in metabolism of coenzymes (Figure 2). Moller and Lin (1986) have been suggested that, the reduction level is modulated by factors such as enzyme kinetics, respiratory state and the concentration of pyridine nucleotides. In fact, the decrease in nicotinamide adenine dinucleotide (NAD) and NADP reduction level observed in this paper or in the previous (Smiri et al., 2010b) may be due to the mitochondrial dysfunction, and the response of mitochondrial respiration is controlled by NAD and NADP levels in pea seeds treated with CdCl$_2$. The respiratory chain of plant mitochondria contains three NAD(P)H dehydrogenases on the matrix surface of the inner membrane. The activities of these dehydrogenases are dependent on the concentration and reduction levels of NAD and NADP in the matrix (Agius et al., 2001). NADP has a multitude of potential roles (Møller and Rasmusson, 1998; Agius et al., 2001). One such role is

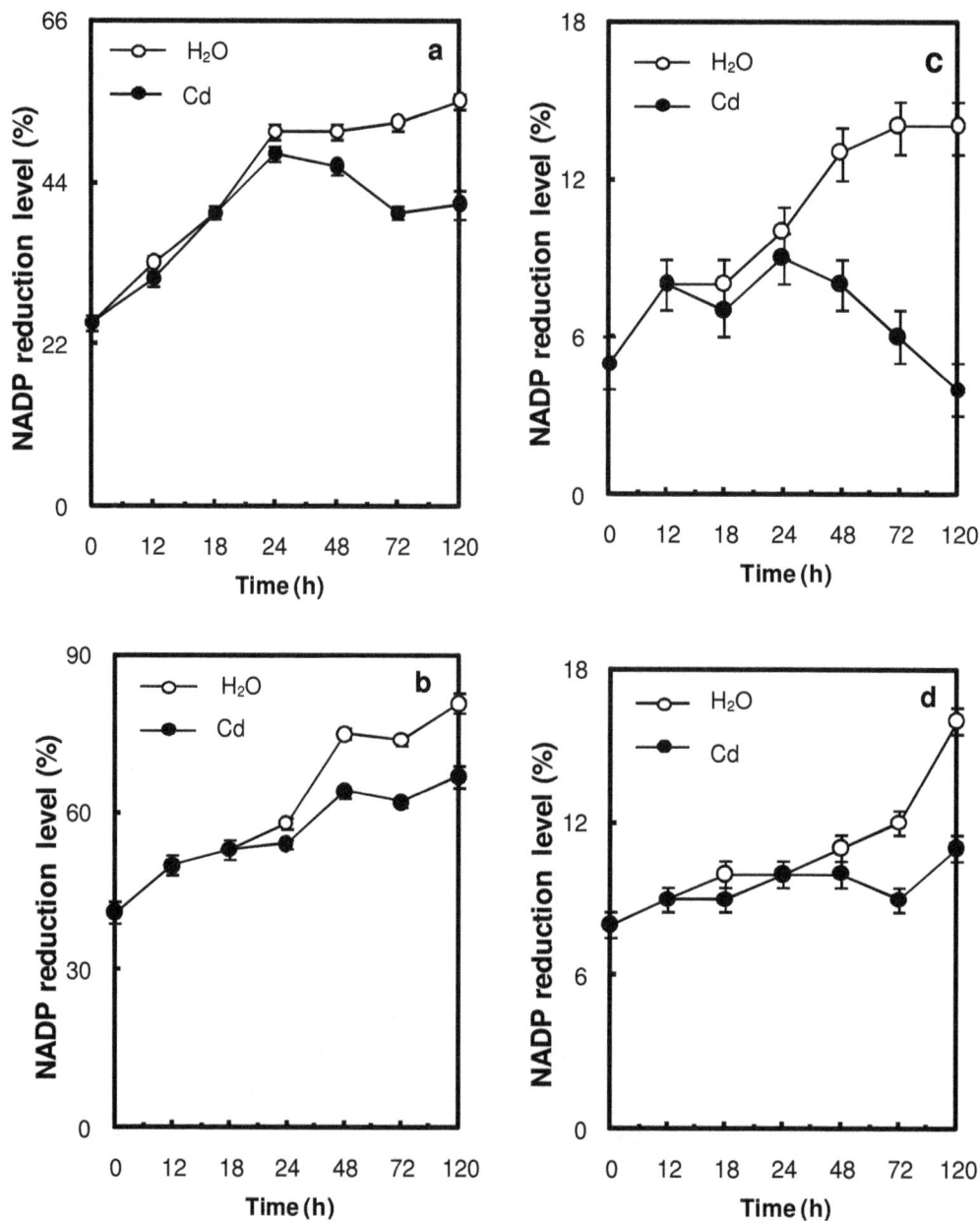

Figure 2. NADP reduction level (evaluated as the percentage of NADPH from total coenzymes (NADPH + NADP$^{+)}$) in peroxisomal fraction (a, cotyledons; b, embryonic axis) and mitochondria (c, cotyledons; d, embryonic axis) of pea seeds during germination after imbibition with H$_2$O or 5 mM CdCl$_2$. Values are the averages of 6 individual measurements (± SE). Each measurement was performed in an extract obtained from several germinating seeds.

contributing to electron transport through the activity of an NADPH dehydrogenase facing the matrix (Rasmusson and Møller, 1991; Melo et al., 1996; Agius et al., 1998). In addition, mitochondrial alternative pathways are expected to provide additional flexibility to mitochondrial metabolism in the situations of stress (Douce and Neuburger, 1989; Rasmusson et al., 2004). When mitochondria respiration capacity decreases, secondary pentose phosphate pathway can be activated to

compensate for the failure in energy supply (Smiri et al., 2009). Reducing power can be supplied via secondary NAD(P)H-recycling dehydrogenase activities, such as G6PDH and 6PGDH (Smiri et al., 2009, 2010c).

The NADPH oxidase activity in the mitochondrial and peroxisomal fractions of cotyledons and embryos imbibed with water for five days was ~50 to 75% lower than that of dry (0 h) tissues. Activity in Cd-treated mitochondria and peroxisomal fractions was ~170 to 250% higher than that

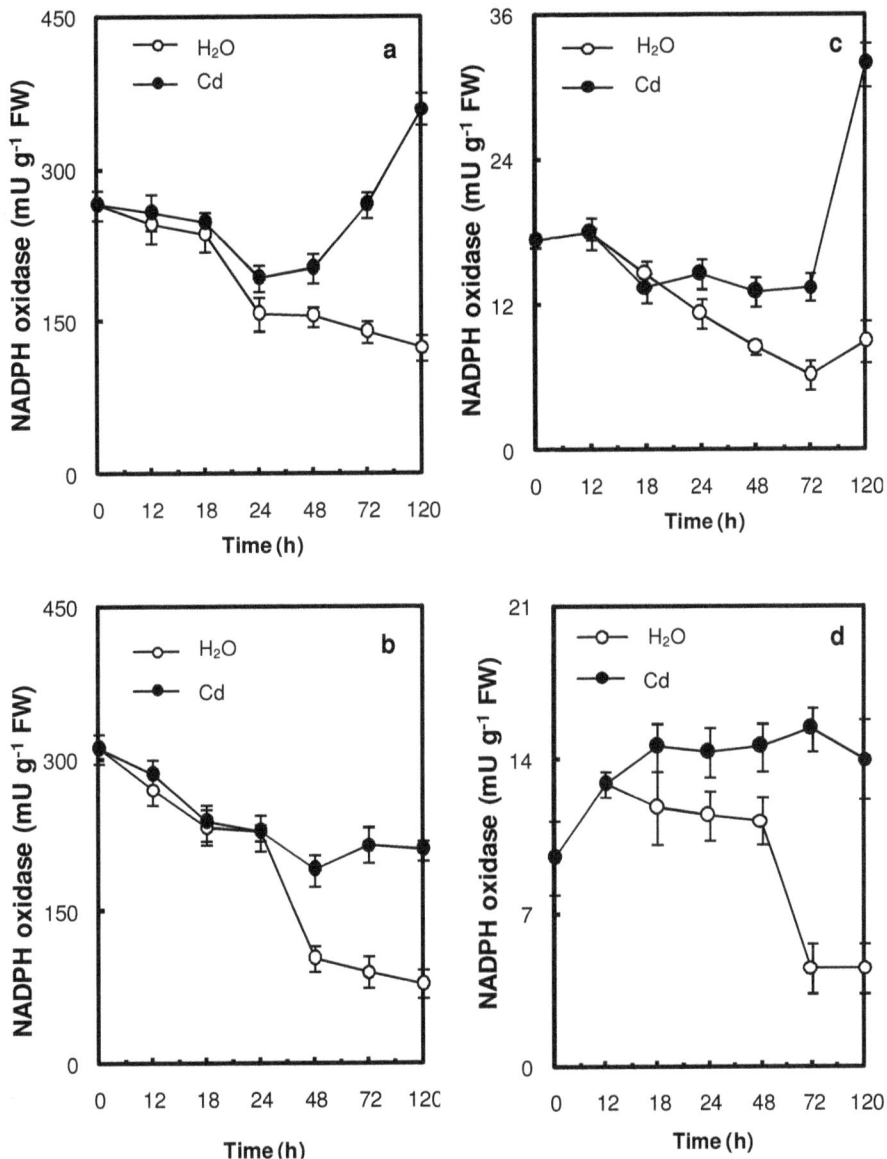

Figure 3. NADPH oxidase activity in peroxisomal fraction (a, cotyledons; b, embryonic axis) and mitochondria (c, cotyledons; d, embryonic axis) of pea seeds during germination after imbibition with H_2O or 5 mM $CdCl_2$. Values are the averages of 6 individual measurements (± SE). Each measurement was performed in an extract obtained from several germinating seeds.

of control (water-treated) from 5 days of exposure (Figure 3). NADPH oxidase transfers electrons from NADPH to O_2 to form superoxide radical (O_2^-), followed by the dismutation of O_2^- to H_2O_2. Cd caused a significant consumption of reduced nicotinamide, as evidenced by the increase in $NADP^+$ balance (Table 1), probably by the stimulation of enzymatic oxidation via NADPH oxidase activities (Figure 3). In adult pea, plant responds to Cd toxicity by increasing the activity of antioxidative enzymes (Chaoui et al., 2004; Rodríguez-Serrano et al., 2006). In plant cells, the production of ROS such as O_2^-, H_2O_2 and the hydroxyl radical takes place in chloroplasts,

mitochondria, peroxisomes, the plasma membrane and the apoplastic space (Navrot et al., 2006).

All biologically relevant macromolecules, that is nucleic acids, membrane lipids and proteins, are susceptible to damage by ROS. Thus, a number of studies have documented the production of ROS during the germination of various species (Bailly, 2004), and the production of reactive oxygen species by germinating seeds, has been regarded as a cause of stress, that might affect the success of germination. We suggest that alteration of redox and oxidative properties in both tissues of pea seeds due to treatment with $CdCl_2$ is highly responsible for the decrease of germination success.

REFERENCES

Agius SC, Bykova NV, Igamberdiev AU, Moller IM (1998). The internal rotenone-insensitive NADPH dehydrogenase contributes to malate oxidation by potato tuber and pea leaf mitochondria. Physiol. Plant, 104(3): 329-336.

Agius SC, Rasmusson AG, Moller IM (2001). NAD(P) turnover in plant mitochondria. Aust. J. Plant Physiol., 28(6): 461-470.

Atkin OK, Tjoelker MG (2003). Thermal acclimation and the dynamic response of plant respiration to temperature. Trends Plant Sci., 8(7): 343-351.

Attucci S, Carde JP, Raymond P, Saint-Ges V, Spiteri A, Pradet A (1991). Oxidative phosphorylation by mitochondria extracted from dry sunflower seeds. Plant Physiol., 95(2): 390-398.

Bailly C (2004). Active oxygen species and antioxidants in seed biology. Seed Sci. Res., 14(1): 93-107.

Berger F, Ramirez-Hernandez MH, Ziegler M (2004). The new life of a centenarian: sinalling functions of NAD(P). Trends Biochem. Sci., 29(3): 111-118.

Bewley JD (1997). Seed germination and dormancy. Plant Cell, 9(7): 1055-1066.

Bewley JD, Black M (1994). Seeds: Physiology of development and germination. Plenum, New York, p.445.

Chai MF, Wel PC, Chen QJ, Rui A, Jia C, Shuhua Y, Wang XC (2006). NADK3, a novel cytoplasmic source of NADPH, is required under conditions of oxidative stress and modulates abscisic acid responses in Arabidopsis. Plant J., 47(5): 665-674.

Chaoui A, Jarrar B, El-Ferjani E (2004). Effects of cadmium and copper on peroxidase, NADH oxidase and IAA oxidase activities in cell wall, soluble and microsomal membrane fractions of pea roots. J. Plant Physiol., 161(11): 1225-1234.

Chugh LK, Sawhney SK (1996). Effect of cadmium on germination, amylases and rate of respiration of germinating pea seeds. Environ. Pollut., 92(1): 1-5.

Douce R, Neuburger M (1989). The uniqueness of plant mitochondria. Annu. Rev. Plant Physiol., 40(1): 371-414.

Ehrenshaft M, Brambl R (1990). Respiration and mitochondrial biogenesis in germinating embryos of maize. Plant Physiol., 93(1): 295-304.

Ernst WHO (1998). Effects of heavy metals in plants at the cellular and organismic level ecotoxicology. In: Bioaccumulation and Biological Effects of Chemicals (Gerrit S, Bernd M, eds.). John Wiley Sons Inc. and Spektrum Akademischer Verlag, pp. 587-620.

Garnier L, Simon-Plas F, Thuleau P, Agnel JP, Blein JP, Ranjeva R, Montillet JL (2006). Cadmium affects tobacco cells by a series of three waves of reactive oxygen species that contribute to cytotoxicity. Plant Cell Environ., 29(10): 1956-1969.

Grataö PL, Polle A, Lea PJ, Azevedo RA (2005). Making the life of heavy metals-stressed plant a little easier. Funct. Plant Biol., 32(6): 481-494.

Hill SA, Grof CPL, Bryce JH, Leaver CJ (1992). Regulation of mitochondrial function and biogenesis in cucumber (Cucumis sativus L.) cotyledons during early seedling growth. Plant Physiol., 99(1): 60-66.

Hunt L, Holdsworth MJ, Gray JE (2007). Nicotinamidase activity is important for germination. Plant J., 51(3): 341-351.

Ishida A, Ookubo K, Ono K (1987). Formation of hydrogen peroxide by NAD(P)H oxidation with isolated cell wall-associated peroxidase from cultured liverwort cells, Marchantia polymorpha L. Plant Cell Physiol., 28(4): 723-726.

Matsumura H, Miyachi S (1980). Cycling assay for nicotinamide adenine dinucleotides. Methods Enzymol., 69: 465-470.

Melo AMP, Roberts TH, Moller IM (1996). Evidence for the presence of two rotenone-insensitive NAD(P)H dehydrogenases on the inner surface of the inner membrane of potato tuber mitochondria. Biochim. Biophys. Acta, 1276(2): 133-139.

Mittler R, Vanderauwera S, Gollery M, Van Breusegem F (2004). Reactive oxygen gene network of plants. Trends Plant Sci., 9(10): 490-498.

Møller IM (2001). Plant mitochondria and oxidative stress. Electron transport, NADPH turnover and metabolism of reactive oxygen species. Annu. Rev. Plant Physiol. Plant Mol. Biol., 52(1): 561-591.

Møller IM, Jensen PE, Hansson A (2007). Oxidative modifications to cellular components in plants. Annu. Rev. Plant Biol., 58: 459-481.

Møller IM, Lin W (1986). Membrane-bound NAD(P)H dehydrogenases in higher plant cell. Annu. Rev. Plant Physiol., 37(1): 309:334.

Møller IM, Rasmusson AG (1998). The role of NADP in the mitochondrial matrix. Trends Plant Sci., 3(1): 21-27.

Moore CS, Cook-Johnson RJ, Rudhe C, Whelan J, Day DA, Wiskich JT, Soole KL (2003). Identification of AtNDI1, an internal non-phosphorylating NAD(P)H dehydrogenase in Arabidopsis mitochondria. Plant Physiol., 133(4): 1968-1978.

Murata Y, Pei ZM, Mori IC, Schroeder J (2001). Abscisic acid activation of plasma membrane Ca2+ channels in guard cells requires cytosolic NAD(P)H and is differentially disrupted upstream and downstream of reactive oxygen species production in abi1-1 and abi2-1 protein phosphatase 2C mutants. Plant Cell, 13(1): 2513-2523.

Navrot N, Collin V, Gualberto J, Gelhaye E, Hirasawa M, Rey P, Knaff DB, Issakidis E, Jacquot JP, Rouhier N (2006). Plant glutathione peroxidases are functional peroxiredoxins distributed in several subcellular compartments and regulated during biotic and abiotic stresses. Plant Physiol., 142(4): 1364-1379.

Nawa Y, Asahi T (1971). Rapid development of mitochondria in pea cotyledons during the early stage of germination. Plant Physiol., 48(1): 671-674.

Noctor G, Queval G, Gakiere B (2006). NAD(P) synthesis and pyridine nucleotide cycling in plants and their potential importance in stress conditions. J. Exp. Bot., 57(8): 1603-1620.

Prasad TK, Anderson MD, Stewart CR (1995). Localization and characterization of peroxidases in the mitochondria of chilling-acclimated maize seedlings. Plant Physiol., 108(4): 1597-1605.

Rahoui S, Chaoui A, El Ferjani E (2008). Differential sensitivity to cadmium in germinating seeds of three cultivars of faba bean (Vicia faba L.). Acta Physiol. Plant, 30(4): 451-456.

Rahoui S, Chaoui A, El Ferjani E (2010). Membrane damage and solute leakage from germinating pea seed under cadmium stress. J. Hazar. Mater., 178(1): 1128-1131.

Rasmusson AG, Møller IM (1991). NAD(P)H dehydrogenases on the inner surface of the inner mitochondrial membrane studied using inside-out submitochondrial particles. Physiol. Plant, 83(3): 357-365.

Rasmusson AG, Soole KL, Elthon TE (2004). Alternative NAD(P)H dehydrogenases of plant mitochondria. Annu. Rev. Plant Biol., 55(1): 23-39.

Rodríguez-Serrano M, Romero-Puertas MC, Zabalza A, Corpas FJ, Gómez M, del Río LA, Sandalio LM (2006). Cadmium effect on oxidative metabolism of pea (Pisum sativum L.) roots. Imaging of reactive oxygen species and nitric oxide accumulation in vivo. Plant Cell Environ., 29(8): 1532-1544.

Romero-Puertas MC, Corpas FJ, Rodríguez-Serrano M, Gomez M, Del Río LA, Sandalio LM (2007). Differential expression and regulation of antioxidative enzymes by cadmium in pea plants. J. Plant Physiol., 164(10): 1346-1357.

Romero-Puertas MC, Rodriguez-Serrano M, Corpas FJ, Gomez M, Del Rio LA, Sandalio LM (2004). Cadmium induced subcellular accumulation of superoxide and H2O2 in pea leaves. Plant Cell Environ., 27(9): 1122-1134.

Sandalio LM, Dalurzo HC, Gomez M, Romero-Puertas MC, Del Rio LA (2001). Cadmium induced changes in the growth and oxidative metabolism of pea plants. J. Exp. Bot., 52(364): 2115-2226.

Sato S, Asahi T (1975). Biochemical properties of mitochondrial membrane from dry pea seeds and changes in the properties during imbibitions. Plant Physiol., 56(1): 816-820.

Smiri M, Chaoui A, El Ferjani E (2009). Respiratory metabolism in the embryonic axis of germinating pea seed exposed to cadmium. J. Plant Physiol., 166(3): 259-269.

Smiri M, Chaoui A, Rouhier N, Gelhaye E, Jacquot JP, El Ferjani E (2010a). Redox regulation of the glutathione reductase/iso-glutaredoxin system in germinating pea seed exposed to cadmium. Plant Sci., 179(5): 423-436.

Smiri M, Chaoui A, Rouhier N, Gelhaye E, Jacquot JP, El Ferjani E

(2010b). NAD pattern and NADH oxidase activity in pea (Pisum sativum L.) under cadmium toxicity. Physiol. Mol. Biol. Plants, 16(3):305-315 In press, DOI 10.1007/s12298-010-0033-7.

Smiri M, Chaoui A, Rouhier N, Jacquot JP, El Ferjani E (2010c). Effect of cadmium on resumption of respiration in cotyledons of germinating pea seeds. Ecotox. Environ. Saf., 73(6): 1246-1254.

Stochs SJ, Bagchi D (1995). Oxidative mechanism in the toxicity of metal ions. Free Rad. Biol. Med., 18(2): 321-336.

Tamoi M, Miyazaki T, Fukamizo T, Shigeoka S (2005). The Calvin cycle in cyanobacteria is regulated by CP12 via the NAD(H)/NADP(H) ratio under light/dark conditions. Plant J., 42(4): 504-513.

Wagner GJ (1993). Accumulation of cadmium in crop plants and its consequences to human health. Adv. Agron., 51(1): 173-212.

Woolhouse HW (1983). Toxicity and tolerance in the responses of plants to metals. In: Encyclopedia of Plant Physiology (Lange OL, Nobel PS, Osmond CB, Ziegler H, eds.). Springer Verlag, Berlin, pp. 245-300.

Xu Y, Feng L, Jeffrey PD, Shi Y, Morel FMM (2008). Structure and metal exchange in the cadmium carbonic anhydrase of marine diatoms. Nature, 452(7183): 56-62.

Zhao Z, Hu X, Ross CW (1987). Comparison of tissue preparation methods for assay of nicotinamide coenzymes. Plant Physiol., 84(4): 987-988.

12

Pesticide residue analysis of fruits and vegetables

Rohan Dasika, Siddharth Tangirala and Padmaja Naishadham*

Department of Chemistry, Osmania University, Hyderabad, India.

This project describes an efficient and effective analytical method to screen pesticides in fruits and vegetable samples using liquid chromatography tandem mass spectrometry (LC-MS/MS). A quick, easy, cheap, effective, rugged, and safe (QuEChERS) method with acetate buffering (AOAC Official Method 2007.01) was used for sample preparation, which has been previously shown to yield high-quality results for hundreds of pesticide residues in foods.

Key words: Liquid chromatography/mass spectrometry (LC/MS), pesticide residues, pesticides, fruits, vegetables.

INTRODUCTION

Plant protection products (more commonly known as pesticides) are widely used in agriculture to increase the yield, improve the quality, and extend the storage life of food crops (Fernández-Alba and García-Reyes 2008). Pesticide residues are the deposits of pesticide active ingredient, its metabolites or breakdown products present in some component of the environment after its application, spillage or dumping. Residue analysis provides a measure of the nature and level of any chemical contamination within the environment and of its persistence. The pesticides must undergo extensive efficacy, environmental, and toxicological testing to be registered by governments for legal use in specified applications. The applied chemicals and/or their degradation products may remain as residues in the agricultural products, which becomes a concern for human exposure. Selected sampling programmes can be used to investigate residual levels of pesticide in the environment, their movement and their relative rates of degradation.

The maximum residue levels (MRLs) (or 'tolerances' in the United States) limit the types and amounts of residues that can be legally present on foods are set by regulatory bodies worldwide. Pesticide residue analysis is tremendously an important process in determining the safety of using certain pesticides. Pesticides polluting the

earth and causing problems in human beings and wildlife, the quantity of pesticide being consumed becomes a necessary knowledge. Analytical quality requirements like trueness, precision, sensitivity and selectivity have been met to suit the need for any particular analysis.

Unfortunately, not all farmers follow legal practices and due to the tremendous number of pesticides and crops in production, a number of analytical methods designed to determine multiple pesticide residues (Food and Drug Administration, 1999; Luke et al., 1975; Specht and Tilkes, 1980; Lee et al., 1991; Andersson and Pålsheden, 1991; Cook et al., 1999; General Inspectorate for Health Protection, 1996; Fillion et al., 2000; Sheridan and Meola, 1999; Lehotay, 2000). In 2003, the QuEChERS method for pesticide residue analysis was introduced Anastassiades et al. (2003), which provides high quality results in a fast, easy, an inexpensive approach. Follow-up studies have further validated the method for greater than 200 pesticides (Lehotay et al., 2007).

Technical developments always follow the way from the primitive via the complicated to the simple. The most common techniques in modern multi-residue target pesticide analysis are gas chromatography, liquid chromatography coupled to mass spectrometry (GC-MS, LC-MS) and/or tandem mass spectrometry (GC-MS/MS, LC-MS/MS) with triple quadrupole mass analysers. The numerous methods available for pesticide analysis show the importance of this application and rapid pace of developments in analytical chemistry. For example, Aguïera et al. (2000) described a method (Splitless large-

---

*Corresponding author. E-mail: tapadmaja@gmail.com.

volume GC-MS injection for the analysis of organophosphorus and organochlorine pesticides in vegetables using a miniaturised ethyl acetate extraction) for the measurement of only ten organophosphorus and organochlorine pesticides by GC-MS, but over the past decade, the number of pesticides typically included in methods has increased dramatically. The sample preparation techniques have also advanced to complement the analytical techniques depending on the types of analytes and matrices monitored.

Anastassiades et al. (2003) described the 'quick, easy, cheap, effective, rugged, low solvent consumption, wide pesticide range (Polar, pH – dependent compounds) and safe' (QuEChERS) method for pesticide residues in food as an example of a method that takes advantage of the powerful features of nearly universal selectivity and high sensitivity of modern GC- and LC-MS(/MS) instruments. The QuEChERS approach has been extensively validated for hundreds of pesticide residues in many types of foods, and has become Association of Analytical Communities (AOAC) Official Method 2007.01 (Lehotay et al., 2007) and CEN (2008). The QuEChERS method has several advantages over most traditional methods of analysis. High recoveries (greater than 85%) are achieved for a wide polarity and volatility range of pesticides, including notoriously difficult analytes. Very rugged because extract clean up is done to remove organic acids.

The most common approach is to use matrix-matched calibration standards. However, it can be difficult to find a blank matrix from which to prepare the calibration standards and compensation from one sample to another (even for the same matrix) may not be the same. A method of standard additions in the sample extract may be an alternative approach.

MATERIALS AND METHODS

Sampling

For the present project random sampling was done from the various markets in the twin cities of Hyderabad, Secunderabad and R. R. District local fields and markets. Depending upon the nature of the vegetation (size, shape, etc.), samples were enclosed in a clean blotting paper and wrapped inside a clean, paper envelope. The addition of a small sachet of silica gel to the envelope helps to reduce the moisture content of the system. However, samples were provided to the analytical laboratory to check for possible co-extractives, which could interfere with the analysis.

Reagents and materials

High-performance liquid chromatography (HPLC)- grade acetonitrile; deionized water was obtained from a Milli-Q reagent water system; dimethyl formamide; anhydrous magnesium sulfate obtained from Merck and primary secondary amine (PSA)-bonded silica was obtained from Supelco, Sigma Aldrich. Each sample was filtered through a 0.45 mm polyvinylidene difluoride (PVDF) filter before injection. Acetic acid and sodium acetate from Merck were used for the sample preparation procedure. Analytical-grade pesticide standards were ordered from Sigma- Aldrich.

Individual standard stock solutions were prepared by dissolving the crystalline standards in acetonitrile (or dimethyl formamide for those insoluble in acetonitrile) to reach the final concentration of 1000 to 4000 mg/ml. For method optimization, individual standard solutions were used, which were prepared by diluting the stock solution to a concentration of 1 to 4 mg/ml. A standard mix solution in acetonitrile for preparation of calibration standards was prepared from the individual stock solutions to yield 10 mg/ml.

Sample preparation

The acetate-buffered QuEChERS sample preparation method for pesticides (AOAC Official Method 2007.01) was applied to all the samples. After homogenization with a house-hold mill (equipped with stainless steel knives), a 15 g portion of the homogenized sample was weighed into a 50 ml polytetra fluoro ethylene (PTFE) tube and 100 ml of 50 mg/ml triphenyl phosphate (TPP) surrogate standard solution in acetonitrile was added followed by 15 ml of acetonitrile containing 1% acetic acid (v/v not accounting for purity). Then, 6 g MgSO4 and 2.5 g sodium acetate trihydrate (equivalent to 1.5 g of anhydrous form) were added, and the sample was shaken forcefully for 4 min. The sample was then centrifuged at 4000 rpm for 5 min and 5 ml of the supernatant were transferred to a 15 ml PTFE tube to which 750 mg MgSO4 and 250 mg PSA were added. The extract was shaken using a vortex mixer for 20 s and centrifuged at 4000 rpm again for 5 min. Approximately 3ml of the supernatant were filtered through a 0.45 mm PTFE filter (13 mm diameter), and 800 ml portions were transferred to autosampler vials. The extracts were evaporated to dryness under a stream of argon and reconstituted in 800 ml acetonitrile/water (20/80, v/v) for the LC-MS/MS analysis.

For the matrix-matched and standard addition calibrations, 4 to 80 ml of reconstituted samples were transferred into autosampler glass inserts and 20 ml portions of 0, 250, 500 and 1250 ng/ml standard mix solutions containing pesticides in 25/75 acetonitrile/ water (v/v) were added to reach the final additional concentrations of 0, 50, 100 and 250 ng/ g equivalents, respectively.

LC-MS analysis

For LC analysis, an Agilent 1100 HPLC system was used. It contained a binary pump, a degasser, column thermostat and an autosampler. A reverse-phase C8 analytical column of 150mm x 4.6mm internal diameter (i.d.) and 5 μm particle size and a guard column of 125 x 4.6 mm particle size were coupled to the LC system. Deionized water containing 0.1% formic acid (mobile phase component A) and acetonitrile (component B) were employed for the gradient programme, which started with 20% B for 3 min and was linearly increased to 100% B in 27 min (held for 3 min). The column was then re-equilibrated forr 12 min back to 20% B.

Thus, the total run time took 45 min. The flow rate was constant at 0.6 ml/min, and injection volume was 10 μl. For the MS/MS analysis, an Applied Biosystems 3200 QTRAP system was used. Applied Biosystems Analyst 1.4.2 software was used for instrument control and data processing. For the determination of pesticides, the commercial method of Applied Biosystems (2005) and its library was used.

Statistical analysis

1. An Analysis Of Variance (ANOVA) test and a t-Test were conducted between the 'no wash' and 'salted lukewarm water wash'

results to check their statistical significance. The data is given in Table 7.

2. The null hypothesis - there is "no correlation" between the pesticide's 'no wash' residue values and 'salted lukewarm water wash' residue values. It means the difference between the mean values of 'no wash' residual values and 'salted lukewarm water wash values' is insignificant".

3. The alternative hypothesis (the opposite of the null hypothesis) - there "is a correlation" between the two values.

4. The α-value – Probability that the null hypothesis is valid. This is assumed as 0.05. This means that 95% of the times the data was not a result of chance (means good data).

5. The 'p-value' is the α-value resulting at the end if statistical test. Should be much lower than 0.05. There is sufficient evidence to conclude that the mean of no wash values is significantly greater than the mean of salted water wash values. The Null hypothesis was hence rejected. The t-Tests showed 'p-value' well under 0.05 for almost all results, thus, indicating that the data is statistically significant.

RESULTS AND DISCUSSION

This study describes the combination of two parallel methods in the qualitative and quantitative screening of pesticide residues: (1) qualitative screening for target pesticides by LC-MS/MS using MRM data and (2) confirmation, quantitative determination of the frequently used /or previously detected pesticides using the MRM method. Compared with other available methods, the QuEChERS method is believe to give the best result. This concept was believed to give the widest scope with the least effort and still give excellent qualitative and quantitative results, particularly when using QuEChERS for sample preparation.

FRUITS

Apples

Residual levels for 15 different pesticides have been tested on apples. Fungicide Pyraclostrobin, Dodine etc, insecticides Pirimicarb, Thiacloprid and Acetamip etc, noted negative presence in comparison to the MRL even before any wash. The data is given in Table 1 and shown in Figure 1.

1. No wash: Pesticides Diphenylamine, Chlorpyrifos, Thiabendazole and Malathion were found to be extremely persistent when tested with "no wash".
a. Chlorpyrifos was lower than MRL on Pink Lady Apples, but was heavily persistent on Granny Smith and Gold Del Yellow apples.
2. Lukewarm water wash: Diphenylamine was low or nearly absent on Pink Lady apple, while others were still present. On Granny Smith and Gold Del apples, Diphenylamine was between 50 to 90% above MRL.
3. Salted Lukewarm water wash: Thiabendazole and Malathion were still persistent on all the 3 varieties. Average percentage deviation from the MRL was 165

and 45% respectively.
a. Chlorpyrifos was still shown high on Granny Smith and Gold Del apples.

Grapes

No Wash: The fungicides Imazalil and Thiabendazole remained approximately 70 and over 100% above the MRL respectively. Insecticide Phosmet, on an average was found to be 50% above MRL. The data is given in Table 2 and shown in Figures 2 and 3.

1. Black grapes had higher initial residual values than green grapes for all pesticides.
2. Lukewarm Water Wash: Phosmet was found to go below MRL, but others were still present, but showed decrease of about 20 to 25% in residual levels.
3. Salted Lukewarm water wash: Green grapes showed Imazalil closer to MRL, but Thiabendazole was still about 50% more than MRL. Black grapes showed much higher residual values of these 2 pesticides.

Pears

All the seven tested pesticides on the green pear were initially present in relatively low quantities compared to the residue levels on the apples. The only chemical where the residue after cleaning with salted water, greater than the MRL was Thiabendazole. After lukewarm water wash, the residue level of Thiabendazole was 78% above the MRL; after a salted lukewarm water wash, it was 43% above. The data is given in Table 3.

Guava

All the pesticide residues on the guava were well under the prescribed level before any wash, excepting Thiabendazole. After the wash with lukewarm water, Thiabendazole residue was still above, while after salted lukewarm water wash, the Thiabendazole residue went down to Non Detectable (ND) in comparison to MRL. The data is given in Table 4.

Vegetables

Egg plant or Brinjal

No Wash: Eggplants showed the most pesticides that have residue levels over the MRL. They ranged from 100 to 500% above MRL.

Lukewarm Water Wash: The overall residue levels of all pesticides tested showed significant improvement with values going down anywhere from 50 to 80% when compared against no wash readings. However, at this

Table 1. Pesticide residues in Apple.

		Granny Smith Green Apple			
S/N	Pesticide	Maximum residue limit (mg/kg)	No wash	Lukewarm water wash	Salted lukewarm water wash
1	Pyraclostrobin	0.3	0.05	ND	ND
2	Dithiocarbamates	3	2.2	1.9	1.7
3	Dodine	5	3.4	2.5	2.2
4	Acetamiprid	0.1	0.07	ND	ND
5	Diphenylamine	5	7.8	3.4	2.8
6	Dithiocarbamates	3	2.2	1.3	1.2
7	Imazalil	5	3.8	2.7	2.4
8	Phosmet	10	9.2	7.5	6.4
9	Pirimicarb	1	0.06	ND	ND
10	Chlorpyrifos	0.5	2.8	1.7	1.1
11	Captan	3	2.2	1.8	1.7
12	Dithiocarbamates	3	2.6	2.3	2.1
13	Malathion	0.5	1.2	1	0.8
14	Thiacloprid	0.3	0.2	0.18	0.18
15	Thiabendazole	10	43.2	37.4	31

		Pink Lady Apple			
S/N	Pesticide	Maximum residue limit	No Wash	Lukewarm water wash	Salted lukewarm water wash
1	Diphenylamine	5	6	3.5	2
2	Phosmet	10	8	5	4
3	Chlorpyrifos	0.5	0.14	0.1	0.08
4	Malathion	0.5	0.9	0.7	0.6
5	Thiabendazole	10	36.8	29	27.3

		Gold Del Michigan Yellow Apple			
S/N	Pesticide	Maximum residue limit	No Wash	Lukewarm water wash	Salted lukewarm water wash
1	Diphenylamine	5	9.6	5.4	4.2
2	Phosmet	10	12	9.5	8.6
3	Chlorpyrifos	0.5	3.4	1.9	0.8
4	Malathion	0.5	1.8	0.9	0.8
5	Thiabendazole	10	46.8	37.7	32.6

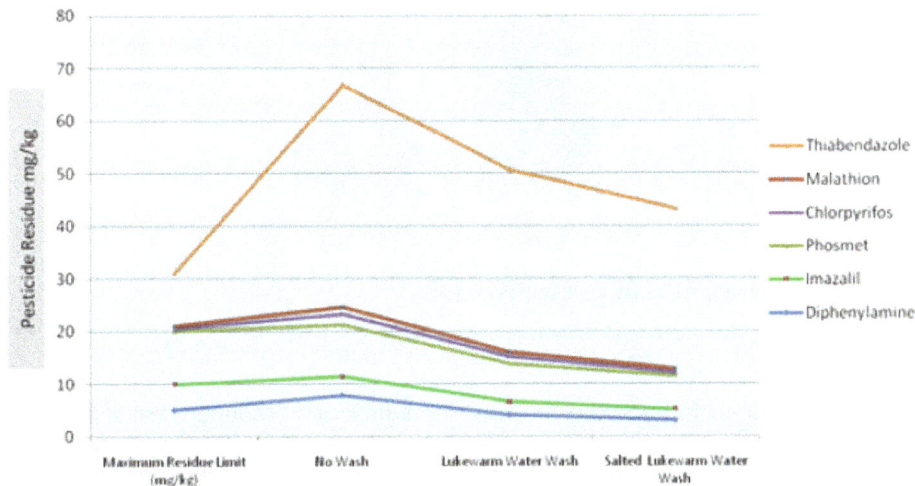

Figure 1. Average pesticide residues (Pink Lady, Granny Smith, Gold Del Apples).

Table 2. Pesticide residues in grapes.

		Green grapes			
S/N	Pesticide	Max residue limit (mg/kg)	No wash	Lukewarm water wash	Salted lukewarm water wash
1	Imazalil	5	8.6	6.9	5.4
2	Thiabendazole	10	21.4	17	14.7
3	Phosmet	10	11.8	8.9	7.6
4	Decamethrin	0.05	0.03	ND	ND

		Black grapes			
S/N	Pesticide	Max residue limit (mg/kg)	No wash	Lukewarm water wash	Salted lukewarm water wash
1	Imazalil	5	9.3	8.2	6.6
2	Thiabendazole	10	32	27.4	21
3	Phosmet	10	16.7	11.3	8.9
4	Decamethrin	0.05	0.05	0.04	0.02

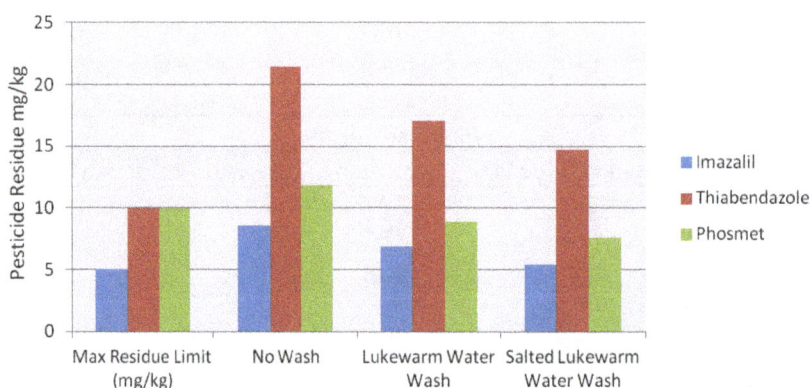

Figure 2. Pesticide residues in green grapes.

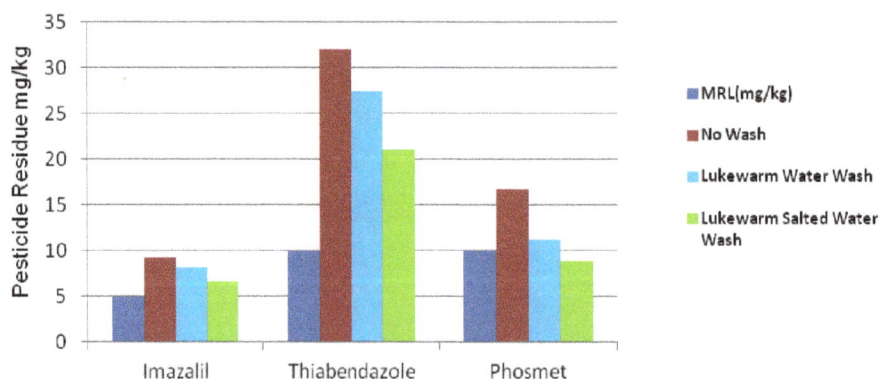

Figure 3. Pesticide residues on black grapes.

stage more washing will be required to bring the residue levels closer to MRL.

Salted Lukewarm water wash: Chlorothalonil residues showed 300% above MRL. Diphenylamine residues significantly reduced, but still 22% above MRL.

Chlorpyrifos, Thiabendazole, Acephate residues were also very high suggesting another round of wash, which was out of scope for the project. The data is given in

Table 3. Pesticide residues in pear.

S/N	Pesticide	Max residue limit (mg/kg)	Pear					
			No wash	Deviation from MRL (%)	Lukewarm water wash	Deviation from MRL (%)	Salted lukewarm Water Wash	Deviation from MRL (%)
1	Diphenylamine	5	6.1	22	4.3	-14	3.4	-32
2	Imazalil	5	3	-40	2.1	-58	1.8	-64
3	Phosmet	10	11	10	8.9	-11	8.8	-12
4	Pirimicarb	1	1.8	80	1.2	20	0.6	-40
5	Chlorpyrifos	0.5	0.6	20	0.3	-40	0.2	-60
6	Captan	3	0.06	-98	ND	ND	ND	ND
7	Thiabendazole	10	26.4	164	17.8	78	14.3	43

Table 4. Pesticide residues in guava.

Name of the pesticide	Max residue limit (mg/kg)	Guava		
		No Wash	Lukewarm water wash	Salted lukewarm water wash
Diphenylamine	5	4.8	2.2	1.2
Imazalil	5	4.2	3.6	3.4
Phosmet	10	9.3	7.9	7.2
Chlorpyrifos	0.5	0.43	0.28	0.18
Thiabendazole	10	0.02	ND	ND

Table 5 and shown in Figure 4.

Bell Peppers

1. No Wash: Bell Peppers, both green and red had Chlorpyrifos, and Diphenylamine residues present from 75 to 100% above MRL. Thiabendazole was found but in comparison to other vegetables, this was relatively in smaller levels.

2. Lukewarm water wash: Chlorpyrifos and Diphenylamine persist and are about 40 to 60% above MRL. Thiabendazole was 23% above MRL on green one, while it was just about the MRL level on Red pepper.

3. Salted Lukewarm water wash: Diphenylamine is still persistent 25 to 44% above MRL, while others have shown great reduction. The data is given in Table 6 and shown in Figures 5 and 6.

All fruits and vegetables had great disparity between the maximum Thiabendazole levels instituted by the EPA and the levels found on the produce. This was especially evident on apples and Indian eggplants. Even after washing the fruit or vegetable, the Thiabendazole residue still remained higher than the tolerable levels. The only exceptions to this pattern are Red bell peppers and guavas.

Conclusions

The developed combination of the two methods described above permitted the fast and easy qualitative screening of target pesticides in a 45 min LCMS/ MS run.

Although the manual evaluation of the given chromatograms increased the analysis time by an additional 15 min per sample, very little time, cost and labour was spent on sample preparation. In the case of dirty samples, some false indications were observed, but these were caught by the use of the MRM confirmatory and quantitative method for the more common pesticides. The construction

Table 5. Pesticide residues in eggplants or brinjal.

		Indian Egg Plant			
S/N	Pesticide	Maximum residue limit (mg/kg)	No Wash	Lukewarm water wash	Salted lukewarm water wash
1	Diphenylamine	5	9.8	5.3	4.4
2	Imazalil	5	4.2	2.5	1.8
3	Phosmet	10	9	8.4	6.8
4	Chlorpyrifos	0.5	1.4	1	0.8
5	Thiabendazole	10	44.1	29.3	14.3
6	Chlorothalonil	0.1	0.5	0.3	0.19
7	Endosulfan	2	2.2	1.9	1.5
8	Acephate	2	5.2	4.2	3.6
		Chinese Egg Plant			
S/N	Pesticide	Maximum residue limit (mg/kg)	No Wash	Lukewarm water wash	Salted lukewarm water wash
1	Diphenylamine	5	9.8	7.3	6.1
2	Imazalil	5	3.8	2.9	2.4
3	Phosmet	10	9.3	6.9	6.4
5	Chlorpyrifos	0.5	2.1	1.7	1.3
6	Thiabendazole	10	33.2	24.1	21
7	Chlorothalonil	0.1	0.6	0.45	0.38
8	Endosulfan	2	2.8	2.3	1.8
9	Acephate	2	7.8	5.7	4.7

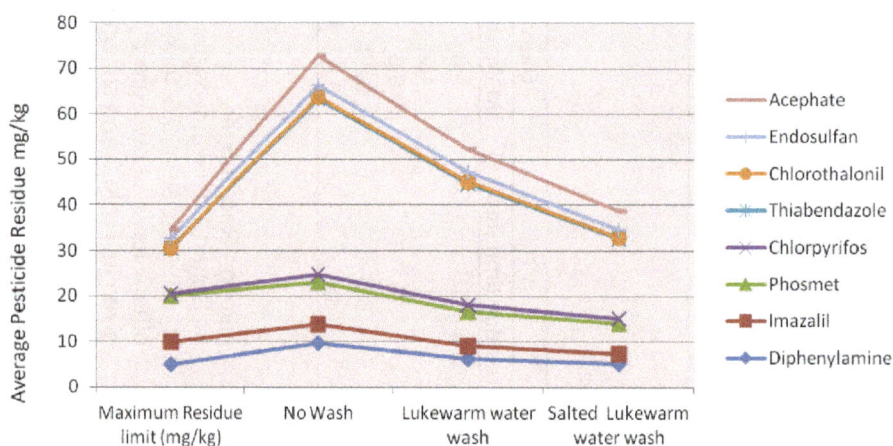

Figure 4. Average pesticide residues (Indian and Chinese eggplants).

of standard addition calibration was carried out with the same extract that was previously injected for screening the compounds. A large calibration database in different matrices was collected to show the consistency of the average calibration slope, which helped us check the accuracy of the calculated results from the method of standard additions.

Apples

Test revealed increased residue levels of Chlorpyrifos, Thiabendazole, Malathion and Diphenylamine. Most

fungicides, insecticides and pesticides have been washed away with the lukewarm water wash.

Pink Lady Apples showed no residue of Chlorpyrifos and showed only Malathion and Thiabendazole before wash and have greatly reduced after lukewarm water wash.

However, Granny Smith Apples and Gold Del Apples showed higher residue of Chlorpyrifos after the lukewarm water wash.

Grapes

Imazalil is among the persistent pesticides on grapes with

Table 6. Pesticide residues in bell peppers.

			Red Bell pepper					
S/N	Pesticide	Max residue limit (mg/kg)	No Wash	Deviation from MRL (%)	Lukewarm water wash	Deviation from MRL (%)	Salted lukewarm water wash	Deviation from MRL (%)
1	Chloropyrifos	2	3.5	75	2.7	35	2.2	10
2	Imazilil	5	4.8	-4	3.7	-26	3.3	-34
3	Thiabendazole	10	11.4	14	10	0	9.7	-3
4	Diphenylamine	5	8.7	74	6.9	38	6.3	26

			Green Bell pepper					
S/N	Pesticide	Max residue limit (mg/kg)	No Wash	Deviation from MRL (%)	Lukewarm water wash	Deviation from MRL (%)	Salted lukewarm water wash	Deviation from MRL (%)
1	Chloropyrifos	2	4.1	105	2.9	45	2.3	15
2	Imazilil	5	3.5	-30	2.4	-52	1.8	-64
3	Thiabendazole	10	14.6	46	12.3	23	11.5	15
4	Diphenylamine	5	9.8	96	7.9	58	7.2	44

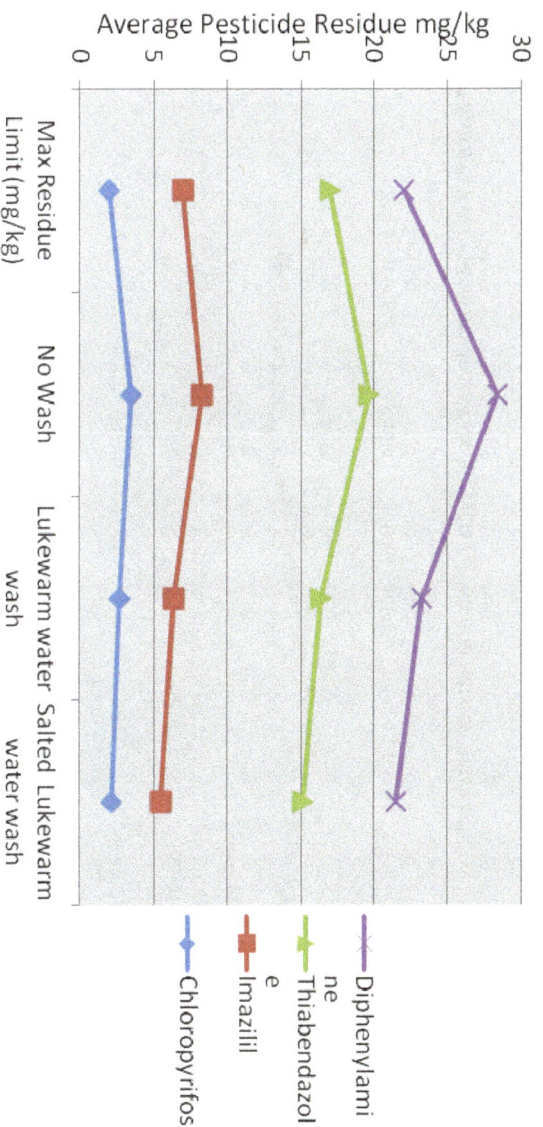

Figure 5. Pesticide residue analysis on red bell pepper.

Figure 6. Average pesticide residue analysis (green and red bell peppers).

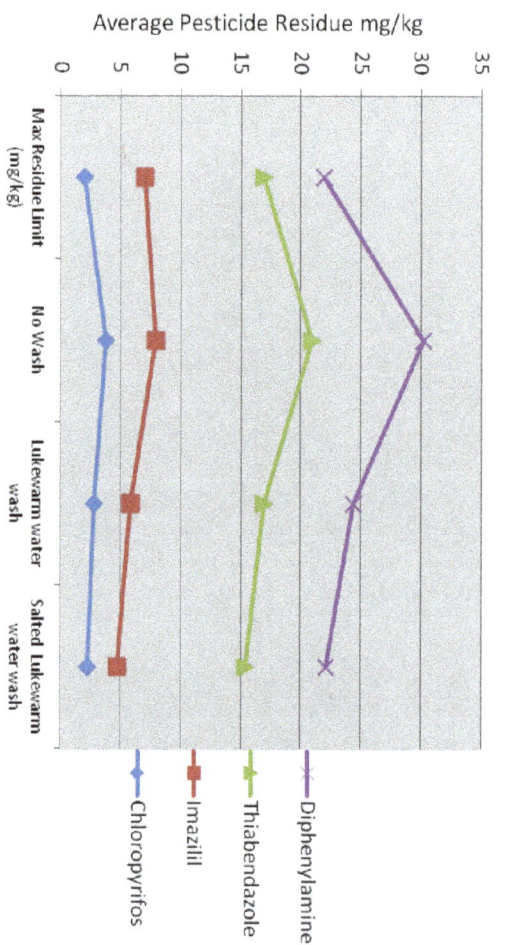

Table 7. Statistical analysis.

Pesticide tested	α-value	Degrees of freedom	t-value	Look-up t-value	p-value	Standard deviation for salted lukewarm water wash	Statistically significant?
				Statistical Tests			
Diphenylamine	0.05	15.86	4.05	1.753	0.00097	2.04	yes
Chlorpyrifos	0.05	12.56	1.88	1.782	0.085	0.83	yes
Imazalil	0.05	18.74	2.13	1.74	0.046	1.64	yes
Thiabendazole	0.05	17.4	2.31	1.74	0.033	8.5	yes
Phosmet	0.05	12.96	3.45	1.771	0.0043	1.55	yes

allowed MRL of 5mg/kg. When consumed over longer periods, i.e. 2 to 5 times (50 mg/kg/day) more than MRL showed some mortality issues and significant brain enzyme depression.

Green pears

Chlorpyrifos, Phosmet, and Diphenylamine are

Guava

The residue levels of most toxic pesticides were

present above MRL values before wash, but go down after lukewarm water wash. Their residual values drop further after salted lukewarm water wash.

Green and red bell peppers

Chlorpyrifos and Diphenylamine are significantly

under MRL even before any wash, except slightly higher Thiabendazole, which is non-toxic. After salted lukewarm water wash, all the pesticides were well under MRL, even Thiabendazole.

present even after salted lukewarm water wash.

Eggplants

There are significant levels of Chlorpyrifos and Diphenylamine without any wash. After salted lukewarm water wash, Diphenylamine was under the MRL, but Chlorpyrifos remained higher. Although the difference might not be much, Chlorpyrifos is very toxic even in low doses.

Farmer's survey report conclusion

In summary, the survey reflected that, no produce can be consumed right after it is purchased from the market. It may be safe to pick produce and consume right away on smaller farms. The survey reports from the farmers indicated the following:

1. Farmers use IPM (Integrated Pest Management)
2. Pest management which relies on common sense and information about pests and pesticides in farm.
3. Used to lessen the pest damage with least harm to humans, plants, and their environment.
4. Big farmers use pesticides heavily, but with a gap of 10 to 14 days before each application.
5. Small farmers do not have the space or need to use heavy pesticides. Most produce can be eaten right after picking without washing.
6. However, large farmers, because of the size rely more heavily on pesticides.
7. Large farmers employ sophisticated cleaning mechanisms before the product is sold.

REFERENCE

Agu¨ era A, Piedra L, Hernando MD, Ferna´ ndez-Alba AR,Contreras M (2000). Splitless large-volume GC-MS injection for the analysis of organophosphorus and organochlorine pesticides in vegetables using a miniaturised ethyl acetate extraction. Analyst., 125:1397–1402.

Anastassiades M, Lehotay SJ, Štajnbaher D, Schenck FJ (2003). Fast and easy multiresidue method employing acetonitrile extraction/partitioning and "dispersive solidphase extraction" for the determination of pesticide residues in produce. J. AOAC. Int., 86: 412-431.

Andersson A, Pålsheden H (1991). Comparison of the efficiency of different GLC multi-residue methods on crops containing pesticide residues. Fresenius J. Anal. Chem., 339: 365-367.

Applied Biosystems (2005). Application Note 114AP38-02.Available from: http://www.absciex.com/LITERATURE/ cms_041987.pdf/ Banerjee K, Oulkar DP, Dasgupta S, Patil SB, Patil SH, Savant R, Adsule PG.2007.

Cook J, Beckett MP, Reliford B, Hammock W, Engel M (1999). Mutiresidue analysis of pesticides in fresh fruits and vegetables using procedures developed by the Florida Department of Agriculture and Consumer Services. J. AOAC Int., 82: 1419-1435.

Fernández-Alba AR, Garcia-Reyes JF (2008). Large-scale multi-residue methods for pesticides and their degradation products in food by advanced LC-MS. Trac-Trend. Anal. Chem., 27 (11): 973-990.

Fillion J, Sauvé F, Selwyn J (2000). Multiresidue method for the determination of residues of 251 pesticides in fruits and vegetables by gas chromatography/mass spectrometry and liquid chromatography with fluorescence detection. J. AOAC Int., 83: 698-713.

Food and Drug Administration (1999). Pesticide Analytical Manual Volume I: Multiresidue Methods, 3rd Edition, U.S. Department of Health and Human Services, Washington, DC.

General Inspectorate for Health Protection (1996). Analytical Methods for Pesticide Residues in Foodstuffs, 6th edition, Ministry of Health Welfare and Sport, The Netherlands.

http://www.cfsan.fda.gov/~frf/pami3.html

Lee SM, Papathakis ML, Hsiao-Ming CF, Carr JE (1991). Multipesticide residue method for fruits and vegetables: California Department of Food and Agriculture. Fresenius J. Anal. Chem., 339, 376-383.

Lehotay SJ (2000). Determination of pesticide residues in nonfatty foods by supercritical fluid extraction and gas chromatography/mass spectrometry: collaborative study. J. AOAC Int., 83, 680-697.

Lehotay SJ, Hiemstra M, van Bodegraven P, de Kok A (2007). Validation of a fast and easy method for the determination of more than 200 pesticide residues in fruits and vegetables using gas and liquid chromatography and mass spectrometric detection. J. AOAC Int., 88 (2005) 595.

Luke MA, Froberg JE, Masumoto HT (1975). Extraction and cleanup of organochlorine, organophosphate, organonitrogen, and hydrocarbon pesticides in produce for determination by gas-liquid chromatography. J. Assoc. Off. Anal. Chem., 58, 1020-1026.

Sheridan RS, Meola JR (1999) Analysis of pesticide residues in fruits, vegetables, and milk by gas chromatography/tandem mass spectrometry. J. AOAC Int., 82, 982-990.

Specht W, Tilkes M (1980). Gaschromatographische bestimmung von rückständen an pflanzenbehandlungsmitteln nach clean-up über gel-chromatographie und mini-kieselgelsäulen- chromatographie. Fresenius J. Anal. Chem., 301, 300-307.

13

Toxicological effects of burrow pit effluent from a waste dump on periwinkle (*Tympanotonus fuscatus linne*)

OGELEKA D. F.[1]* and TUDARARO-AHEROBO L. E.[2]

[1]Department of Chemistry, Western Delta University, Oghara, Delta State, Nigeria.
[2]Department of Environmental Sciences, Federal University of Petroleum Resources, Effurun, Delta State, Nigeria.

Burrow pit effluent collected from a waste dump site in the Niger Delta area of Nigeria was subjected to sublethal test using periwinkle (*Tympanotonus fuscatus linne*). This was to ascertain if the heavy metals and organic constituents in the burrow pit effluent bioaccumulated in the tissues of the organisms. The test was conducted using the Organization for Economic Cooperation and Development (OECD) protocol #218 in a sediment medium with varying concentrations of the test effluent. Low bioaccumulation potentials were observed for the metals in the tissues of the organisms. The determination of sixteen polycyclic aromatic hydrocarbons (PAHs) revealed only three components at relatively low concentrations in the tissues of the organisms at test termination of 28 days for test effluents concentrations (3.125, 12.5 and 50%). Concentrations for benzo [b] fluoranthene were (0.0017 – 0.0039 ppm), phenanthrene, (0.0021 – 0.0049 ppm) and pyrene, (0.0035 - 0.0081 ppm). However, the concentration of PAHs in the tissues of the organisms at test initiation was <0.0001 ppm. There was significant ($p < 0.05$) difference in the PAHs concentrations in the organisms exposed to the test effluent and the controls. This could lead to adverse ecological imbalance on a variety of aquatic species including bottom dwelling organisms inhabiting such environment if the release of untreated effluent is not controlled in the Niger Delta area of Nigeria.

Key words: Periwinkle (*Tympanotonus fuscatus linne*), sediment toxicity, effluent, sublethal toxicity (Bioaccumulation).

INTRODUCTION

The environment is perceived to be at risk from thousands of toxic substances and chemicals of both anthropogenic and natural origin. When hazardous substances are released into the environment, an evaluation is necessary to determine the possible impact of these substances on human health and other biota (Adams et al., 1992; Ogeleka et al., 2010). An important process through which chemicals substances can affect living organisms is bioaccumulation. Bioaccumulation is a process by which chemicals or substances are taken up by an organism either directly from exposure to a contaminated medium or by consumption of food containing the chemical or substance. Bioaccumulation

means an increase in the concentration of a chemical or substance in a biological organism over time, compared to its concentration in the environment (Corl, 2001; Relyea and Diecks, 2008). Thus understanding the dynamic process of bioaccumulation is very important in protecting human beings and other organisms from the adverse effects of chemical exposure and has become a critical consideration in the regulation of chemicals (DPR, 2002; OECD, 2003).

The natural aquatic systems may extensively be contaminated with heavy metals released from industrial and other anthropogenic activities. Metals are non-biodegradable and are considered as major environmental pollutants causing cytotoxic, mutagenic and carcinogenic effects in animals (Ajayi and Osibanjo, 1981; Hayat et al., 2007; Hussain et al., 2011). Heavy metal contamination may have devastating effects on the ecological balance of the recipient environment and a diversity of aquatic

*Corresponding author. E-mail: dorysafam@yahoo.com.

organisms (Ashraj, 2005; Waqar, 2006; Farombi et al., 2007; Tawari-Fufeyin and Ekaye, 2007; Yilmaz et al., 2007).

Among animal species, fish and bottom dwelling organisms are the inhabitants that cannot easily escape from the detrimental effects of these pollutants (Vinodhini and Narayanan, 2008; Ezemonye et al., 2009). The mechanisms of heavy metal and organic excretion, deposition and detoxification in aquatic organisms are not capable of handling heavy metals in short time frames; thus these chemicals tend to accumulate specifically in metabolically active tissues and organs (Langston, 1989; Cicik and Engin, 2005; Ezemonye et al., 2007 a; Vinodhini and Narayanan, 2008). It is also known that physiological and biochemical parameters in aquatic organisms' blood and tissues could change when exposed to heavy metals and other toxic substances exerting an extra stress on the organisms (Sastry and Rao, 1984; Cicik and Engin, 2005; Davies et al., 2006).

Many aquatic pollutants such as polyaromatic hydrocarbons (PAHs) and their halogenated forms are chemically quiet stable; owing to their lipophilic nature, they can easily penetrate biological membrane and accumulate in organisms. Polyaromatic hydrocarbons (PAHs) consist of hydrogen and carbon arranged in the form of two or more fused benzene rings. They are important environmental pollutants because of their ubiquitous presence and carcinogenicity and are the most toxic of the hydrocarbon families (Tuvikene, 1995). The United State Environmental Protection Agency (USEPA) and the World Health Organisation (WHO) have identified 16 PAHs as priority pollutions while some of these e.g. benzo (a) anthracene, chrysene and benzo (a) pyrene are considered potential human carcinogens (EPA, 1980; Kanchanamayoon and Tatrahun, 2008).

Bioconcentration refers to the absorption or uptake of a chemical from a medium to concentrations in the organism's tissues that are higher than that in the surrounding environment. The degree to which a contaminant would concentrate in an organism is expressed as a bioconcentration factor (BCF). BCF refers to the concentration of a chemical in an organism's tissues divided by the exposure concentration (McGeer et al., 2003). It has been found that chemicals or substances displaying a half-life greater than 30 days, a bioconcentration factor (BCF) greater than 1000 or an octanol/water partition coefficient, log Kow value greater than 4.2 tend to be persistent and bioaccumulate (EPA, 2000; Ezemonye et al., 2007).

The aim of this study was to assess the effects of burrow pit effluent from a waste dump site in the Niger Delta area of Nigeria on bottom dwelling organism, periwinkle (*Tympanotonus fuscatus linne*). The toxicological end point of assessment was growth, bioaccumulation of heavy metals (copper, zinc, total iron, lead, chromium) and PAHs. The test species were chosen because they are abundant, sensitive and available all the year round in the Niger Delta ecological

zone of Nigeria. They are of great economic importance since they play a role in the coastal food web serving as a source of nutrition for humans and other organisms (Beeby, 2001; Ciarelli et al., 1997).

MATERIALS AND METHODS

Area description

The waste dumpsite is in Ughelli area of the Niger Delta ecological zone of Nigeria. Ughelli is located in Ughelli North local Government Area of Delta State, Nigeria. The site boundary georeferences are latitude 05°28'55.6"N and longitude 005°48'34.0"E. The study site is a fresh water environment with temperature ranging between 21.5 and 36.9°C and relative humidity between 55 and 94%. An annual rainfall of 2900 mm is normal for the area. Freshwater swamp forest zone dominates but most of it has been cleared due to high human population density and rural to urban drift.

Collection and acclimation of test organisms

The effluent sample was collected from a waste dump site in Ughelli in the Niger Delta area of Nigeria and stored at 4°C before starting the experiment. Healthy test species of periwinkle were collected from a fresh water cultured farm in Ekrheranwhen in the Niger Delta. Ekrheranwhen is located in Ughelli North Local Government Area of Delta State, Nigeria (Latitude 05°32'43.6"N and longitude 005°55'04.6"E). The test organisms were acclimated under laboratory conditions for a period of seven days. The size of the organisms used for the sediment bioassay was 2.67 ± 0.20 g.

Experimental sublethal bioassay procedure

The 28 day experiment was carried out using the OECD #218 sediment toxicity bioassay protocol with spiked sediment (OECD, 2004). The test began with a range-finding test to determine the concentrations to be used in the definitive test. Approximately 24 hours before the test, the sediment samples were acclimated at 25°C and weighed. Triplicate treatment tanks for each concentration of the test effluent were prepared with 1 kg of the sediment. The prepared test solutions (3.125, 12.5, 50%) were added to the treatment tanks, homogenized and allowed to settle for 2 to 3 hours. The test organisms were placed in the dilution water to rinse off debris that may interfere with the test and ten (10) organisms were weighed to obtain the initial weight. The organisms were then gently transferred into each amber-coloured glass tank containing the sediment and test effluent. The controls were maintained in clean sediment and habitat water without the test effluent (Environment Canada, 1992). The treatment tanks were gently aerated using oil-free low whisper aerators. After 28 days the sediment was sieved and the periwinkles were rinsed in the dilution water and weighed to obtain the final weight.

Analysis of metals and polyaromatic hydrocarbons (PAHs)

For metal analysis, soft tissues were collected after the shells were removed. These were dried at 480°C for 24 h. The dried tissues were ground and known weights were digested in a mixture of nitric acid (5 ml), sulphuric acid (3 ml) and perchloric acid (3 ml). Five (5) trace metals namely lead, copper, total iron, chromium and zinc were determined. The heavy metal contents of the digest (habita

Table 1. Concentrations of heavy metals and organics in periwinkle after 28 days.

Parameter (ppm)	Concentration of effluent			
	Variable (%)			Control (%)
	3.125	12.5	50	0
Total iron	37.15 ± 0.11	43.09 ± 0.14	49.52 ± 0.16	34.01 ± 0.10
Zinc	0.34 ± 0.07	0.35 ± 0.04	0.44 ± 0.03	0.30 ± 0.04
Phenanthrene	0.0021 ± 0.002	0.0039 ± 0.007	0.0049 ± 0.010	<0.001
Pyrene	0.0035 ± 0.001	0.0065 ± 0.002	0.0081 ± 0.002	<0.001
Benzo[b] fluoranthene	0.0017± 0.001	0.0027 ± 0.001	0.0039 ± 0.001	<0.001

water, burrow pit effluent, soft tissues of test species and controls) were tested using atomic absorption spectrophotometer (AAS, Shimazu 6701 F model).

The samples collected for PAH analysis were crushed, dried with sodium sulphate and extracted with dichloromethane (DCM) for 4 h in a ratio of 1:10 sample: solvent. The subsequent extract was then concentrated to 1 ml by evaporation in a secured fume hood and passed through a fractionating column. The resulting extract was dried with sodium sulphate and placed in clean amber coloured vials rinsed with DCM. An appropriate volume (1 μl) was injected into a GC/MS (Agilent 5975C) for the analysis of the 16 different PAHs components. The GC/MS was calibrated using specific PAHs standards. Standard stock solutions (1 mg/ml) were prepared by dissolving 10 mg of the desired PAH in 10 ml DCM and stored at 20 °C. All working solutions were freshly prepared by serial dilution with DCM.

Water chemistry

The determination of physico-chemical parameters of the dilution water was carried out to provide relevant information on possible changes that could result in potential hazards to the biological indicators. Physico-chemical constituents determined include; pH, temperature, dissolved oxygen (DO), salinity and conductivity.

Physiological effect

The physiological endpoint used in this assessment was growth. Mean weight of the total number of organisms used for the test was taken at initiation (day 0) and termination (day 28) of the test. It was computed using the formula:

$$\text{Mean weight} = \frac{\text{weight of organism at day 0 + weight of organism at day 28}}{2}$$

RESULTS

The analysis of the concentration of contaminants in the test organisms is an important approach to assessing the bioavailability of substances and to evaluate their behaviour in the environment. The concentrations of physico-chemical parameters, heavy metals and PAHs in the burrow pit effluent, dilution water (control), and periwinkle soft tissues at days 0 and 28 are presented in Table 1, Figure 1 and 2.

The control and the burrow pit effluent recorded 5.60 ±0.28 and 7.21 ± 0.43 pH units respectively. Salinity was

0.04 ± 0.004 and 0.73 ± 0.03 ppt while conductivity was 290 ± 23 μS/cm and 2420 ± 82 μS/cm in the same order. The mean temperature during the experimental period in all the test was 27 ± 2 °C with a 16:8 h light: darkness photoperiod. Dissolved oxygen concentrations recorded was 6.4 ± 0.4 mg/l. Concentrations in the control for copper, zinc and total iron were 0.02 ± 0, 0.05 ± 0.01 and 0.06 ± 0.02 mg/l respectively while in the burrow pit effluent <0.01, 0.06 ± 0.02 and 0.36 ± 0.02 mg/l respectively were recorded. Lead and chromium were not detected in the control and burrow pit effluent. Three components of PAHs (benzo [b] fluoranthene, phenanthrene, pyrene) out of the sixteen components determined, were found in the burrow pit effluent. There were no PAHs in the control sample; however, the PAHs concentrations for benzo [b] fluoranthene, phenanthrene and pyrene in the burrow pit effluent were 0.008 ± 0.001, 0.010 ± 0.001 and 0.059 ± 0.004 ppm respectively.

The pH of the native sediment (control) was 4.98 ± 0.5 pH units. It had an organic carbon content of 3.3 ± 0.13% while PAHs were < 0.001 ppm. Concentrations recorded for zinc, copper, iron, chromium and lead were 41.92 ± 1.2, 9.47 ± 0.89, 14213 ± 123, 6.32 ± 0.46 and < 0.01 ppm respectively.

Concentrations reported in the tissue of the control organism for copper, zinc and total iron were 0.07 ± 0.01, 0.30 ± 0.04 and 34.01 ± 0.10 ppm respectively while chromium and lead were not detected. Zinc and total iron concentrations in the test organisms for 3.125, 12.5 and 50% test effluent were 0.34 to 0.44 ppm and 37.15 to 49.52 ppm respectively (Table 1). Concentrations for PAHs were obtained mainly from three components namely benzo [b] fluoranthene, phenanthrene and pyrene. The concentration of PAHs in the tissues of the organisms at day 0 was <0.001 ppm while the concentrations of benzo [b] fluoranthene, phenanthrene and pyrene in the tissues of the organisms after 28-day exposure to the test effluent are given in Table 1. The lowest PAHs concentration was recorded in the organisms exposed to test effluent concentration of 3.125% (benzo [b] fluoranthene, 0.0017 ± 0.001 ppm) while the highest concentration was recorded in exposure concentration of 50% (pyrene, 0.0081 ± 0.002 ppm).

The mean weight of the periwinkle was higher in the

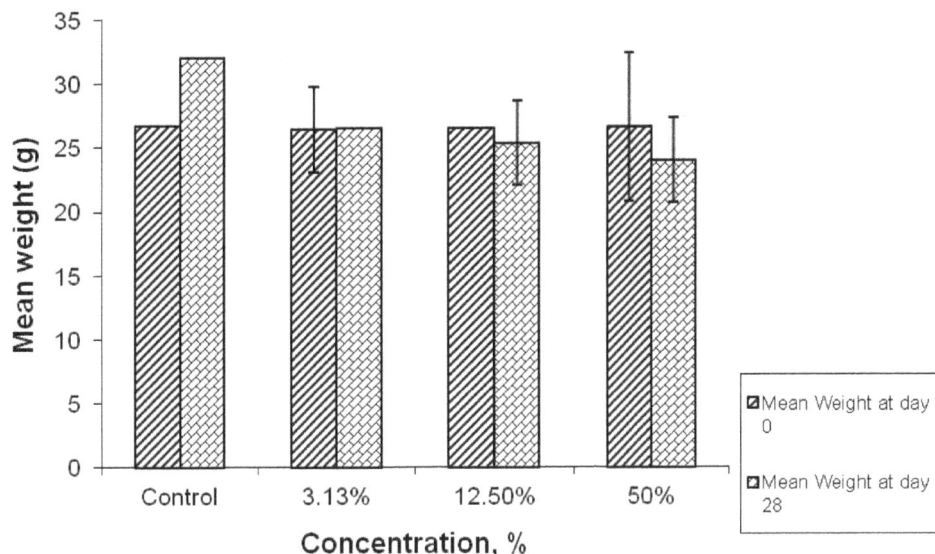

Figure 1. Mean weight ± SEM of periwinkle exposed to effluent at day 0 and 28.

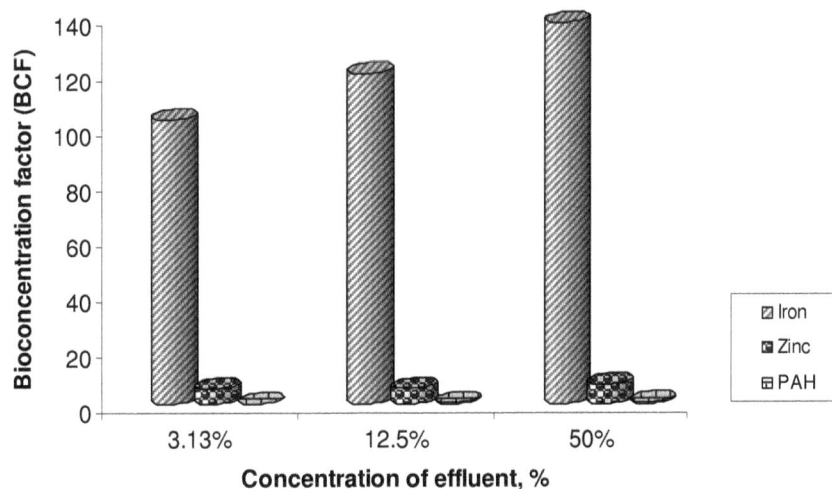

Figure 2. Mean Bioconcentration Factor (BCF) of burrow pit effluent at day 28.

control and the 3.125% test effluent than 12.5% and 50% test effluents concentrations (Figure 1). Bioconcentration factor (BCF) for total iron in the test organisms in the test effluent concentrations (3.125, 12.5 and 50%) were 103, 120 and 138 respectively. Zinc recorded a BCF ranging from 5.67 to 7.33. The test organisms showed levels of accumulation of PAHs, which varied from 0.22 in 3.125% to 0.63 in 50% test effluent (Figure 2).

DISCUSSION

Knowledge of heavy metal concentrations in aquatic organisms is important due to the nature of management

and human consumption of these species. Bioaccumulation of metals in periwinkles can be considered an index of metal pollution in aquatic bodies (Davies et al., 2006). The baseline concentrations of metal components in the unexposed test organisms at test initiation indicate generally low levels than observed for the tissues of the organisms after uptake of the effluent. However, the observed accumulation of iron and zinc in the test organisms could be from the sediment since the values of these metals in the effluent were relatively very low in this study.

Certain factors may have influenced the differential uptake of metals in the test organisms. The chemical form of the metal, metal kinetics, route of exposure, concentration and exposure duration may have affected the slight accumulation observed (Joel and Amajuoyi, 2009). Benthic organisms are the most directly impacted by contaminants, thus the removal of these aquatic components from the food web, due to pollution, may indirectly affect an ecosystem as a direct toxic effect since they serve as food for humans and other animals (Beeby, 2001). Invertebrates are generally more sensitive to pollutants than either fish or algae. Benthic species are particularly sensitive to water soluble contaminants, affecting the ability of the organisms to reproduce, avoid predators or feed (Bat et al., 1999; Chindah et al., 2001; Ezemonye et al., 2007 b). Consequently, when periwinkles are exposed to elevated metal levels in an aquatic environment, they can absorb the bioavailable metals directly from the environment via the tissue or through the ingestion of contaminated water, sediment and food (Burton, 1992; Trefry et al., 1996; Alexander, 2000). Uptake routes include: absorption primarily by passive diffusion with some active transport; the digestive system supports uptake of essential metals. Respiratory and body surfaces of aquatic organisms also support uptake of essential metals which are used for the maintenance of hyper (fresh) or hypo (marine) osmotic conditions. Rate of excretion / regulation determines toxicity / tolerance among populations. The results from this study compares favourably with works of Davies et al. (2006) and Chindah et al. (2009).

Very low concentrations of PAHs were found in the test organisms exposed to the contaminated effluent at test termination on the 28 day. The levels of PAHs in the organisms' tissues could have been induced by uptake from the burrow pit effluent since PAHs was not detected in the sediment at day 0. Polyaromatic hydrocarbons (PAHs) are readily absorbed by aquatic organisms during exposure to contaminated food, water and sediment, reaching levels higher than those in the ambient medium. However, PAHs do not accumulate in the same manner as some other lipophilic organic compounds, instead, they are converted to more water-soluble forms, which facilitate their subsequent slow excretion from the organism (Neff, 1985).

However, PAHs with a high molecular weight may cause sublethal effects; such as growth reduction, chronic diseases, reproductive impairment, at very low concentrations in biota: (5 to 100 ppb) in the tissue of the animal. Concentrations which are considered toxic (0.2 to 10 ppm) have been found in biota from heavily polluted areas (EPA, 2000; Kanchanamayoon and Tatrahun, 2008). Concentrations of PAHs in the aquatic environment are generally highest in sediment, intermediate in biota and lowest in the water column (CCME, 1992). In water, PAHs attach to sediment, impacting bottom-dwelling organisms like periwinkle, shrimp, oysters and plankton. As these organisms spend time in or near contaminated sediments, they accumulate PAHs in their tissues leading to harmful effects. A wide range of PAH-induced ecotoxicological effects in a diverse suite of biota, including microorganisms, aquatic biota, amphibians and terrestrial mammals have been reported (Delistraty, 1997; Ekundayo and Benka-Coker, 1994; Khan and Law, 2005). These studies assessed the effects of PAHs on survival, growth, metabolism and tumor formation, i.e. acute, developmental, reproductive toxicity, cytotoxicity, genotoxicity and carcinogenicity.

According to Canterford et al. (1978) it is useful to express results in terms of bioconcentration factor (BCF) when comparing the order of uptake of metals. In this study bioconcentration potentials were very low and varied with the test effluent concentrations. Although there was accumulation of zinc, total iron and PAHs in the tissue of the test organisms, BCF for these parameters were relatively very low i.e. less than 1000, thus are not persistent and could be eliminated from the test organisms with time if exposure is not continuous (EPA, 2000). However, the observed BCF indicates that periwinkles have potential to concentrate contaminants in their soft tissues (Ademoroti, 1996; Eja et al., 2003). The results from this study agreed with those obtained by Davies et al. (2006). The other parameters tested did not indicate any level of bioaccumulation in the tissues of the organisms.

In this study, mean weight of the periwinkle was higher in the control and the 3.125% test effluent than 12.5 and 50% test effluents concentrations. This indicates that the effect of the burrow pit effluent on the growth of periwinkle showed a slight reduction for organisms exposed to both 12.5 and 50% when compared to the control and 3.125%. Polycyclic aromatic hydrocarbons affect organisms through various toxic actions and they are being recognized with increasing frequency as major contributors to the hazard to aquatic life of contaminated sediments, particularly near areas of intense human activity (Neff and Burns, 1996; Neff et al., 2005; Pies et al., 2008).

Some authors suggested that PAHs detected in river sediments and water may contribute to endocrine disruption, which is supported by numerous studies on the physiological effects of PAH exposure on fish and other aquatic organisms (Lintelmann et al., 2003). Other studies have reported evidence of PAH exposure among fish populations (van den Heuvel et al., 1999), and cited PAHs as potential stressors causing increased mortality, growth and malformations in fish (Colavecchia et al., 2004; Heintz, 2007). In spite of the relatively low bioaccumulation levels recorded for metals and PAHs in the test organisms, it is important to caution that if the concentration of the components in the effluent increases, the amount inside the organism would also increase, as was observed in the organisms in the higher concentration, until it reaches a new equilibrium. Exposure to large amounts of the effluent for a long period of time, however, may overwhelm the equilibrium potentially causing harmful effects.

Conclusion

Information concerning effluent and chemical accumulation is fundamental in determining environmental quality guidelines by regulatory bodies. This will further assist in grouping substances that are potential hazards as well as qualifying the risk of chemicals and substances on the ecosystem and human health. Periodic surveillance is therefore essential to ensure ecosystem balance for these economically viable bottom dwelling organisms.

REFERENCES

Adams WJ, Kimerle RA, Barnett WJ (1992). Sediment quality and aquatic life assessment. Environ. Sci. Technol., 26(10): 1864- 1874.

Ademoroti CMA (1996). Environmental chemistry and toxicology. Foludex Press Ltd, Ibadan. pp. 217.

Ajayi SO, Osibanjo O (1981). Pollution studies on Nigerian Rivers. 11 Water quality of some Nigeria Rivers. Environ. Pollut., (Series B), 2: 87-95.

Alexander M (2000). Aging, bioavailability and overestimation of risk from environmental pollutants. Environ. Sci. Technol., 34: 4259-4265.

Ashraj W (2005). Accumulation of heavy metals in kidney and heart tissues of Epinephelus microdon fish from the Arabian Gulf. Environ. Monit. Assess., 101 (1-3): 311-316.

Bat L, Gundogdu A, Sezgin M, Culha M, Gonlugur G, Akbulut M (1999). Acute toxicity of zinc, copper and lead to three species of marine organisms from the Sinop Peninsula, Black Sea. Tr. J. Biol., 23: 537–544.

Beeby A (2001). What do sentinels stand for? Environ. Pollut., 112: 285-298.

Burton AG Jr (1992). Assessing contaminated aquatic sediments. Environ. Sci. Technol., 26 (10): 1862 – 1863.

Canadian Council of Ministers of the Environment, CCME (1992). Canadian water quality guidelines, prepared by the task force on water quality guidelines of the Canadian Council of Ministers of the Environment, Eco-health branch, Ottawa, Ontario, Canada.

Canterford GS, Bichanan AS, Ducker SC (1978). Accumulation of heavy metals by the marine diatom Ditylum brightrelli (West) Grunow. Aust. J. Mar. freshwater Res., 29: 611-622.

Chindah AC, Braide SA, Nduaguibe UM (2001). Tolerance of periwinkle (Tympanotonus fuscatus linne) and shrimp (Palemonetes fricanus balss) to waste water from Bonny light crude oil tank farm. Polish J. Environ. Protect. Nat. Resour., 21: 61-72.

Chindah AC, Braide SA, Amakiri J, Chikwendu SON (2009). Heavy metal concentrations in sediment and periwinkle –Tympanotonus fuscastus in the different ecological zones of Bonny River system, Niger Delta, Nigeria. The Open Environ. Pollut. Toxicol. J., 1: 93-106.

Ciarelli S, Vonck WA, Straalen PMA, Van NM (1997). Reproducibility of spiked-sediment bioassays using the fresh and brackish benthic amphipod, Corophium volutator; fresh and brackish. Environ. Res., 43(4): 329-343.

Cicik B, Engin K (2005). The effects of cadmium on levels of glucose in serum and glycogen reserves in the liver and muscle tissues of Cyprinus carpio (L., 1758). Turk. J. Vet. Anim. Sci., 29: 113-117.

Colavecchia MV, Backus SM, Hodson PV, Parrott JL (2004). Toxicity of oil sands to early life stages of fathead minnows (Pimephales promelas). Environ. Toxicol. Chem., 23(7): 1709–1718.

Corl E (2001). Bioaccumulation in the ecological risk assessment (ERA) process. Issue Papers pp. 1- 12.

Davies OA, Allison ME, Uyi HS (2006). Bioaccumulation of heavy metals in water, sediment and periwinkle (Tympanotonus fuscatus var radula) from the Elechi Creek, Niger Delta. Afr. J. Biotechnol., 5(10): 968-973.

Delistraty D (1997). Toxic equivalency factor approach for risk assessment of polycyclic aromatic hydrocarbons. Toxicol. Environ. Chem., 64: 81-108.

Department of Petroleum Resources, DPR (2002). Environment

guidelines and standards for the petroleum industry in Nigeria (EGASPIN) Revised Edition.

Eja ME, Ogri ORA, Arikpo GE (2003). Bioconcentration of heavy metals in surface sediments from the Great Kwa Rivers Estuary, Calabar, South Eastern Nig. J. Nig. Environ. Soc., 2: 247-256.

Ekundayo JA, Benka-Coker MO (1994). Effects of exposure of aquatic snails to sublethal concentrations of waste drilling fluid. Environ. Monitoring Assess., 30: 291-297.

Environment Canada (1992). Biological test method: Acute test for sediment toxicity using fresh and brackish or estuarine amphipods. conservation and protection, Ottawa, Ontario Report EPS1/RM/26.

Environmental Protection Agency, EPA. U. S Environmental Protection Agency (1980). Ambient water quality criteria for polynuclear aromatic hydrocarbons. U.S. Environ. Protection Agency. Rep. 440/5-80-069. pp 193.

Environmental Protection Agency, EPA. U.S. Environmental Protection Agency (2000). Bioaccumulation testing and interpretation for the purpose of sediment quality assessment: status and needs. Bioaccumulation analysis workgroup, Washington D.C. EPA/823/R-00/001.

Ezemonye LIN, Ogeleka DF, Okieimen FE (2007a). Biological alterations in fish fingerlings (Tilapia guineensis) exposed to industrial detergent and corrosion inhibitor. Chem. Ecol., 23(5): 1-10.

Ezemonye LIN, Ogeleka DF, Okieimen FE (2007b). Desmoscaris Tripsinosa and Palamonetes Africanus response to concentrations of Neatex and Norust CR 486 in sediment. J. Surfactants Deterg., 10 (4): 301 - 308.

Ezemonye LIN, Ogeleka DF, Okieimen FE (2009). Lethal toxicity of industrial detergent on bottom dwelling sentinels. Int. J. Sediment Res., 24: 478-482.

Farombi EO, Adelowo OA, Ajimoko YR (2007). Biomarkers of oxidative stress and heavy metal levels as indicators of environmental pollution in African cat fish (Clarias gariepinus) from Nigeria Ogun River. Int. J. Environ. Res. Pub. Health, 4(2): 158-165.

Hayat S, Javed M, Razzaq S (2007). Growth performance of metal stressed major carps viz. Catla catla, Labeo rohita and Cirrhina mrigala reared under semi-intensive culture system. Pakistan Vet. J., 27(1): 8-12.

Heintz RA (2007). Chronic exposure to polynuclear aromatic hydrocarbons in natal habitats leads to decreased equilibrium size, growth, and stability of pink salmon populations. Integr. Environ. Assess. Manag., 3(3): 351–363.

Hussain SM, Javed M, Javid A, Javid T, Hussain N (2011). Growth responses of Catla catla, Labeo rohita and Cirrhina mrigala during chronic exposure of iron. Pak. J. Agri. Sci., 48: 239-244.

Joel OF, Amajuoyi CA (2009). Evaluation of the Effect of Short-Term Cadmium Exposure on Brackish Water Shrimp-Palaemonetes Africanus. J. Appl. Sci. Environ. Manage., 13(4): 23 – 27.

Kanchanamayoon W, Tatrahun N (2008). Determination of polycyclic aromatic hydrocarbons in water samples by solid phase extraction and gas chromatography. World J. Chem., 3(2): 51-54.

Khan ZM, Law FCP (2005). Adverse effects of pesticides and related chemicals on enzyme and hormone systems of fish, amphibians and reptiles: A review. Proc. Pakistan Acad. Sci., 42(4): 315-323.

Langston RW (1989). Toxic effects of metals and the incidence of marine ecosystem. In: Furness, RW, Rainbow, PS, Eds. Heavy metals in the marine environment. CRC Press, New York, pp. 128-142.

Lintelmann J, Katayama A, Kurihara N, Shore L, Wenzel A (2003). Endocrine disruptors in the environment (IUPAC Technical Report). Pure Appl. Chem., 75(5): 631–681.

McGeer JC, Brix KV, Skeaff JM, Deforest DK, Brigham SI (2003). Inverse relationship between bioconcentration factor and exposure concentration for metals: Implication for hazard assessment of metals in the aquatic environment. Environ. Toxicol. Chem., 22: 1017-1037.

Neff JM (1985). Polycyclic aromatic hydrocarbons. pp 416-454 in Rand GM, Petrocelli SR (eds.). Fundamentals of aquatic toxicology. Hemisphere Publ. Corp., New York.

Neff JM, Burns WA (1996). Estimation of polycyclic aromatic hydrocarbon concentrations in the water column based on tissue residues in mussels and salmon: An equilibrium partitioning approach. Environ. Toxicol. Chem., 15: 2240–2253.

Neff JM, Stout SA, Gunster DG (2005). Ecological Risk Assessment of

Polycyclic Aromatic Hydrocarbons in Sediments: Identifying Sources and Ecological Hazard. Integrated Environ. Assess. Manage., 1(1): 22–33.

Ogeleka DF, Ezemonye LIN, Okieimen FE (2010). Sublethal effects of industrial chemicals on fish fingerlings (*Tilapia Guineensis*). Afr. J. Biotechnol., 9(12): 1839-1843.

Organisation for Economic Co-operation and Development, OECD (2003). Environment, health and safety publications series on pesticides persistent, bioaccumulative, and toxic pesticides in OECD Member countries results of survey on data requirements and risk assessment approaches No. 15 1 – 67.

Organization for Economic Cooperation and Development, OECD (2004). OECD Guidelines for the Testing of Chemicals No. 218: "Sediment-water chironomid toxicity test using spiked sediment".

Pies C, Hoffmann B, Petrowsky J, Yang Yi, Ternes TA, Hofmann T (2008). Characterization and source identification of polycyclic aromatic hydrocarbons (PAHs) in river bank soils. Chemosphere, 72 (10): 1594-1601.

Relyea RA, Diecks N (2008). An unforeseen chain of events: lethal effects of pesticides on frogs at sublethal concentrations. Ecol. Appl., 18(7): 1728-1742.

Sastry KV, Rao DR (1984). Effects of mercuric chloride on some biochemical and physiological parameters of the freshwater murrel Channa punctatus. Environ. Res., 34: 343-350.

Tawari-Fufeyin P, Ekaye SA (2007). Fish species diversity as indicator of pollution in Ikpoba river, Benin City, Nigeria. Rev. Fish Biol. Fisheries, 17: 21-30.

Tuvikene A (1995). Response of fish to polyaromatic hydrocarbons (PAHs). Ann. Zool. Fennici., 32: 295-309.

Trefry JH, Trocine RP, Naito KL, Metz S (1996). Assessing the potential for enhanced bioaccumulation of heavy metals from produced water discharges to the Gulf of Mexico. Produced water 2. Environmental issues and mitigation technologies. Edited by Mark Reed and Stale Johnsen, Plenum Press, New York and London. pp. 339-354.

Van den Heuvel M, Power M, MacKinnon MD, Dixon DG (1999). Effects of oil sands related aquatic reclamation on yellow perch (*Perca flavescens*) I: Water quality characteristics and yellow perch physiological and population responses. Can. J. Fish Aquat. Sci., 56(7): 1213-1225.

Vinodhini R, Narayanan M (2008). Bioaccumulation of heavy metals in organs of fresh water fish *Cyprinus carpio* (Common carp). Int. J. Environ. Sci. Tech., 5(2): 179-182.

Waqar A (2006). Levels of selected heavy metals in Tuna fish. Arab. J. Sci. Eng., 31(1A): 89–92.

Yilmaz F, Ozdemir N, Demirak A, Tuna AL (2007). Heavy metal levels in two fish species *Leuciscus cephalus* and *Lepomis gibbosus*. Food Chem., 100: 830-835.

14

Toxicological effects of methomyl and remediation technologies of its residues in an aquatic system

Ismail I. El-Fakharany, Ahmed H. Massoud, Aly S. Derbalah* and Mostafa S. Saad Allah

Pesticides Department, Faculty of Agriculture kafr-EL-Shiekh University, 33516 Egypt.

This study was carried out to evaluate toxicological effects of methomyl at low concentration level with respect to some biochemical parameters (acetylcholinesterase [AChE], alkaline phasphates [ALP], glutamic-pyrovic transaminase [GPT], glutamic-oxaloacetic transaminase [GOT] and glutathion-S-transferase [GST]) and histopathological changes of treated rats organs (kidney and liver). Furthermore, to evaluate the efficacy of different remediation techniques (advanced oxidation processes [AOPs] and bioremediation) for removing the tested insecticide in aquatic system. The tested insecticide at dose level of 10 mg kg^{-1} induced significant toxicity against the treated rats relative to control with respect to biochemical parameters and histopathological changes in treated rats. Photo-Fenton like reagent was the most effective chemical remediation treatment for methomyl removal in aquatic system followed by than Fe^{3+}/UV, H_2O_2/UV, Fe^{3+}/H_2O_2 and UV only systems, respectively. Bioremediation of methomyl using *Pseudomonas sp.* (EB20) isolate removed 77% of its initial concentration. This study concluded that, methomyl at the tested concentration level in water is expected to induce side effects on human health. Bioremediation using *Pseudomonas* sp. (EB20) can be regarded as a safe remediation technology of methomyl in drinking water. However the photo-Fenton like reagent would be more preferable as effective treatment of methomyl in wastewater.

Key words: Methomyl, residues, toxicity, remediation.

INTRODUCTION

Wide spread use and disposal of organophosphorus and carbamates compounds that have been used as an alternative to organochlorine compounds for pest control (Muller and Schwack, 2001) resulted in the release of their residue into natural water, thus inducing an environmental problem (Derbalah et al., 2004b). Dimethoate, malathion and methomyl considered to be priority pollutants in water due to the wide range use of these pesticides against different pests (Lasarm et al., 2009).

Pesticide pollution of surface waters and wastewaters has increased sharply and it constitutes a major pollutant problem and health hazards due to an extensive use of these substances (Derbalah et al., 2004b; Evgenidou et al., 2007). Therefore, evaluation of its side effects on human health considered a source of major concern.

Toxicity of organophosporus and carbamates insecticides used compounds against human and animals were always evaluated by assessment of such biochemical parameters alterations and histopathological changes in tissues and organs (Cronelius et al., 1959; Ghanem et al., 2006; Massoud et al., 2010). However, there is lack of evaluating the toxicity of these pesticides at low concentration levels near the environmental level. Since most previous studies were using high doses of the tested pesticides to expect significant toxicological effect of these insecticides, however, this is did not reflect real situation.

Due to the great environmental and human risk of pesticide residues in water resources, their removal becomes a very important task for human being. Thus, advanced methods are in demand for effective treatment of pesticides-polluted water to achieve complete mineralization of target pesticides and to avoid the formation of toxic end products (Derbalah, 2009). Advanced oxidation processes (AOPs), which are

*Corresponding author. E-mail: aliderbalah@yahoo.com.

constituted by the combination of several oxidants, are characterized by the generation of very reactive and oxidizing free radicals in aqueous solution such as hydroxyl radicals, which posses a great destruction power to the organic pollutants (Benitez et al., 2002).

The photo assisted-Fenton reaction process as advanced oxidation process proved to be very powerful in destroying persistent pesticides in the wastewater (Penuela and Barcelo, 1998; Fallmann et al., 1999; Derbalah et al., 2004). In this respect, it is necessary to apply this method in water with low concentration level of pesticides (near to the environmental level), which would be able to generalize photo-Fenton reaction process for pesticides removal from water (Derbalah et al., 2004a). Bioremediation of chemo-pollutants becomes the method of choice because it is economically feasible and safer than chemical remediation technologies (Derbalah et al., 2008). *Pseudomonas* sp is known for their versatility in degradation of xenobiotic compounds such as pesticides in water (Abd El-Razik, 2006; Massoud et al., 2007b; Derbalah et al., 2008).

Therefore, the present study aimed to evaluate the toxic effects of the most frequently detected compound in water resources (methomyl), at low dose near the environmental levels with the respect to some biochemical's (AChE, ALP, GOT, GPT and GST) parameters in blood and histological changes in treated rats organs (liver and kidney) and finally to evaluate the efficacy of different remediation technologies (advanced oxidation processes and bioremediation) for the removal of the tested insecticide residue in the aquatic system.

MATERIALS AND METHODS

Toxicity experiment

Animals' treatment

For Toxicity assessments 8-week-old 80-100g Wistar male rats (*Rattus norvegicus*) obtained from Faculty of Medicine, Tanta University were used. Wister rats were housed in wire cages under standard conditions with free access to drinking water and food. The rats were kept in temperature-controlled room with 14 h Light and 10 h dark cycles and given standard diet consist. Before treatment, rats were made adaptation for two weeks during feeding. The animals were randomly divided into four groups each comprising of three animals. Two groups of tested insecticide (24 h and 21 days), group for control (without methomyl) and group for ethanol control.

Rats were treated with methomyl (99%) that obtained from Kafr El Zyat for Chemicals and Pesticides Company Limited, Kafr-El-Zayat, Egypt. The tested insecticide was dissolved in ethanol and gave to rats by oral dose at level of 10 mg kg^{-1} (volume 1ml). Rats were scarified under anesthesia. Then tissue and blood samples were taken after 24 h (acute toxicity) and 21 days (sub-chronic toxicity). Blood samples were taken by cardiac puncture in vials containing heparin. Blood samples were centrifuged at 4500 rpm for 20 min and serum was collected for enzymes activity determination. For histopathological test, rat organs (liver and kidney) were taken and kept in formalin 10% for histopathological test (Derbalah, 2009).

Enzymes assays

The colorimetric methods of Ellman et al. (1961), Gornal et al. (1949), Belfield and Goldberg (1971), Reitman and Frankel (1957), and Rose and Wallbank (1986) were used for determining the activity of AChE (acetylcholinesterase), (ALP) alkaline phasphates, GPT (Glutamic-Pyrovic Transaminase), GOT (Glutamic-Oxaloacetic Transaminase) and GST (glutathion-S-transferase) in blood , respectively.

Histopathological tests

The histopathology test was carried out at Dep. of Histopathology, Fac. of Veterinary Medicine, Kafr El-Sheikh Univ. Egypt. This experiment was conducted to study the histopathological lesions of the organs (liver and kidney) in treated rats with the tested insecticide; these organs were removed and prepared for histopathological examination according to the method described by Bancroft and Stevens (1996).

Chemical remediation of tested insecticide in aqueous system

A UV mercury lamb (model VL-4 LC (80W) was employed for the irradiation of the tested insecticide (methomyl). Ferric chloride was used as a source of iron catalyst because it remains unchanged before and after oxidation and this made the study of the reaction and the future engineering scale-up simpler, because the system remained homogeneous (Derbalah et al., 2004a). The solution was prepared by addition of desired amounts of methomyl technical grade (10 ppm) in distilled water. Then freshly prepared ferric chloride, $FeCl_3$, at concentration level of 1 mM as ferric ion was added followed by addition of H_2O_2 at 20 mg/l and the total volume was reached 100 ml by distilled water. The initial pH of the prepared solution was adjusted at 2.8 using hydrochloric acid 1 Molar for all experiments (Derbalah et al., 2004a; Derbalah, 2009) using pH meter Jenway (Model 3510, PH/mV/Temperature Meter). All degradation experiments were carried out at room temperature. The solution was transferred from standard flask to glass cell and exposed to irradiation of UV lamp (the distance between the lamp and pesticides solution 15 cm)with a wave length of 265 nm (Derbalah, 2009). Illumination times were 10, 20, 40, 80, 160 and 320 min. Samples were removed at these regular intervals for HPLC analysis. Moreover, three experiments were carried out, the first in the absence of hydrogen peroxide (Fe^{3+}/UV) to account for the degradation of methomyl under iron, the second in the absence of iron to account for the degradation of methomyl under hydrogen peroxide (H_2O_2/UV) and finally the third in the absence of iron and hydrogen peroxide to account the degradation under UV light only. Moreover, to account the effect of light on Fenton degradation ability, one experiment was carried out in the presence of Fenton components under dark conditions. The irradiated samples were analyzed directly by HPLC system in the Central Laboratory of Pesticides, Agriculture Research Center, El-Dokey, Egypt. A mixture of acetonitrle and distilled water (20:80) was used as mobile phase under the isocratic olution mode. The flow rate of mobile phase was maintained at 0.7 ml /min. The used detector was UV and the wavelength was 231 nm (Tamimi et al., 2008).

Bioremediation of tested insecticide in aqueous system

Pseudomonas sp. (EB20) was isolated from El-Hamoul water at Kafr-El-Sheikh Governorate, which polluted by persistent organic pollutants (POPs) (Ashry et al., 2006) and identified according to its morphological and physiological parameters as described by Holt (1984). The bioremediation test was carried out at Microbiology

Table 1. Effect of methomyl at dose level of 10 mg/kg on activity of some biochemical' parameters in rats after 24 h of treatment.

Treatments	AChE (U/L)	ALP (U/L)	GPT (Units/ml)	GOT (Units/ml)	GST (Units/ml)
Control	$5.51 \times 10^{-1} \pm 0.035bc$	$1.28 \times 10^{-1} \pm 0.0002b$	$8.493 \times 10^{-1} \pm 0.007d$	$10.58 \times 10^{-1} \pm 0.007c$	$1.15 \times 10^{-2} \pm 0.003b$
Ethanol	$5.86 \times 10^{-1} \pm 0.023b$	$4.67 \times 10^{-2} \pm 0.0006d$	$9.58 \times 10^{-1} \pm 0.009c$	$14.47 \times 10^{-1} \pm 0.002a$	$2 \times 10^{-3} \pm 0.0005c$
Methomyl	$6.92 \times 10^{-1} \pm 0.012a$	$1.455 \times 10^{-1} \pm 0.0002a$	$13.12 \times 10^{-1} \pm 0.006a$	$14.02 \times 10^{-1} \pm 0.002b$	$2.9 \times 10^{-2} \pm 0.0043a$

*a, b and c letters shows the significance and non-significance between the means at p value of 0.05 using Duncan's multiple range test.

Table 2. Effect of methomyl at dose level of 10 mg/kg on activity of some biochemical's parameters in rats after 21 days of treatment.

Treatments	AChE (U/L)	ALP (U/L)	GPT (Units/ml)	GOT (Units/ml)	GST (Units/ml)
Control	$9.73 \times 10^{-1} \pm 0.012b$	$1.128 \times 10^{-1} \pm 0.0007a$	$26.07 \times 10^{-1} \pm 0.009a$	$21.93 \times 10^{-1} \pm 0.005a$	$1.73 \times 10^{-2} \pm 0.002a$
Ethanol	$14.56 \times 10^{-1} \pm 0.012a$	$4.77 \times 10^{-2} \pm 0.0003c$	$10.27 \times 10^{-1} \pm 0.002d$	$4.19 \times 10^{-1} \pm 0.002d$	$1.3 \times 10^{-3} \pm 0.0002d$
Methomyl	$8.44 \times 10^{-1} \pm 0 \ c$	$4.95 \times 10^{-2} \pm 0.0004b$	$16.39 \times 10^{-1} \pm 0.0045c$	$14.70 \times 10^{-1} \pm 0.006c$	$6 \times 10^{-4} \pm 0.0005c$

*a, b and c letters shows the significance and non-significance between the means at p value of 0.05 using Duncan's multiple range test.

laboratory, Dep. of Agric. Botany, Fac. of Agric. Kafr El-Sheikh Univ. The selected microbial isolate *Pseudomonas* sp. (EB20) was cultured onto Mineral Slat Medium (MSL) spiked with the tested insecticide (methomyl) separately for 7 days and then the growing colonies was washed with three ml sterilized MSL liquid medium. The cell suspension of 10^8 cfu/ml (colony forming unit) was used to inoculate 100 ml MSL liquid medium containing 10 ppm of the tested insecticide. The cultures were incubated at 30°C, pH (7) and 150 rpm as optimum conditions for the growth of the tested microbial isolate (Derbalah et al., 2008) for 14 days. Samples were collected at 0, 3, 7, 10 and 14 days for monitoring methomyl degradation. Control flasks of equal volume of MSL liquid medium and the tested insecticide without the selected microbial isolate were run in parallel at all intervals to asses a biotic loss. The collected water samples of the tested insecticide were filtered using syringe filter (Derbalah et al., 2008) followed by HPLC analysis as mentioned before.

Statistical analysis

Data from the enzymes experiments were statistically analyzed using one-way repeated measurement analysis of variance. Duncan's multiple range test were used to separate means using SAS program (Version 6.12, SAS Institute Inc., Cary, USA).

RESULTS AND DISCUSSION

Toxicity of methomyl on some biochemical parameters in rats

The obtained data in Table 1 showed that, the activity of ALP, GOT, GPT and GST were increased after 24 h of treatment with methomyl at dose level of 10 mg/kg comparing with control treatment. On the other hand, the same liver functions enzymes activities were decreased after 21 days of treatment with methomyl at the same dose relative to control treatment as shown in Table 2. The increase of hepatic enzymes after 24 h of treatment with methomyl was due to the fact that the liver is often

primary target organ for the toxic effect of xenbiotics and the elevation of these defense enzymes is expected due to the early damage in the hepatic cells (Massoud et al., 2010). Moreover, the histopathological changes found in liver tissue elsewhere in this study confirmed this approach. However, the decrease of hepatic enzymes in rats treated methomyl after 21 days with ethanol may be due to the presence of ethanol as a solvent which acts as free radical producer, increasing the enzyme inhibition (Sivapiriya et al., 2006). Furthermore, the detoxification of the tested insecticide with the time after treatment increased by these defense enzymes and subsequently their activity with time gradually decreased.

Data in Tables 1 and 2 showed that, the activity of GST and ALP enzymes were decreased either after 24 h or 21days of treatment with ethanol relative to control treatment. On the other hand, the activity of GPT and GOT enzymes were increased after 24 h of treatment with ethanol while after 21 days of treatment the same enzymes activity were decreased comparing with control as shown in Tables 1 and 2. The decrease of hepatic enzymes in rats treated with ethanol may be due to the reason mentioned elsewhere in this study (Sivapiriya et al., 2006).

Referring to acetylcholinesterase activity, the obtained results revealed that, the activity decreased after 24 h or after 21 days of treatment with methomyl at dose level of 10 mg/kg relative to control treatment as shown in Tables 1 and 2. The inhibition of acetylcholinesterase activity in treated rats relative to control treatment due to that methomyl and other carbamates known as acetylcholinestrase inhibitors (Derbalah 2009). Moreover, the obtained results indicated that, the activity of AChE enzyme was increased either after 24 h or after 21 days of treatment by oral administration with ethanol comparing with control treatment (Tables 1 and 2). The increase in AChE activity in rat treated with ethanol

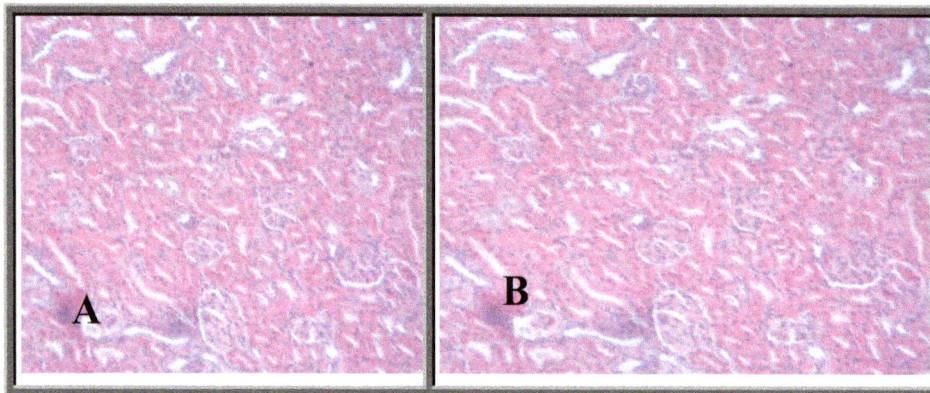

Figure 1. Sections from kidney of rats after 24 h of treatment with methomyl at dose level of 10 mg/kg (B) relative to control (A).

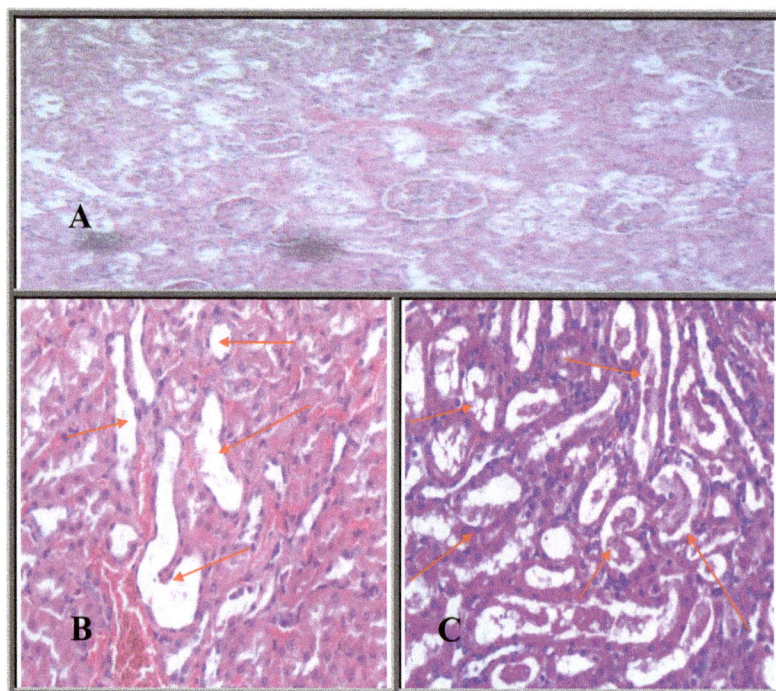

Figure 2. Sections from kidney of rats after 21 days of treatment with methomyl at dose level of 10 mg/kg (B and C) relative to control (A).

relative to methomyl may be due to that ethanol known to significantly reduce the inhibition of the AChE inhibitors (Sivapiriya et al., 2006).

Histological changes in different rat organs after treatment with methomyl

Histopathological changes in kidney

In normal histologic structure of the kidney, the cortex contains glomerular tufts scattered in between proximal and distal convoluted tubules (Figures 1A and 2A). Kidney of rats treated with methomyl at dose level 10 mg/kg after 24 h of treatment showed no changes (Figure 1B). However, for rats treated with methomyl at the same dose after 21 days of treatment showed cystic tubular degeneration (Figure 2B) and advanced tubular degeneration with cystic dilatation of the lumens (Figure 2C).

Histopathological changes in liver

Normal liver structure appeared in the form of hepatic

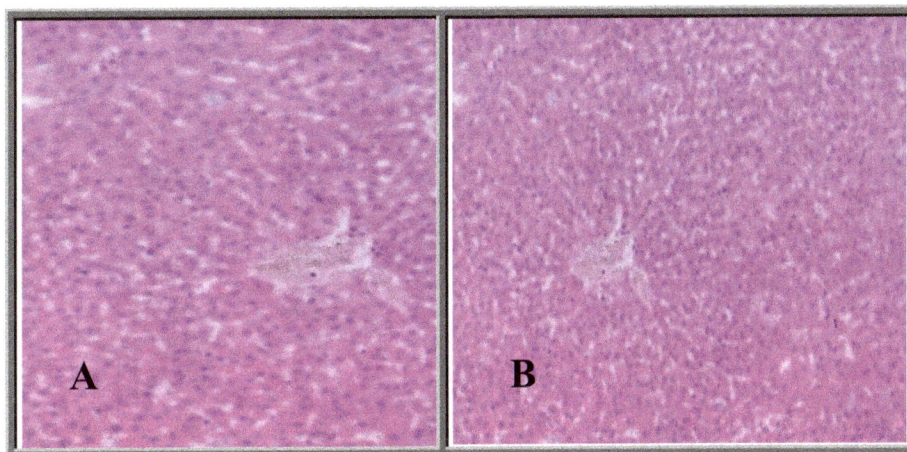

Figure 3. Sections from liver of rats after 24 h of treatment with methomyl at dose level 10 mg/kg (B) relative to control (A).

lobules in which there were centrally located central veins, which were surrounded by hepatocytes arranged in the form of hepatic cords separated from each other by hepatic sinusoids (Figures 3A and 4A). The liver of rats treated with methomyl at dose level 10 mg/kg after 24 h of treatment was normal as control (Figure 3B). However, the liver of rats treated with methomyl at dose level 10 mg/kg after 21 days of treatment showed advanced scarring of the hepatocytes cytoplasm (Figure 4B) and advanced sinusoidal congestion (Figure 4C). The increase of hepatic enzymes (GPT and GOT) after 24 h of treatment with methomyl mentioned before support the histopahtological changes recorded in liver. Since the liver is often primary target organ for the toxic effect of xenbiotics and the elevation of these defense enzymes is expected due to the early damage in the hepatic cells (Roganovic, 1998; Massoud et al., 2010).

Chemical remediation of the tested insecticide in aqueous solution

The loss in methomyl initial concentration with the irradiation time under UV, H_2O_2/UV, Fe_{3+}/UV, Fe_{3+}/H_2O_2 and Fe_{3+}/H_2O_2/UV systems was evaluated. The results in **Figure 5** showed that, the degradation rate of the tested insecticide were greatly enhanced by irradiation under Fe_{3+}/H_2O_2/UV (Fenton like reaction) relative to UV, H_2O_2/UV, Fe_{3+}/UV, Fe_{3+}/H_2O_2 systems. More than 90% of methomyl initial concentration was degraded under Fe_{3+}/H_2O_2/UV system within 320 min of irradiation time compared with 52, 46 , 45 and 38% of the same insecticide in the presence of, UV, H_2O_2/UV, Fe_{3+}/UV, Fe_{3+}/H_2O_2 and Fe_{3+}/H_2O_2/UV systems, respectively within the same irradiation time (Figure 5).

The photodegradation of methomyl under UV light is due to the direct absorbance of UV light (photolysis).

(Derbalah et al., 2004a). While the degradation of methomyl under H_2O_2/UV system due to firstly, the direct photolysis and secondly due to the generation of hydroxyl radicals from hydrogen peroxide equation (1) (Benitez et al., 2002; Derbalah et al., 2004a; Derbalah, 2009).

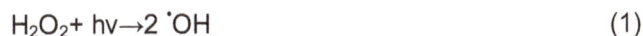

$$H_2O_2 + hv \rightarrow 2\ ^\cdot OH \qquad (1)$$

However, the photodegradation of methomyl under Fe^{3+}/UV system is due to the direct photolysis of the tested compound by absorbance of UV light. Moreover, due to the indirect photolysis of this compound by the hydroxyl radicals generated from $Fe(OH)^{2+}$ in the presence of UV light as shown in Equation (2)

$$Fe(OH)^{2+} + hv \rightarrow Fe^{2+} + ^\cdot OH \qquad (2)$$

The degradation of methomyl under Fe^{3+}/H_2O_2 system is due to the generation of hydroxyl radicals under this system in the absence of the light as shown in equation 3

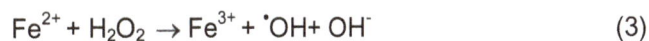

$$Fe^{2+} + H_2O_2 \rightarrow Fe^{3+} + ^\cdot OH + OH^- \qquad (3)$$

The degradation rate of methomyl under H_2O_2/UV system was slightly slower than its degradation under Fe^{3+}/UV system which may be due to the lower formation rate of hydroxyl radicals under H_2O_2/UV system relative to Fe^{3+}/UV system (Derbalah et al., 2004a). This low gene-ration rate of hydroxyl radicals under H_2O_2/UV system compared to Fe^{3+}/UV system can attributed to the fact that hydrogen peroxide absorbs weakly above 300 nm and that hydroxyl radicals are only generated through direct photolysis of hydrogen peroxide (Benitez et al., 2002). On the other hand, under Fe^{3+}/UV system the $Fe(OH)^{2+}$ complex is the predominant species for generating hydroxyl radicals absorbed light at wavelengths up to 410 which may lead to the generation

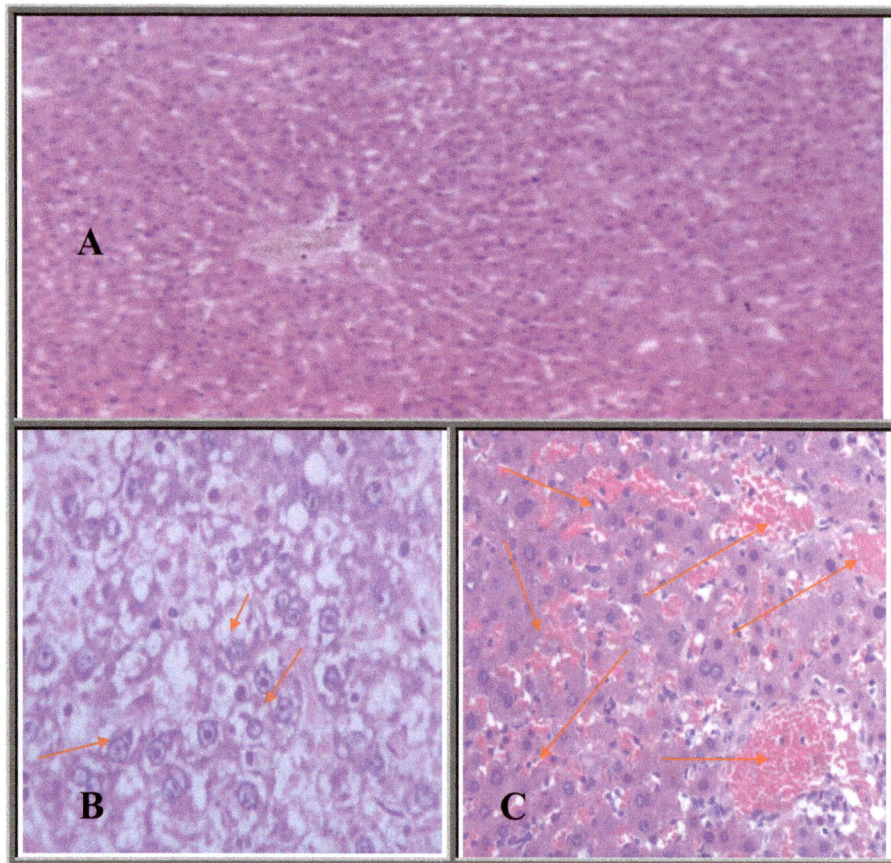

Figure 4. Sections from liver of rats after 21days of treatment with methomyl at dose level 10 mg/kg (B and C) relative to control (A).

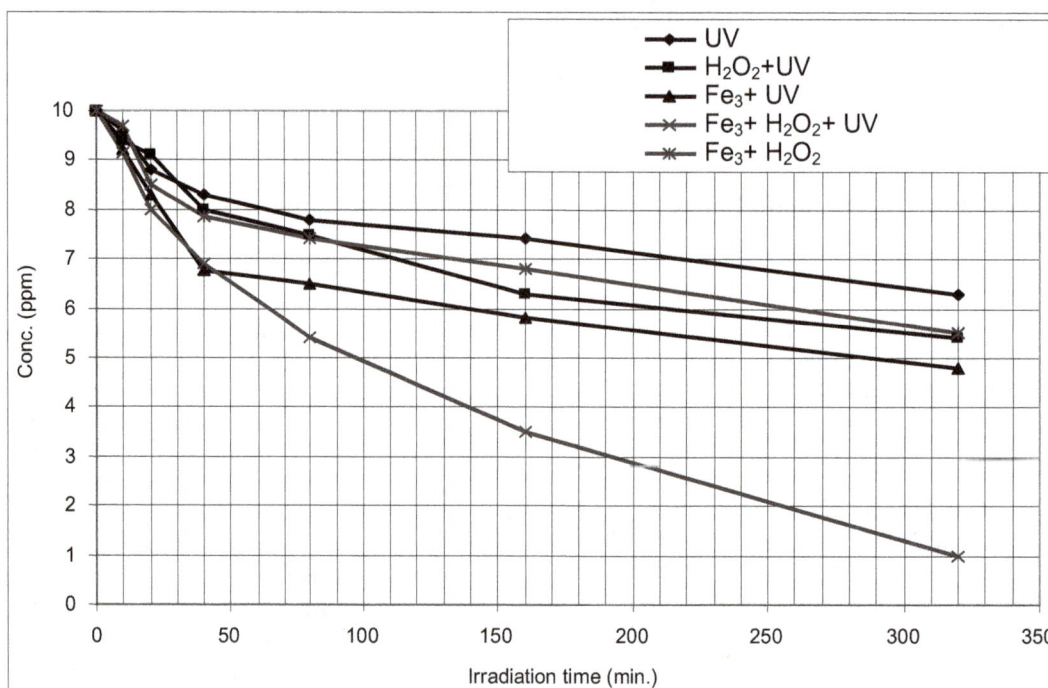

Figure 5. Degradation of methomyl at initial concentration of 10 mg/L under $Fe^{3+}/H_2O_2/UV$, H_2O_2/UV, Fe^{3+}/UV system and Fe^{3+}/H_2O_2 in distilled water. H_2O_2=20mg/l, $FeCl_3$=1mmol, pH=2.8.

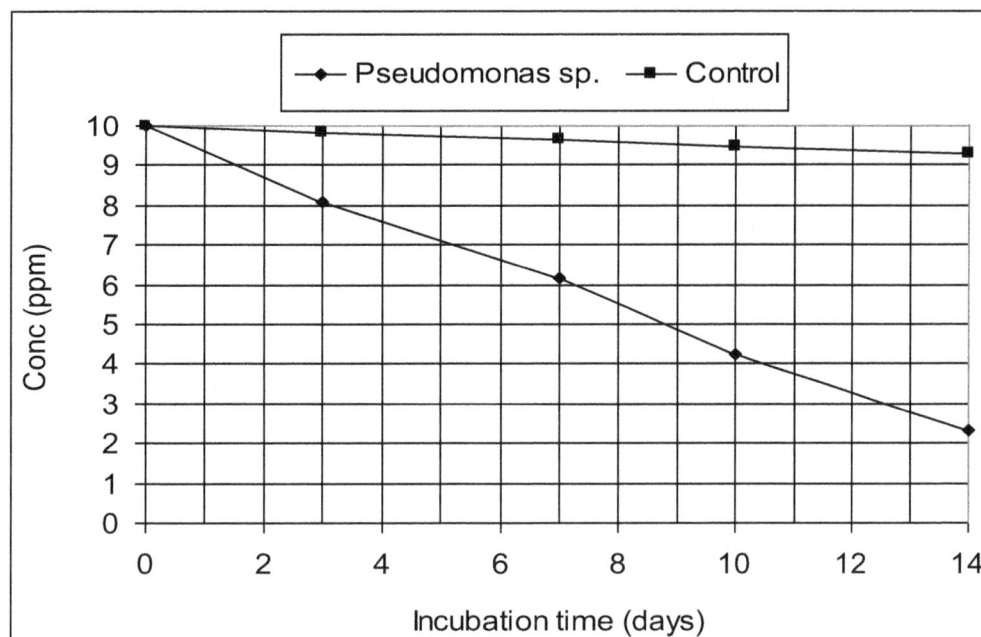

Figure 6. Biodegradation of methomyl at concentration level of 10 ppm by *Pseudomonas* sp. (EB20) isolate in aquatic medium.

of hydroxyl radicals under this system more than H_2O_2/UV system (Derbalah et al., 2004a).

The great enhancement in tested insecticide degradation rate under photo-Fenton's like reagent system ($Fe^{3+}/H_2O_2/UV$) relative to the other advanced oxidation processes is due to the higher generation rate of hydroxyl radicals under this system ($Fe^{3+}/H_2O_2/UV$) than the other systems (Derbalah et al., 2004a; Derbalah, 2009). This high generation rate of hydroxyl radicals under photo-Fenton like reagent due to many reasons. Firstly, the photolysis of hydrogen peroxide itself, which leads to the formation of hydroxyl radical's Equation (1) (Benitez et al., 2002). Secondly, the photolysis of Fe $(OH)^{2+}$ complex the predominant specie of Fe^{3+} for generating hydroxyl radicals Equation (2) (Larson et al., 1991).

Thirdly, the photodecarboxylation of ferric carboxylate complexes generate ferrous ion (Equation 3), which can in turn react with hydrogen peroxide to generate additional hydroxyl radicals (Equation 4) (Pignatello and Sun, 1995).

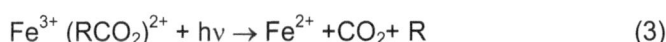

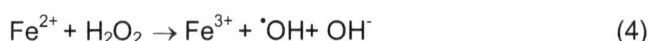

$$H_2O_2 + h\nu \rightarrow 2\ ^{\cdot}OH \tag{1}$$

$$Fe(OH)^{2+} + h\nu \rightarrow Fe^{2+} + \ ^{\cdot}OH \tag{2}$$

$$Fe^{3+}(RCO_2)^{2+} + h\nu \rightarrow Fe^{2+} + CO_2 + R \tag{3}$$

$$Fe^{2+} + H_2O_2 \rightarrow Fe^{3+} + \ ^{\cdot}OH + OH^- \tag{4}$$

The degradation of methomyl and other carbamates insecticides have been reported before (Benitez et al., 2002; Oller et al., 2007; Tamimi et al., 2008; Derbalah, 2009).

Bioremediation of the tested insecticide in aqueous solution

The ability of the selected microbial isolate *Pseudomonas* sp (EB20) for the biodegradation of the tested insecticide was illustrated in Figure 6. The results in Figure 6 indicated that, *Pseudomonas* sp. (EB20) showed high potential in the degradation of the tested insecticide. Since around 77% of methomyl of initial concentration level (10 ppm) was degraded within two weeks of incubation with *Pseudomonas* sp (EB20). On the other hand, the tested insecticide degradation percentages reached to 6.6% at the end of incubation time in control or non-inoculated samples. This is implied that the quote of tested insecticide decay due to temperature effect, photodecomposition and volatilization is very slight or negligible. The degradation of methomyl may be attributed to the secretion of enzymes which are capable of degrading pesticides (Bollag and Liu, 1990). The genus *Pseudomonas* showed the highest rate of methomyl degradation and has considerable potential for the biotransformation and biodegradation of acetylcholineeaterase inhibitors insecticides with widely differing chemical structures (Massoud et al., 2008; Derbalah et al., 2008).

Conclusions

Methomyl induced toxicological effects in treated rats relative to control with the respect to enzymes activity and histological changes in treated rats' organs. More extensive studies are needed to evaluate the toxicity of methomyl at concentration level more close the environmental level which in return helps to evaluate its toxicity under real environmental conditions. Photo-Fenton like reagent was the most effective treatment for the removal of methomyl residues in aquatic system and may be preferable in wastewater treatment. *Pseudomonas* sp. (EB20) could be regarded as a safe removal treatment of methomyl in drinking water.

REFERENCES

Abd El-Razik MAS (2006). Toxicological studies on some agrochemical Pollutants. M.SC. Thesis, Fac. of Agric. Kafrelsheikh Univ. 1-167.

Ashry MA, Bayoumi OC, El-Fakharany II, Derbalah AS, Ismail AA (2006). Monitoring and removal of pesticides residues in drinking water collected from Kafr El-Sheikh governorate. Egy. J. Agric. Res. Tanta Univ., 32: 691-704.

Bancroft JD, Stevens A (1996). Theory and Practice of Histological Techniques. Fourth edition.

Belfield A, Goldberg DM (1971). Alkaline phosphates colorimetric method. Enzyme, 12:561.

Benitez FJ, Acero JL, Real FJ (2002). Degradation of carbofuran by using ozone, UV radiation and advanced oxidation processes. J. Hazardous Mat. B., 89: 51-65.

Bollag JM, Liu SY (1990). Biological transformation process of pesticides. Pesticides in the Soil Environment (ed. Chang H.H.) pp. 169–211 Soil Science Society of America, Madison, WI.

Cronelius CE, Charles W, Arhode E (1959). Serum and tissue transaminase activities in domestic animals. Cornell Vet., 49: 116-121.

Derbalah AS (2009). Chemical remediation of carbofuran insecticide in aquatic system by advanced oxidation processes. J. Agric. Res. Kafr El-Sheikh Univ., 35: 308-327.

Derbalah AS, Massoud AH, Belal EB (2008). Biodegrability of famoxadone by various microbial isolates in aquatic system. Land Contamination Reclamation, 16: 13-23.

Derbalah AS, Nakatani N, Sakugawa H (2004a) Photocatalytic removal of fenitrothion in pure and natural waters by photo-Fenton reaction.Chemosphere, 57: 635-644.

Derbalah AS, Wakatsuki H, Yamazaki T, Sakugawa H (2004b). Photodegradation kinetics of fenitrothion in various aqueous media and its effect on steroid hormones biosynthesis. Geochem. J., 38: 201-213.

Ellman GL, Courtney KD, Andres V, Featherstone RM (1961). A new and rapid calorimetric determination of acetylcholinesterase activity. Biochem. Pharmacol., 7: 88-95.

Evgenidou E, Konstantinou I, Fytianos K, Poulios I (2007). Oxidation of tow organophosphorus insecticides by the photo-assisted Fenton reaction. Water Res., 41: 2015-2027.

Fallmann H, Krutzler T, Baue R, Malato S, Balanco J (1999). Applicability of the photo-Fenton methods for treating water containing pesticides. Catal. Today 54: 309-319.

Ghanem NF, Hassan NA, Ismail AA (2006). Biochemical and histological changes induced in rats fed on diets and byproducts of fumigated wheat grains with phostoxin. Pro. 4th Int. Con. Biol.Sc. (zool) 259-268.

Gornal AC, Bardawill CJ, David MM (1949). Protein–Biuret colorimetric method. J. Biol. Chem., 177:751.

Holt GH (1984). Ordinary gram negative bacteria. Bergy,s Manual of Systematic Bacteriology, (ed. Krieg N.R) Williams and Wilkins, Baltimore.

Larson RA, Schlauch MB, Marley K (1991). Ferric ion promoted photodecomposition of triazines. J. Agric. Food Chem., 39: 2057-2062.

Lasram MM, Annabi AB, El-Elj N, Kamoun A, El-Fazaa S, Gharbi N (2009). Metabolic disorders of acute exposure to malathion in adult wistar rats. J. Hazardous Mat., 163:1052-1055.

Massoud AH, El-Fakhrany II, Abd El-Razik MAS (2007a). Monitoring of some agrochemical pollutants in surface water in Kafr El-Sheikh Governorate. J. Pest. Cont. Environ. Sci., 15: 21-41.

Massoud AA, Derbalah AS, Iman A, Abd-Elaziz IA, Ahmed MS (2010) Oral Toxicity of Malathion at Low Doses in Sprague-Dawley Rats: A Biochemical and Histopathological Study. Monofyia Vet. J., 7: 1:183-196.

Massoud AH, Derbalah AS, Belal EB (2008) Microbial Detoxification of metalaxyl in aquatic system. J. Environ. Sci., 20: 262–267.

Muller H, Schwack W (2001). Photochemistry of organophosphorus insecticides. Rev. Environ. Contam. Toxicol., 172: 129-228.

Oller I, Malato S, Sanchez-Perez JA, Maldanonads MI, Gasso R (2007). Detoxification of waste water containing five common pesticides by solar AOPs-biological coupled system. Catal. Today, 129: 69-78.

Penuela GA, Barcelo D (1998). Photodegradation and stability of chlorothalonil in water studied by solid phase extraction followed by gas chromatographic techniques. J. Chromat. A., 823: 81-90.

Pignatello JJ, Sun Y (1995). Complete oxidation of metholachlor and methyl parathion in water by the photoassisted Fenton reaction. Water Res., 29 (8):1837-1844.

Reitman A, Frankel S (1957). GPT (ALT) Glutamic–Pyruvic Transaminase and GOT (AST) Glutamic –Pyruvic Transaminase colorimetric method. Am. J. Clin. Path., 28:56.

Roganovic M (1998). "Liver lesions in bleak (*Alhurnus alburnus alborella Filippi*) collected from some contaminated sites on lake Ohrid. A histopathological evidence," Ekol. Zast. Zivot. Sred., 6: 11-18.

Rose HA, Wallbank BE (1986). Mixed-function oxidase and Glutathion-s-transferase activity in asussceptible and Fenitrothion-resistant strain of *Oryzaephilus surimamensis*. J. Econ. Entmol., 79: 896-899.

Sivapiriya V, Jayan T, Venkatraman S (2006). Effects of dimethoate (*O,O*-dimethyl *S*-methyl carbamoyl methyl phosphorodithioate) and Ethanol in antioxidant status of liver and kidney of experimental mice. Pesticide Bioch. Physiol., 85: 115–121.

Tamimi M, Qourzal S, Barka N, Assabbane A, Ait-Ichou Y (2008). Methomyl degradation in aqueous solution by Fenton's reagent and photo-Fenton system. Separation Purif. Technol., 61:103-108.

15

Potential climate effects on nitrogen eco-toxicology of freshwater Lake, Victoria

Opio Alfonse

Department of Biology, Faculty of Science, Gulu University, Gulu-Uganda. E-mail: alfonseopio@gmail.com.

Lake Victoria has experienced changes that include introduction of alien species, over exploitation of fish, eutrophication and climate change. This review is a scenario of nitrogen cycle acceleration resulting in retention above the total inflow and the potential effect of climate on the N cycle. Excess nitrogen is attributed to nitrogen fixation, algal proliferation and decomposition. The nitrogen transformation like ammonia conversion to nitrate is enhanced over horizontal distance at higher temperature during dry season. Nitrogen (N) concentration in the vertical profile is related to climate variability of water temperature, lake water movement and differences in nitrogen loads from the catchment. Despite all, the effect of eddy currents or heat transfer caused by solar radiation on nitrogen processes is unknown. However, annual cycle of vertical oxygen distribution caused by stratification seems to provide potential condition for nitrous oxide production throughout the lake as compared to nitrogen gas. Therefore, understanding the relationships between organisms' diversity and community structure particularly of autotrophic and heterotrophic nitrogen bacteria, and their ecosystem functions in the entire freshwater lake is important for nitrogen budget. Due to scarcity of information, it is not possible to ascertain projection of climate influence on N dynamics in the Lake Victoria ecosystem. The ultimate suggestions on mitigation measures are to enforce policies that reduce both point and non-point sources of N into the lake and maintain riparian forests and wetlands.

Key words: Climate-effects, freshwater, Lake Victoria, nitrogen-ecotoxicology.

INTRODUCTION

Although Lake Victoria supports one of the largest fisheries, it has experienced changes that include introduction of alien species, over exploitation of fish, eutrophication and climate change (Hecky, 1993; Verschuren et al., 2002). Climate change is expected in an increased global temperature (IPCC, 2007). Already global mean surface temperatures have risen by 0.74°C over the past 100 years (1906 - 2005) and the warming rate of the last 50 years is almost double that over the previous 100 years (Trenberth et al., 2007). Africa countries in equatorial African region are warming at a slightly slower rate of about 1.4°C with respect to the 1961-1990 average (IPCC, 2001). Precipitation is also simulated to increase over Africa by 2050 (Hudson and Jones, 2002). Further, ozone depletion occurring over the latitudes that include much of Africa has potential effect on biogeochemical cycles such as alteration of sources and sinks of greenhouse gases and ozone

(http://www.epa.gov/ozone/science/sc_fact.html). All these changes are likely to affect aquatic systems in Africa in complex ways (Lovejoy and Hannah, 2005). The changes may influence the water quality and biological processes through runoff from the catchment and hydrology processes within the lake system.

The global nitrogen cycle could come under increasing pressure, not only from direct anthropogenic perturbations but also from the consequences of climate change (Gruber and Galloway, 2008). In addition, the interactions of nitrogen with carbon and how these interact with the climate system is less emphasized (Falkowski et al., 2000), despite the climate change effects that may have started in tropical ecosystems including aquatic systems (Wandiga, 2003). The human impact on the dynamics is yet unknown. However, general responses to supply of nutrients and changing climate have been reported (Regier et al., 1990; Benke, 1993; Lovejoy and Hannah, 2005).

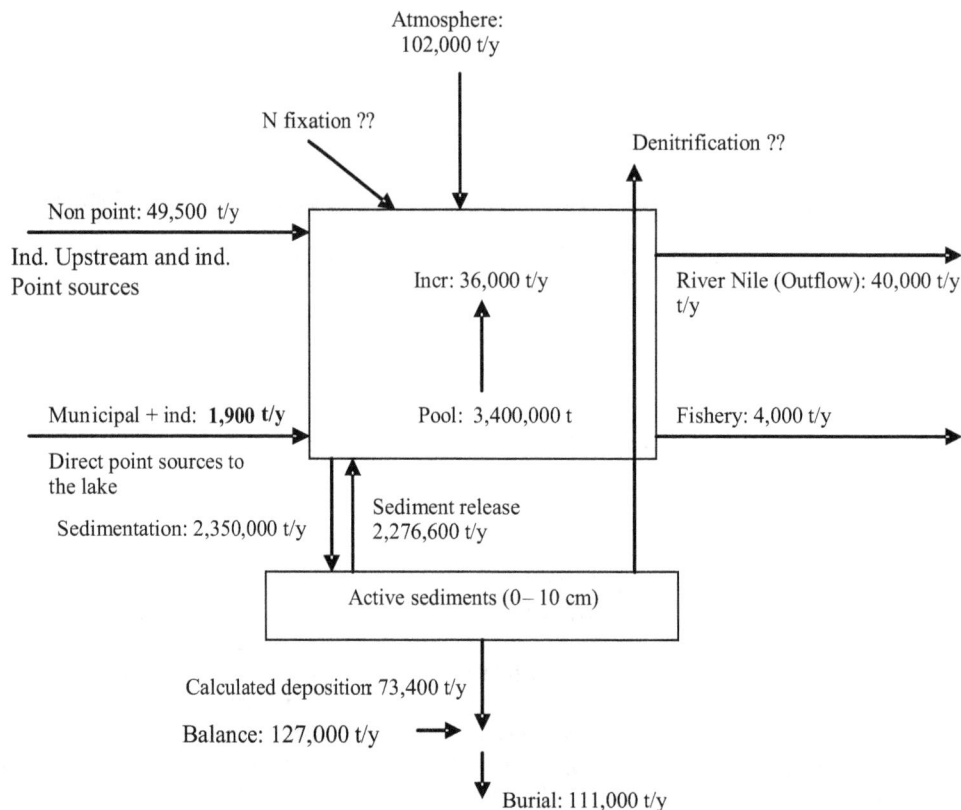

Figure 1. Nitrogen mass balance for Lake Victoria (Kayombo and Jorgensen, 2006).

The incorporation of ecosystem functions, as the rates of certain biological processes into regular monitoring program is becoming important (Bunn, 1995; Young, 2007). Many studies are conducted on linkages between climatic variables and effects on socio-economic status (Opera et al., 2007), sustainable use of natural resources (Niang et al., 2007) and biogeochemical cycles are now becoming of interest.

Nitrogen cycle acceleration affects the environment through the eutrophication of terrestrial and aquatic systems, and global acidification (Gruber and Galloway, 2008). In Lake Victoria, there are different catchment characteristics and there is uneven temporal and spatial nitrogen loads and distribution in the Lake system (Rutagemwa et al., 2005; Kayombo and Jorgensen, 2006; Pascal et al., 2007). Overall TN pollution loading (t/y) of the riparian countries (Uganda, Kenya and Tanzania) considering 50% load reduction by other treatment systems before discharge into the environment is 767 (21.88%), 2,019 (57.60%) and 719 (20.5) for urban wastewater and runoff, and 33 (7.97%), 57 (13.77%) and 324 (78.26) from industrial activities respectively (Kayombo and Jorgensen, 2006). However, the effect of nitrogen loads (inflow) into the Lake over a short period of time may not be visible (Biswas, 1976); although models indicate nutrients dispersion within the system (Banadda

et al., 2011; Bongomin and Opio, Unpublished). The concentration of nitrogen compounds in aquatic systems are derived from allochthonous and autochthonous sources, while, in the system, nitrogen is either retained or released. The sources and the amount of nitrogen retained in freshwater, Lake Victoria are indicated in Figure 1.

The summary of the total nitrogen budget of the lake excludes nitrogen fixation and denitrification. The atmospheric deposition (102,000 t/y) is expected to increase with increasing nitrogen compounds in the atmosphere. Millennium Ecosystem Assessment (2005) estimated an increase of total reative nitrogen deposition from the atmosphere into the Lake Victoria surface water region to a threshold between 1000 - 2000 mg N m^{-2} y^{-1} in 2050. Nitrogen removed by fish harvesting is estimated at 4, 000 t/y. A total of 73,400 t/y of nitrogen which is above the total input (51, 400 t/y) is deposited in the sediment. This is attributed to nitrogen fixation, algal proliferation and decomposition (Bugenyi and Balirwa, 1998) and also explains the relatively higher concentrations of inorganic nitrogen (IN) (0.477mg l^{-1}) and dissolved organic nitrogen (DON) (0.406 mg l^{-1}) at the bottom of the lake (Pascal et al., 2007). The high values are contribution of accretion process (0.131 g m^{-3} d^{-1}) of nitrogen transformation or settling of nitrogen into the sediments (Pascal et al.,

2007).

Accretion in the lake has not been quantified into the varied contribution of the dead organisms (i.e. plants, algae, plankton and bacteria) and wastes from the fishes. However, it depicts nitrogen 'top-down' effects. Mugidde (1993) however reported 'bottom-up' effects which is a result of eutrophication dynamics.

The inversion and prolific production of water hyacinth in Lake Victoria attributed to lack of natural enemies, ample space, optimal temperature and abundant nutrients (Opande et al., 2004) and caused significant changes in N dynamics of the lake. Water hyacinth is generally mobile except in lagoons and beaches that have little external interferences from wind actions. Open water is always clear of water hyacinth due to the frequent wave actions. The impact of water hyacinth on reducing fisheries harvests has been reported (Kateregga and Sterner, 2009). In addition, decomposition of the sunken water hyacinth cause prolong depression of dissolved oxygen to even anoxic levels close to the lake bottom and increase diversity and abundance of phytoplankton, macro invertebrates and fishes. Increasing infestation by water hyacinth is correlated with both *Chironomid* and mollusk densities (Bugenyi and Balirwa, 1998). Succession of the plant caused the disappearance of other free-floating macrophytes like *Pistia stratiotes* while providing substrate for emergent *Vossia cuspidate*. The weed is believed to have led to extinction of *Azolla nilotica* in the lake. This means a major change in the contribution of N retention and stock in the different floral and faunal components of the lake system.

The total organic nitrogen released into the lake undergoes a hydrolytic reaction, producing ammonia which provides a food source for the nitrifying bacteria that converts ammonia (NH_3) to nitrite (NO_2^-) and then nitrate (NO_3^-) (Equations 1 and 2).

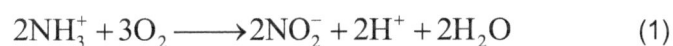

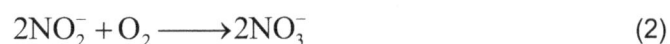

$$2NH_3^+ + 3O_2 \longrightarrow 2NO_2^- + 2H^+ + 2H_2O \qquad (1)$$

$$2NO_2^- + O_2 \longrightarrow 2NO_3^- \qquad (2)$$

The process of nitrification is mediated by *Nitrosomonas* (1) and *Nitrobacters* (2) that produce nitrite and nitrate respectively. Nitrification is however inhibited by free ammonia, nitrous acid and nitrite at low pH (Anthonisen et al., 1976; Metcalf and Eddy, 2003; Arceivala and Asolekar, 2008). Despite the significance of these processes in Lake Victoria, only values for Tanzania section of the lake is available (Pascal et al., 2007).

Although nitrification reduces toxicity of ammonia and contributes to biological oxygen demand (BOD), in Lake Victoria, more conversion of ammonia to nitrate in the horizontal distance occurs during dry season due to increased temperature, and at this time there is virtually

no non-point allochthonous organic input (Banadda et al., 2011). The same study also reported similar concentrations for ammonia, nitrites, and nitrates at various vertical depths of the lake section in Gabba area in Uganda in the rainy season. However, an increasing concentration in the vertical profile (depth) was reported in the Tanzania section of the lake (Pascal et al., 2007). The differences in the N concentrations could be a result of the depth and accretion process at the different sites. In addition, shielded bays tend to have less mixing unless there is strong seiche from the open waters that cause dilution of the nitrogen compounds (Larsson et al., 2008). Moreover, nitrogen is introduced into the lake during the wet season due to non-point sources of pollution such as atmospheric deposition; precipitation and land run off through rivers from agricultural activities (Kayombo and Jorgensen, 2006; Pascal et al., 2007; Banadda et al., 2011). Therefore, the reportedly 10 - 20% increase in runoff as a result of climate change for most of Uganda (MLWE, 2002) is likely to affect nutrient inflow into the lake and their dispersion within the lake system.

Denitrification process within aquatic systems lowers nitrogen concentration (Equations 3 and 4).

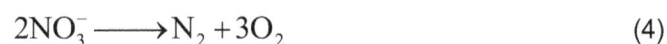

$$2NO_2^- \longrightarrow N_2 + 2O_2 \qquad (3)$$

$$2NO_3^- \longrightarrow N_2 + 3O_2 \qquad (4)$$

In Lake Victoria, denitrification estimation has been reported within the general range of 250 - 500 kg N km^{-2} y^{-1} (Seitzinger et al., 2006). The daily average value for denitrification in the Tanzania section of the lake is presented in Figure 2. These processes affect the overall nitrogen budget estimation in the lake system.

In aquatic systems, aerobic nitrification may be followed by anaerobic denitrification and for that reason; there is high potential of nitrous oxide production as compared to nitrogen gas (Takaya et al., 2003). Such conditions may accumulate nitrous oxide ($N_2O - N_i$) in the lake although this has never been estimated. Nitrogen budget of the lake indicates that there is a lot of nitrogen accumulating in the lake bottom through accretion (Kayombo and Jorgensen, 2006; Pascal et al., 2007). The nitrogen would be lost or locked up in the sediment if it were not the 'bottom-up' effect reported by Mugidde (1993). The general process affect the total nitrogen (TN) and total phosphorus (TP) ratio (Van Ginkel, 2002) with ultimate reflection in the type, distribution and abundance of plant species in the lake. TN:TP ratio of greater than 10:1 favors green algae, which may be less problematic to manage while any ratio less than 10:1 encourages *Cyanobacteria* that is able to fix atmospheric nitrogen. Cyanobacterial blooms have been reported in the inshore of Lake Victoria and is attributed to light as the limiting factor since dissolved organic matter (DOM) attenuate most of the light in the

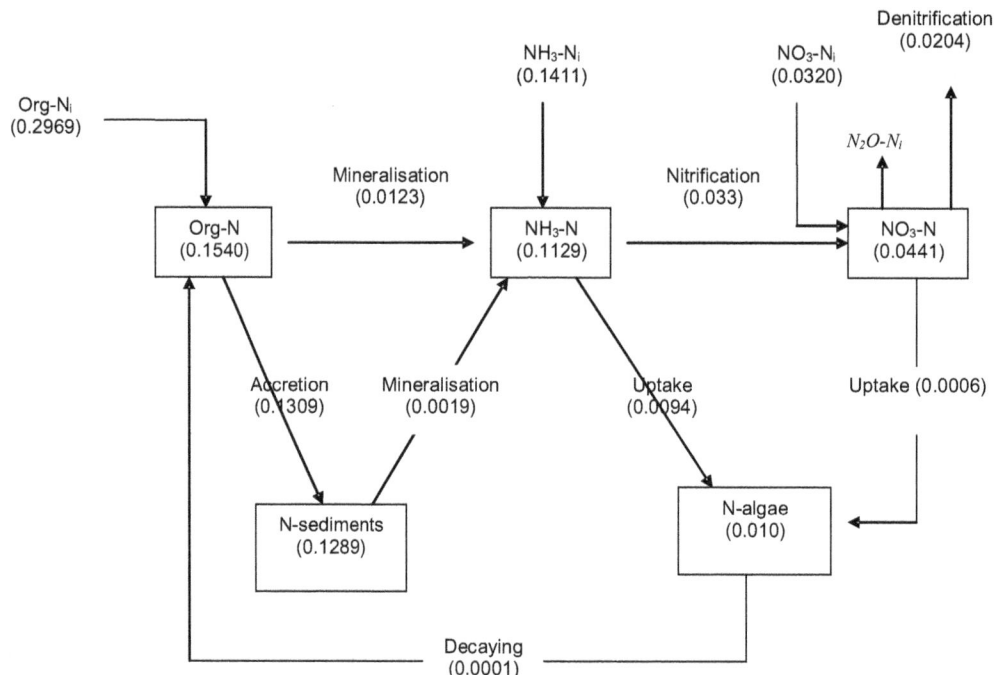

Figure 2. Nitrogen mass balance for Tanzania section of Lake Victoria (Pascal et al., 2007). All values are in g m^{-3} d^{-1}.

inshore zone (Verschuren et al., 2002). This also corresponds with the positive correlation of nitrogen bacteria to nitrate-N concentration in aquatic systems (Chia and Bako, 2008). TN:TP ratios and especially DIN : SRP ratios in Lake Victoria decrease with the wetland presence along the coastline, showing a higher probability of N limitation in the inshore waters where large wetlands are present (Co´zar et al., 2007). The results therefore points to denitrification processes in the wetland ecotones as the cause of this trend.

Solar radiation effects on nitrogen related factors

The visible solar radiation range is important in aquatic productivity. The shorter wavelength of solar radiation (ultraviolet (UV)) has potential effects on aquatic biogeochemical cycles. The general effect of solar radiation in Lake Victoria waters is the buildup of thermal differences resulting in stratification of the waters. In addition, the radiation triggers off conventional movement of water that transfer heat from the southeastern area to the northeastern colder region of the lake (Song et al., 2004). The wave of heat transfer is thought to vary on the intra-seasonal, inter-annual, inter-decadal and palaeo-climate time scales causing eddy currents. Therefore, the role of the eddy currents in the transport of energy and momentum may impact on the spatial and temporal nitrogen processes. The role of eddy currents could be similar to the water movement in the great lakes or tides

in marine environment that organizes the structure and function of the ecosystems (Keough et al., 1999) particularly the nitrifying and denitrifying bacteria which are sensitive to temperature changes.

Solar radiation intensity within water column is strongly related to water quality and therefore vary from inshore to the open waters (Dattilo et al., 2001). General increase in UV radiation in aquatic systems has been reported (Gies et al., 2004; Mckenzie et al., 2003) corresponding with stratospheric ozone depletion (Solomon, 2004). The increasing intensity affects aquatic food chain (Häder et al., 2007). Many factors influence the depth of penetration of radiation into natural waters i.e. dissolved organic compounds (Dattilo et al., 2001; Häder et al., 2007) which are likely to be influenced by future climate change. Owing to the high input of inorganic and organic decaying materials, coupled with the high level of eutrophication, Lake Victoria ecosystems has a enormous potential of UV absorption (Loiselle et al., 2001; Bracchini et al., 2004). The inshore area has high phytoplankton compared to open water zones and is attributed to nutrients load and less stable water column stratification in the inshore areas (Loiselle et al., 2001). Meanwhile the open water zone is characterized by atmospheric nitrogen deposition and seasonal mixing. In addition, variation of DOM from the fringing wetlands directly influences phytoplankton biomass and productivity (Loiselle et al., 2001). The effect is usually enhanced during the wet season as a result of flushing out of DOM that reduces the deleterious effects of UV radiation. Changes in phytoplankton composition and species diversity in the lake therefore impact on the nitrogen processes particularly the fishery and sediment retention.

Exposure to solar UV radiation has been shown to affect both orientation mechanisms and motility in phytoplankton, resulting in reduced survival rates (http://www.epa.gov/ozone/science/effects/index.html).

The effect with other stress factors has been studied in bacteria and *Cyanobacteria* and other primary producers. Excessive visible radiation, non-optimal temperature, toxic heavy metal, and changes in salinity synergistically increase the inhibitory effects of UV on growth, reproduction, ecosystem structure and food dynamics (UNEP, 2000). Lake Victoria catchment is dominated by industrial activities and the chances of heavy metals increasing into the lake cannot be ruled out. This notwithstanding, a number of new compounds absorbing UV have been identified in *Cyanobacteria*, phytoplankton and macro-algae and their role as photoprotectants during evolution is now recognized (UNEP, 2000). The effect of such changes on nitrogen processes in the open Lake Victoria waters is unknown. Analysis of the inshore areas indicates protection/attenuation from high irradiances of UV compared to open waters (Dattilo et al., 2001).

Thermal conditions and nitrogen changes in Lake Victoria

Water temperature usually falls with water depth, because less sunlight penetrates the water at a greater depth. However, temperature change attributed to climate variation in Lake Victoria has been reported (Bugenyi and Magumba, 1996; Wandiga, 2003) and thermal gradients weakness over the last decade in the Lake water column is also documented (Marshall et al., 2009). The sensitivity of fresh water Lakes to climate warming over a narrow range of high water temperature have been reported (Ndebele-Murisa et al., 2010). Therefore, moderate climate warming may destabilize nitrogen cycles and nutrients distribution (Spigel, and Coulter, 1996), as well as occurrence of nitrogen fixation, nitrifying and denitrifying bacteria. For example in Nyanza area of Winam gulf in Lake Victoria, high abundance of nitrogen fixing (*Cylindriospermopsis africana*) and potentially toxic (*Anabaena sporoides*) species during uniformly mixed period is attributed to low availability of dissolved inorganic nitrogen (Gikuma-Njuru et al., 2011). The mixed condition results into uniform oxygen distribution in lower depths; thereby promoting conversion of low oxidized forms of nitrogen to nitrate that is easily assimilated into organic growth. Mathematical models have also indicated the role played by the assumption of uniform mixing factor in nitrification process. The series model for nitrogen dispersion revealed shorter distance for ammonia-N and nitrite-N, and longer distance for nitrate-N (Bongomin and Opio, Unpublished). The information

obtained by using principle of conservation of mass modeling indicated greater distance for all the nitrogen compounds (Banadda et al., 2011). In the later modeling result, there is assumption of uniform mixing and the results from the models also show the effect of thermal stratification on the nitrification process.

Like any other aquatic systems, the general organic nitrogen decrease in Lake Victoria agrees with first order kinetics (Kadlec and Knight, 1996) and temperature impact on the processes like decomposition (Webster and Benfield, 1986). Oxygen consumption and dissolution rate during decomposition depends on temperature (Boyd, 1998). In this regards, nitrogen loading from organic substances is expected to increase with rising temperature range. Ammonia fraction also dominates over ammonium at high temperature values (Boyd, 1998). Therefore, in eutrophic freshwater lakes such as Victoria, daily temperature rise increasing pH value to more than nine (9), shifts the balance of the total ammonia towards un-ionized ammonia (Balirwa, 1998).

Extreme pH values are detrimental to many micro-organisms that influence nitrogen cycle. High pH values may imply high NH_4^+ ion concentration (Boyd, 1998). Research using molecular analysis have also revealed novel nitrifier sub-strains of *Nitrosospira briensis* (3b cluster) being tolerant to high NH_4^+ concentration while activities of *Nitrosospira NpAV* (3a cluster) strain are inhibited at high concentration (Webster et al., 2005). However, the pH ranges suitable for denitrifier growth and optimal ammonification process is 7 - 8 and 6.5 - 8.5 respectively and usually low pH increase the generation of N_2O and NO (Princic et al., 1998).

The survival of aquatic also organisms depend on temperature and warm waters are naturally productive with species that flourish being harmful sometimes (Poff et al., 2002). For example 'nuisance' blooms of algae that occur in warm and nutrient rich conditions are expected to increase in frequency. Large fish predators that require cool water temperatures may also be lost resulting into more blooms of 'nuisance' algae, reduced water quality and pose potential health problems. However, in some cases, i.e. the abundance of nitrogen fixing bacteria such as *Dactyllococcpsis* and *Gomphosphaeria* species are reported to be negatively correlated with temperature (Chia and Bakia, 2008). Optimal nitrification also occurs at the range of 25 - 35°C while temperature range suitable for denitrifies is estimated between 5 °C to 25°C (Kadlec and Knight, 1996). Changes in thermal gradients in the Lake Victoria water column were reported (Marshall et al., 2009) and the effect on spatial and temporal nitrogen processes is yet unknown.

In addition, ammonia and DO concentrations, water mixing and light attenuation are known to influence nitrification process in water hyacinth systems (Webster and Tchnobanoglous, 1985; Todd and Josephson, 1996). However, nitrification process is most sensitive to DO concentration and temperature only affects the rates of physiological processes and vitality of bacterial attachment sites in the water hyacinth (Webster and Tchnobanoglous, 1986). Therefore, nitrification in poorly oxygenated water hyacinth system is limited by oxygen transport through the plant. Elevated temperature only affects nitrification rates when it increases or decreases oxygen transport capacity.

Dissolved oxygen concentration and nitrogen processes

Oxygen flux into aquatic system is driven by mass transfer from the atmospheric sources constituting the first order process (Kadlec and Knight, 1996) and during autotrophic production of aquatic plants. In Lake Victoria, the annual cycle of oxygen vertical distribution in 1990 - 1991 was compared with data of 1960 - 1961 (Hecky et al., 1994). The results showed high oxygen concentration in the mixed layer for 1990 - 1991 with nearly continuous super-saturation in surface waters. The hypolimnetic waters had lower concentration for a longer period; the values were < 1 mg/l at 40 m compared to the shallowest occurrences of > 50 m in 1961. The dissolved oxygen (DO) trend was attributed to increased nutrient loading, altered climate and food web changes (Verschuren et al., 2002). In such a situation of high BOD levels, growth of heterotrophic bacteria that competes with nitrifying bacteria is favored (Tiedje, 1988). DO concentration of less than 0.50 mg l^{-1} is thought to limit nitrification (Metcalf and Eddy, 2003). Low DO has been reported in the lake sediment area since 1950s (Talling, 1957, Hecky et al., 1994). The general condition seems to have worsened with increased eutrophication. However, lack of complete mixing and redistribution of oxygen enhance anoxic denitrification in the sediment layer of aquatic systems (Mortimer et al., 2004), as oxygen acts as a better electron acceptor and there is high amount of energy produced when DO is used. For that matter, decreased DO favors denitrification process since the only source of oxygen is that bound to the nitrate or nitrite. Breaking of nitrogen-oxygen strong bonds in nitrate requires high amount of energy, which the denitrifiers tend to avoid (Kadlec and Knight, 1996). In this regards, denitrification process in bulk water of Lake Victoria will cease under high oxygen condition but the process continues to occur in the microscopic anoxic zones, for instance, at the sediments particularly when the overlying water column is highly productive. Laws (2000) reported denitrification occurring when oxygen concentration drops to below 0.2 gm^{-3}. Aerobic denitrifiers (*Paracoccus denitrificans, Microvirgula aerodenitrificans, Thaurea mechernichesis* and *P. denitrificans*) reduce nitrate even at oxygen saturation levels (Loyd et al., 1987; Robertson and Kuenen, 1990; Takaya et al., 2003). Whether the rate of denitrification is the same irrespective of the conditions is yet to be investigated for freshwater,

Lake Victoria. Otherwise, nitrogen reduction has the potential of occurring throughout the lake waters since stratification plays a major role in DO profile of the lake. Depending on the lake conditions, aerobic nitrification may be followed by anaerobic denitrification. Under such conditions, there is high potential of nitrous oxide production as compared to nitrogen gas. Typical aerobic denitrifiers as *P. denitrificans* produce more N_2O than nitrogen gas under aerobic conditions meanwhile more of the nitrogen is produced by *Pseudomonas stutzeri* TR2 and *Pseudomonas species* (K50 strain) under the same conditions (Takaya et al., 2003). Therefore, the observed changes in invertebrate community structure and species abundance could also be caused by DO changes for example in the case of *Caridina and chironomids* (McMahon et al., 1974).

Influence of aquatic carbon and nitrogen ratio on nitrogen processes

Increasing atmospheric CO_2 due to industrial activities, burning and climate change is bound to result into more CO_2 dissolving into aquatic systems. CO_2 uptake changes the chemical equilibrium and increase the lowering of pH. However, decreasing pH and rising temperature act to reduce CO_2 aquatic buffer capacity and the rate at which aquatic systems take up CO_2 (IPCC, 2007). Such changes therefore influence the C/N ratio that in turn affects the rate of organic matter mineralization in aquatic systems (Boyd, 1998). Carbon affects the activities of autotrophic nitrifying bacteria and denitrifying bacteria (Kadlec and Knight, 1996). When C:N ratio increases, bacteria and other micro-organisms of decay remove nitrate and ammonia (immobilization of nitrogen) from water for use in decomposition; rather less is added to the water (Boyd, 1998). Utilization of some carbon sources such as methanol by denitrifiers is coupled with production of alkalinity (Kadlec and Knight, 1996). In addition, bacterial decomposition of organic matter is also known to depend on both biological parameters such as synthesis of enzymes capable of hydrolyzing the organic substance and chemical characteristics such as compound structure (Wetzel, 2001; Sangkyu and Kang - Hyun, 2003). Lake Victoria waters receiving substances with high carbon will therefore farvor nitrifying autotrophic bacteria. The carbon and nitrogen elemental ratio is also correlated with growth and fecundity of secondary trophic levels in aquatic system (McMahon et al., 1974).

Conclusion and recommendations

N_2O in Lake Victoria is not characterized though large N inputs and DO changes occurring due to stratification create the potential for the production. The nitrogen and carbon cycle, particularly nitrogen fixation and denitrification are reported as processes that need special attention (Gruber and Galloway, 2008). At the moment, no climate and nitrogen projection for the Lake Victoria ecosystem has been done therefore; it is not possible to ascertain climate change effect on the dynamics of N cycle in the lake. The ultimate suggestions on mitigation measures are for the riparian countries to enforce policies that reduce both point and non-point sources of N into the lake and maintain riparian forests and wetlands. This is because aquatic systems have limited ability to adapt to climate change. Reducing the likelihood of impacts to aquatic systems will therefore depend on human activities so as to reduce N sources, aquatic ecosystem stress and enhance adaptation capacity.

REFERENCES

Anthonisen AC, Loehr RC, Prakasam TBS, Srinth EG (1976). Inhibition of nitrification by ammonia and nitrous-acid. Journal Water Pollution Control Federation, 48(5): 835-852.

Arceivala SJ, Asolekar SR (2008). Wastewater Treatment for Pollution Control and Re-use, Third edition, McGraw-Hill Publishers, New Delhi, p. 140.

Balirwa JS (1998). Lake Victoria wetlands and the ecology of the Nile tilapia, Oreochromis niloticus. PhD Thesis. Agricultural University of Wageningen, The Netherlands.

Banadda N, Nhapi I, Wali UG (2011). Determining and modeling the dispersion of non point source pollutants in Lake Victoria: A case study of Gaba Landing site in Uganda. Afr. J. Environ. Sci. Technol., 5(3): 178-185.

Benke AC (1993). Concepts and pattern of invertebrate production in running waters. Verhandlungen der Internationalen Vereinigung für Theoretishe und Ange – wandte Linologie, 25: 15 -38.

Biswas AK (1976). Systems approach to water management, McGraw-Hill, inc.

Boyd CE (1998). Water quality for pond aquaculture. Research and Development series No. 43. International Center for Aquaculture and Aquatic Environments Alabama Agricultural Experiment Station, Auburn University, Alabama.

Bracchini L, Loiselle S, Dattilo AM, Mazzuoli S, C´ozar A, Rossi C (2004). The spatial distribution of optical properties in the ultraviolet and visible in an aquatic ecosystem, Photochem. Photobiol. Sci., 80: 139-149.

Bugenyi FW, Magumba KM (1996). The present physico-chemical ecology of Lake Victoria, Uganda. In: Johnson, T. C. and Odada, E. O. (eds.). The Limnology, Climatology and Paleoclimatology of East African Lakes: Gordon and Breach, Toronto

Bugenyi FWB, Balirwa JS (1998). East African species introductions and wetland management: Sociopolitical dimensions. In: Science in Africa: Emerging water management issues. Symposium proceedings, Philadelphia, PA.

Bunn SE (1995). Biological monitoring of water quality in Australia. Workshop Summary and future directions. Australia J. Ecol., 20: 220-227.

Chia AM, Bako SP (2008). Seasonal variation of Cynobacteria in relation to physico-chemical parameter of some freshwater ecosystems in the Nigerian Guinea Savana. In: Sengupta, M. and Dalwani, R. (eds). Proceedings of Taal 2007: The 12th world Lake conference: pp. 1383-1387.

Co´zar A, Bergamino N, Mazzuoli S, Azza N, Bracchini L, Dattilo AM, Loiselle SA (2007). Relationships between wetland ecotones and inshore water quality in the Ugandan coast of Lake Victoria. Wetlands Ecol. Manag., 15: 499-507.

Dattilo AM, Bracchini L, Tognazzi A, Mazzuoli S, Gichuki J, Rossi C (2001). Penetration and potential impacts of solar radiation in inshore areas of Lake Victoria. In: EC RTD INCO-DEV Programme, 2001:

Tools for wetland ecosystem resource management in Eastern Africa. Scientific results of the Ecotools project: Lake Victoria wetlands and inshore area. ICA4-CT-2001-10036.

Falkowski P, Scholes RJ, Boyle E, Canadell J, Canfield D, Elser J, Gruber N, Hibbard K, Högberg P, Linder S, Mackenzie FT, Moore B, Pedersen T, Rosenthal Y, Seitzinger S, Smetacek V, Steffen W (2000). The Global Carbon Cycle: A Test of Our Knowledge of Earth as a System. Sciences, 290 (5490): 291-296.

Gies P, Roy C, Javorniczky J, Henderson S, Lemus-Deschamps L, Driscoll C (2004). Global solar UV index: Australian measurements, forecasts and comparison with the UK. Photochemem. Photobiol. Sci., 79: 32-39.

Gikuma-Njuru P, Mwirigi P, Okungu J, Hecky R, Abuodha J (2011). Spatial-temporal variability of phytoplankton abundance and species composition in Lake Victoria, Kenya: Implication for water management. Edocfind.com, file name: WLCK-155-159.pdf.

Gruber N, Galloway JN (2008). With humans having an increasing impact on the planet, the interactions between the nitrogen cycle, the carbon cycle and climate are expected to become an increasingly important determinant of the Earth system. Nature, 451: 293-296.

Häder DP, Kumar HD, Smith RC, Worrest RC (2007). Effects of solar UV radiation on aquatic ecosystems and interactions with climate change. Photochem. Photobiol. Sci., 6: 267-285.

Hecky RE (1993). The eutrophication of Lake Victoria. Verh. Internat. Verein. Limnol., 25: 39-48.

Hecky RE, Bugenyi FWB, Ochumba P, Talling JF, Mugide R, Gophen M, Kaufman L (1994). Deoxygenation of the deep water of Lake Victoria, East Africa. Limnol. Oceanogr., 39(6): 1476-1481.

http://www.epa.gov/ozone/science/effects/index.html. United States Environmental Protection Agency. Health and Environmental effects of Ozone layer depletion. In: Ozone layer protection science. Retrieved on 2/09/2011.

Hudson DA, Jones RG (2002). Regional climate model simulations of present day and future climates of Southern Africa. Technical Note No. 39, 41. London: UK Met Office.

IPCC (2001). Third Assessment Report (TAR) of the Intergovernmental Panel on Climate Change: Synthesis report and policymakers summaries. Cambridge: Cambridge University Press.

IPCC (2007). Summary for policymakers. In: Solomon, S., Qin, D., Manning, M., Chen, Z., Marquis, M., Averyt, K, B., Tignor, M. and Miller, H. L. (eds), Climate change 2007. The physical science basis. Working group 1 Contribution to the fourth assessment report of the Intergovernmental Panel on Climate Change. Cambridge University Press, UK.

Kadlec RH, Knight RL (1996). Treatment wetlands. Lewis, Publishers, Boca, Raton.

Kateregga E, Sterner T (2009). Lake Victoria fish stocks and the effects of water hyacinth. J. Environ. Dev., 18(1): 62-78.

Kayombo S, Jorgenson SE (2006). Lake Victoria. Experiences and lessons learned brief. In: Managing Lake basin for sustainable use-Lake basin management initiative final report, 300 State Street, Annapolis, Maryland 21403 USA.

Keough JR, Thompson TA, Guntenspergen GR, Wilcox DA (1999). Hydrogeomorphic factors and ecosystem responses in coastal wetlands of the Great Lakes. Wetlands, 19: 821-834.

Larsson P, Haande S, Luyiga S, Semyalo R, Kizito YS, Miyingo-Kezimbira A, Brettum P, Solheim AL, Odong R, Asio SM, Jensen KH (2008). Surface seiches mediate pollution in autrophic bay of Lake Victoria. In: Haande, S., on the ecology, toxicology and phylogeny of Cynobacteria in Murchison bay of Lake Victoria, Uganda. PhD Dissertation, University of Bergen.

Laws EA (2000). Nutrient enrichment experiments. In: Aquatic pollution: An introductory text, 3rd edition. Pg 23-34.

Loiselle S, Cozar A, Bergamini N, Tognazzi A, Rossi C (2001) Controlling factors in the eutrophication process of Lake Victoria, focus on the inland waters. In: EC RTD INCO-DEV Programme, 2001: Tools for wetland ecosystem resource management in Eastern Africa. Scientific results of the Ecotools project: Lake Victoria wetlands and inshore area. ICA4-CT-2001-10036.

Lovejoy ET, Hannah L (2005). Climate change and biodiversity. Yale University press, New Haven and London.

Loyd D, Boddy L, Davies KJP (1987). Persistence of bacterial denitrification capacity under aerobic conditions: the rule rather than expectation. FEMS Microbiol. Ecol., 45: 185-190.

Marshall B, Ezekiel C, Gichuki J, Mkumbo O, Sitoki L, Wanda F (2009). Global warming is reducing thermal stability and mitigating the effects of eutrophication in Lake Victoria (East Africa). Available from Nature Proceedings, http://hdl.handle.net/10101/npre.2009.3726.1

McKenzie RL, Bj¨orn LO, Bais A, Ilyas M (2003). Changes in biologically active ultraviolet radiation reaching the Earth's surface. Photochem. Photobiol. Sci., 2: 5-15.

McMahon RF, Hunter RD, Russel-Hunter WD (1974). Variation in aufwuchs at six freshwater habitats in terms of carbon biomass and of carbon:nitrogen ratio. Hydrobiologia, 45:391-404.

Metcalf and Eddy (2003). Wastewater engineering. Treatment and Reuse. Tchobanoglous, G., Burton, F. L. and Stensel, H. D. (Eds). 4th Ed. McGraw Hill, Inc., USA.

Millennium Ecosystem Assessment (2005). Ecosystems and human well - being: biodiversity synthesis. World Resources Institute, Washington, D.C., USA.

Ministry of Lands, Water and Environment (MLWE) (2002). Initial national communication on climate change. Uganda.

Mortimer RJG, Harris SJ, Krom MD, Freitag TE, Prosser JI, Barnes J, Anschutz P, Hayes PJ, Davies IM (2004). Anoxic nitrification in marine sediments. Marine Ecology Progress Series, 276: 37 - 51.

Mugidde R (1993). The increase in phytoplankton primary productivity and biomass in Lake Victoria (Uganda). Verh. Internat. Verein. Limnol., 25: 846-849.

Ndebele - Murisa M, Musil CF, Raitt L (2010). A review of phytoplankton dynamics in tropical African lakes. South Afr. J. Sci., 106(1-2). ISSN 0038-2353.

Niang I, Nyong A, Clark B, Desanker P, Din N, Githeko A, Jalludin M, Osman B (2007). Vulnerability, impacts and adaptation to climate change. In: Otter, L., Olago, D. O. and Niang, I. (eds), Global change processes and impacts in Africa: A synthesis. Creative Print House. Kenya.

Opande GO, Onyang JC, Wagai SO (2004). The water hyacinth (Eichhornia crassipes [MART)] SOLMS), its socio-economic effect, control measures and resurgence in the Winam gulf. Limnologica,- Ecol. Manage. Inland Waters, 34(1-2): 105-109.

Opera A, Saayman I, Githuri F (2007). Water resources and global change. In: Otter L, Olago DO and Niang I (eds), Global change processes and impacts in Africa: A synthesis. Creative Print House. Kenya.

Pascal E, Mwanuzi F, Kimwaga R (2007). Study of Nitrogen Transformation in Lake Victoria. Catchment and Lake Research. Universität Siegen, DAAD, GTZ and Geothe-Institut, Gebrekristos Desta Center, Addis Abeba.

Poff NL, Brinson MM, Day JW (2002). Aquatic ecosystems and global climate change. Potential impact on inland freshwater and coastal wetland ecosystems in the United States. Prepared for the Pew Center on Global Climate Change.

Princic A, Mahne I, Megusar F, Eldor PA, Tiedje JM (1998). Effects of pH and oxygen and ammonium concentrations on community structure of nitrifying bacteria from wastewater. Appl. Environ. Microbiol., 64(10): 3584-3590.

Regier HA, Holmes JA, Pauly D (1990). Influence of temperature change on aquatic ecosystems: An interpretation of empirical data. Trans. Am. Fisheries Soc., 119: 374 - 389.

Robertson LA, Kuenen JG (1990). Physiological and Ecological aspects of aerobic denitrification, a link with heterotrophic nitrification? In: N. P. Revsbech and J. Serensen (eds), Denitrification in soils and sediments. Plenum Press, New York.

Rutagemwa DK, Myanza OI, Mwanuzi F (2005). Water Quality Synthesis Report-Lake Monitoring, Lake Victoria, Mwanza, Tanzania.

Sangkyu P, Kang - Hyun C (2003). Nutrient leaching from leaf litter of emergent macrophyte (Zizania latifolia) and the effects of water temperature on the leaching process. Korean J. Biol. Sci., 7: 289-294.

Seitzinger S, Harrison JA, Böhlke JK, Bouwman AF, Lowrance R, Peterson B, Tobias C, Van Drecht G (2006). Denitrification across landscapes and waterscapes: a synthesis. Ecol. Appl., 16: 2064-2090.

Solomon S (2004). The hole truth. What's news (and what's not) about the ozone hole. Nat., 427: 289-291.

Song Y, Semazzi FHM, Xie L, Ogallo LJ (2004). A coupled regional climate model for the Lake Victoria basin of East Africa. Int. J. Climatol., 24(1): 57 - 75.

Spigel RH, Coulter GW (1996). Comparison of hydrology and physical limnology of the East African Great Lakes: Tanganyika, Malawi, Victoria, Kivu and Turkana (with reference to some North American Great Lakes). In: Johnson TC and Odada E. The limnology, climatology and paleo-climatology of the east African lakes. Toronto: Gordon and Breach. Toronto: Gordon and Breach.

Takaya N, Catalan - sakairi MAB, Sakaguchi Y, Kato I, Zhou Z, Shoun H (2003). Aerobic denitrifying bacteria that produce low levels of nitrous oxide. Appl. Environ. Microbiol., 69(6): 3152-3157.

Talling JF (1957). Some observations on the stratification of Lake Victoria. Am. Society Limnol. Oceanogr., 2(3): 213-221.

Tiedje JM (1988). Ecology of denitrification and dissimilatory nitrate reduction to ammonia. In: Zehnder, A. J. B. (ed).

Todd J, Josephson B (1996). The design of living technologies for waste treatment. Ecol. Eng., 6: 109-136.

Trenberth KE, Jones PD, Ambenje P, Bojariu R, Easterling D, Klein-Tank A, Parker D, Rahimzadeh F, Renwick JA, Rusticucci M, Soden B, Zhai P (2007): Observations: Surface and Atmospheric Climate Change. In: Solomon, S., D. Qin, M. Manning, Z. Chen, M. Marquis, K.B. Averyt, M. Tignor and H.L. Miller (eds.), Climate Change 2007: The Physical Science Basis. Contribution of Working Group I to the Fourth Assessment Report of the Intergovernmental Panel on Climate Change. Cambridge University Press, Cambridge, United Kingdom and New York, NY, USA.

UNEP (2000). Environmental effect of ozone depletion: Interim summary. US Global Change Research Information Office, Suite 250, 1717 Pennsylvania Ave, NW Washington, DC 20006.

Van Ginkel CV (2002). Trophic status assessment. Executive summary. Institute for Water Quality Studies, Department of Water Affairs and Forestry, Private Bag X313 Pretoria, South Africa, 0001.

Verschuren D, Johnson TC, Kling HJ, Edington DN, Leavitt PR, Brown ET, Talbot MR, Hecky RE (2002). The chronology of human impact on Lake Victoria, East Africa. Proc. R. Soc. London, 269: 289-294.

Wandiga SA (2003). Lake Basin Management Problems in Africa: Historical and Future Perspectives. http://www.worldlakes.org/uploads/Lake%20Basin%20Problems%20in%20Africa_12.16.03.pdf. Accessed on 16/09/2011.

Weber AS, Tchobanoglous G (1985). Nitrification in water hyacinth treatment systems. J. Environ. Eng., pp 699-713.

Weber AS, Tchobanoglous G (1986). Predicting nitrification in water hyacinth treatment system. J. Water Pollut. Control Federation, 58(5): 376-380.

Webster G, Embley TM, Freitag TE, Smith Z, Prosser JI (2005). Links between ammonia oxidizer species composition, functional diversity and identification kinetics in grassland soils. Environ. Microbiol., 7: 676-684.

Webster JR, Benfield EF (1986). Vascular plant break down in freshwater ecosystems. Annual Rev. Ecol. Systematic, 17: 567-594.

Wetzel RG (2001). Limnology-Lake and river ecosystems. 3rd ed. Academic Press, San Diego, USA.

Young RG (2007). A trial of wood decomposition rates as an ecological assessment tool in large rivers. Prepared for West Coast Regional Council. Cawthron Report No. 1339.

16

Geoelectric investigation of groundwater resources and aquifer characteristics in Utagba-Ogbe kingdom Ndokwa land area of Delta State, Nigeria

Julius Otutu Oseji

Department of Physics, Delta State University, Abraka, Delta State, Nigeria. E-mail: oseji2002@yahoo.com.

Vertical electrical sounding data were acquired from 10 locations evenly distributed within Utagba-Ogbe Kingdom. The apparent resistivity values obtained in the field were plotted against half current electrode spacing. Interpretation of data were done both quantitatively and qualitatively and bringing in to bare the knowledge of the local geology of Utagba-Ogbe and her environs. Based on the geoelectric section, which shows the relationship between the drillers log and the resistivity measurements at a common depth of penetration, 4 prominent geoelectric layers of near surface aquifer that are not confined were identified in Utagba-Ogbe Kingdom. The study revealed Utagba-Ogbe Kingdom as an extensive sandy unit. The best layer of the aquifer in Utagba-Ogbe kingdom for groundwater development is at a depth between 35.00 - 45.00 m within the second layer. This layer consist medium-grained sand to coarse-grained sand formations, which is the best formation to obtain an appreciable quantity of water for sustainable groundwater development. The results when correlated with lithologic log from a producing borehole in second Owessei Street Umusadege were found to be consistent.

Key words: Vertical electrical sounding, groundwater potential, aquifer, Umusadege, Umuseti, Umusedeli, Umusam, Owessei, drillers log and geoelectric section.

INTRODUCTION

When rain falls to the ground, the water does not stop moving. Some of it flows along the surface in streams or lakes, plants use some while some evaporates and return to the atmosphere, others sink into the ground. Imagine putting a glass of water onto a pile of sand. Where does the water go? The water moves into the spaces between the particles of sand.

Groundwater is the water that is found in cracks and spaces within the soil, sand and rocks. The area where water fills the space is called the saturated zone. The top of this zone is the water table. Assume the top of water to be a table. The water may be only a foot below the ground surface or it may be hundreds of feet down.

Groundwater can be found almost everywhere. The water table may be deep or shallow and may rise or fall depending on many factors. Heavy rains or melting snow may cause the water table to rise or an extended period of dry weather may cause the water table to fall.

Groundwater is stored in, and moves slowly through layers of soil, sand and rocks called aquifers. The size of the spaces in the soil or rock and how well the spaces are connected determine the speed at which groundwater flows.

Aquifers typically consist of gravel, sand, sandstone, or fractured rock, like limestone. These materials are permeable because they have large connected spaces that allow water to flow through. Aquifers are also known as underground reservoirs otherwise called underground flood and the water that reached this chamber is usually much cleaner than the water or reservoirs at the earth surface. Aquifer could be confined or unconfined.

Unconfined aquifers lie very near the water table, with little or no overlying rock or sediment and their water is usually at atmospheric pressure.

Most local groundwater comes from unconfined aquifers made of loose slope materials, sands, gravels, and floodplain deposits left by stream and rivers.

Confined aquifers are sandwiched between rock layers that are either effectively impermeable or have very low permeability. However, a combination of the two can occur and that aquifer is called Leaky or a semi-confined aquifer. The local occurrence of groundwater is the consequence of a finite combination of climatic, hydrologic geologic, topographic, and ecological and soil forming factors.

Resistivity of a material therefore is defined as the opposition to the flow of current in Ohms between opposite faces of a unit cube of the material.

Electrical method utilizes direct current or low frequency alternating current to investigate the electrical properties of the subsurface. It is a technique used to study the shallow layer of the earth by sending direct electric current through a pair of electrodes and studying the potential distribution it produces.

From Ohms law, we can then deduce the resistance and hence the resistivity. Apart from current electrodes, the region we are probing sets a limit on the applicability since we need long lengths of cables and strong power source.

Electrical resistivity of earth materials varies over a wide range. Hence it is possible to determine resistivity measurement.

The method is particularly useful for soil testing, engineering purposes or hydrological checks. Generally, the electrical resistivity method involves the use of artificially sourced current, which is introduced into the ground through a pair of electrodes (current electrodes) while the resulting potential difference is measured by another pair of electrodes called potential electrodes which may or may not be located within the current electrodes (Kearey and Brooks, 1991). Potable water is not only commonly found and its provision limits the setting up of villages and towns to places where there is an existence of supply (Shanker, 1994).

Hence the need for Environmental Geophysical Surveys cannot be overemphasized in getting background information on the distribution, formation and type of the near surface aquifer.

The electrical resistivity method is based on the variable resistance in subsurface materials to the conductance of electrical current, depending on variations in fluid content, density and chemical composition of the material (Parasnis, 1986).

This work was carried out to establish a baseline geophysical data and hydrological characteristics using the Schlumberger arrangement (a Vertical Electrical Sounding) and drillers log from the study area. The vertical electrical method was chosen for this study because the instrumentation is simple; field logistics are easy and straightforward and the analysis of data is less tedious and economical (Ekine and Osobonye, 1996; Ako and Olorunfemi, 1989; Etu-Efeotor and Akpokodje, 1987 and 1979; Okolie et al., 2005). The resistivity method has been used successfully in investigating groundwater potential. (Oseji et al 2005) used the method to investigate the aquifer characteristics and groundwater potential in Kwale, Delta State, Nigeria. (Oseji et al., 2006) also used the method to determine the groundwater potential in Obiaruku and environs. (Emenike, 2000) used the same method to explore for groundwater in a sedimentary environment. (Ako and Osunde, 1982) used the method to delineate aquifer units and established the thickness and

Table 1. Observed (Field) and computed (Theoretical) data for Umusedeli (VES1).

AB/2 Values (m)	Observed Values (ohm-m)	Computed Values (ohm-m)
1.00	392.20	407.52
1.47	454.70	466.46
2.15	580.00	547.73
3.16	625.10	634.82
4.64	703.00	701.59
6.81	732.20	727.45
10.00	675.10	716.26
14.70	711.70	714.26
21.50	784.40	785.21
31.60	1030.30	957.98
46.40	1285.60	1202.87
68.10	1559.50	1468.55
100.00	1697.10	1710.72
147.00	1901.60	1897.24
215.00	2033.90	2010.43
316.00	2102.20	2059.75

depth of water bearing formation. (Okwueze, 1996) used the method to determine the groundwater potential at Obudu basement area. (Oseji and Ujuanbi, 2009) used the method to investigate the groundwater potential in Emu Kingdom.

FIELD PROCEDURES

Ten Vertical Electrical Sounding data were acquired from 5 locations evenly distributed within Utagba-Ogbe Kingdom. The data acquired from VES (1 - 5) were similar to those acquired from VES (6 - 10) hence only five of the data were displayed (Tables 1 - 5). Interpretation of data were done quantitatively and qualitatively and bringing in to bare the knowledge of the local geology of the area. The apparent resistivity values were plotted against half the current electrode spacing on a log-log graph. The curves of best fit were then traced and the data obtained from the smooth curve (Smoothed values) were noted. Qualitative and quantitative interpretations of the field curves were carried out by inspection to obtain the type of curves and by partial curve matching respectively.

The resistivity and thickness obtained from the partial curve matching were improved upon by employing an iterative computer program following the main ideas of (Zohdy, 1974; Zohdy et al., 1974) to obtain the layers parameter (resistivity, thickness and depth) (Tables 6 - 10).

Here, the number of geoelectric layers and their corresponding specific resistivities were first taken to be equal to the number of measurable points and difference of adjacent current electrode spacing respectively. Layer parameters were consequently modified in iterative manner until subsequent iteration yields no improvement on the root mean square (rms) error values in percentage.

The numerous layers that were generated by the computer were grouped into relevant geologic depth intervals called geoelectric sections and the resulting layer parameters were then given geologic interpretation. The type of curves, the resistivity of the

Table 2. Observed (Field) and computed (Theoretical) data for Umuseti (VES 2).

AB/2 Values (m)	Observed Values (ohm-m)	Computed Values (ohm-m)
1.00	351.20	369.29
1.47	324.80	333.23
2.15	292.00	288.15
3.16	285.00	258.88
4.64	279.40	274.16
6.81	350.00	345.99
10.00	450.10	469.94
14.70	614.40	635.27
21.50	813.40	824.97
31.60	999.00	1017.82
46.40	1195.40	1159.39
68.10	1190.30	1180.06
100.00	1005.70	1041.75
147.00	747.10	799.58
215.00	580.80	578.05
316.00	470.60	446.09

Table 4. Observed (Field) and computed (Theoretical) data for 2ND Owessei Street Umusadege (VES 4).

AB/2 Values (m)	Observed Values (ohm-m)	Computed Values (ohm-m)
1.00	543.96	535.44
1.47	611.19	603.68
2.15	643.61	538.08
3.16	601.06	598.43
4.64	467.09	467.08
6.81	293.38	293.83
10.00	168.23	166.50
14.70	125.48	120.57
21.50	131.35	124.49
31.60	150.87	144.83
46.40	177.28	175.84
68.10	213.11	219.69
100.00	255.28	270.19
147.00	301.40	316.19
215.00	356.07	349.43
316.00	427.22	370.03

Table 3. Observed (Field) and computed (Theoretical) data for Umusam (VES 3).

AB/2 Values (m)	Observed Values (ohm-m)	Computed Values (ohm-m)
1.00	259.10	252.87
1.47	284.00	278.29
2.15	320.00	317.14
3.16	375.70	362.00
4.64	414.60	398.41
6.81	429.80	411.99
10.00	408.20	395.66
14.70	373.40	360.16
21.50	338.90	331.50
31.60	345.20	333.73
46.40	365.40	380.85
68.10	485.80	470.27
100.00	597.10	586.27
147.00	698.70	709.15
215.00	798.60	820.15
316.00	909.90	910.46

Table 5. Observed (Field) and computed (Theoretical) data for Oseji Estate Umusadede (VES 5).

AB/2 Values (m)	Observed Values (ohm-m)	Computed Values (ohm-m)
1.00	1256.50	1206.52
1.47	1300.00	1291.13
2.15	1383.00	1372.44
3.16	1437.30	1429.74
4.64	1500.10	1454.63
6.81	1524.80	1462.60
10.00	1529.70	1488.65
14.70	1600.00	1574.56
21.50	1719.00	1735.48
31.60	1882.80	1946.57
46.40	2097.00	2138.23
68.10	2186.20	2224.29
100.00	2011.40	2098.08
147.00	1695.90	1688.24
215.00	1103.50	1108.75
316.00	596.10	602.73

sediments and the lithologic logs from nearby boreholes were used in conjunction with the knowledge of the local geology of the study area as guides in the interpretation and analysis of the geologic section in terms of probable and sustainable water supply.

RESULTS AND DISCUSSION

Utagba-Ogbe is within the Sombriero Warri deltaic plain deposit invaded by mangrove. The curve types are KA for Umusedeli and Umusam, AK for Oseji estate, HK for Umuseti and KH for second Owessei Street Umusadege.

The interpreted sounding curves from the locations at Umusedeli, Umusam and 2nd Owessei Street Umusadege, Umuseti and Oseji estate by Ozoro/ Ogwashi-Uku express road revealed 4 prominent geo-electric layers. However, they appears to be a very thin clay layer between the first and the second geoelectric layers at Umuseti and second Owessei street Umusadege

Table 6. Model parameters VES 1.

Geoelectric Layer	Resistivity (ohm-m)	Thickness (m)	Cumulative Thickness(m)
1	357.21	0.82	0.82
2	956.00	2.98	3.80
3	513.00	9.95	13.75
4	2907.45	23.53	37.28
5	2096.65	16.10	53.38
6	2238.20	55.01	108.39
7	2043.45	infinity	Infinity
RMS error	1.65%		

Field measurements and data interpretations by: Oseji Julius Otutu.

Table 7. Model parameters VES 2.

Geoelectric Layer	Resistivity (ohm-m)	Thickness (m)	Cumulative Thickness(m)
1	398.37	0.90	0.90
2	177.00	2.73	3.63
3	2537.83	19.03	22.66
4	889.00	27.06	49.72
5	416.60	58.06	107.78
6	363.00	infinity	Infinity
RMS error	1.76%		

Field measurements and data interpretations by: Oseji Julius Otutu.

Table 8. Model parameters VES 3.

Geoelectric Layer	Resistivity (ohm-m)	Thickness (m)	Cumulative Thickness (m)
1	234.00	0.94	0.94
2	518.03	3.93	4.87
3	265.47	24.63	29.50
4	1285.02	36.35	65.85
5	1071.00	61.12	126.97
6	1055.43	infinity	Infinity
RMS Error	1.20%		

Field measurements and data interpretations by: Oseji Julius Otutu.

Table 9. Model parameters VES 4.

Geoelectric Layer	Resistivity (ohm-m)	Thickness (m)	Cumulative Thickness(m)
1	401.00	0.48	0.48
2	1056.30	0.67	1.15
3	655.00	1.46	2.61
4	66.60	3.82	6.43
5	132.30	21.33	27.76
6	726.92	29.90	57.66
7	347.10	61.12	118.78
8	395.31	infinity	Infinity
RMS error	2.01%		

Field measurements and data interpretations by: Oseji Julius Otutu.

Table 10. Model parameters VES 5.

Geoelectric Layer	Resistivity (ohm-m)	Thickness (m)	Cumulative Thickness (m)
1	1088.41	0.56	0.56
2	1548.50	2.04	2.60
3	1360.40	2.99	5.59
4	1502.66	6.42	12.01
5	2800.60	56.77	68.78
6	1050.00	41.93	110.71
7	342.00	36.14	146.85
8	236.55	infinity	Infinity
RMS error	1.05%		

Field measurements and data interpretations by: Oseji Julius Otutu.

thereby given rise to a fifth geoelectric layer. The lithologic information from a producing borehole within the study area in conjunction with the knowledge of the local geology of the area was used in constructing an earth model. The litholog indicates broadly that within the depth penetrated, the succession is lateritic topsoil, thin clay layer, fine grained sand and medium to coarse-grained sands at various depths.

The data obtained from the field revealed high resistivity values in areas along Umusedeli. This was attributed to both the dried nature of the soil and the effects of buried petrol tanks at Agip Petrol Station that exists below the sounding point and along the spread line of the VES.

The various layers obtained from the iterated results were reduced to 4 relevant geoelectric depth intervals called geoelectric sections as shown in Figure 1.

The first geoelectric layer corresponds to the topsoil with resistivity values ranging from 200.00 - 400.00 Ω m reflecting the various compositions and moisture content of the weathered lateritic layers. The thickness of this layer varies from 0.40 - 1.00 m.

The second geoelectric layer at locations 7 and 9 in Umuseti and second Owessei Street Umusadege has resistivity values of between 66.00 and 170.00 Ω m. This is diagnostics of clay formation, which may act as a confining bed. However, because of the thickness, the area may be susceptible to contamination in the event of pollution.

The second geoelectric layer at Umusedeli, Umusam and Oseji estate is composed of medium/coarse sand formation with resistivity ranging from 100.00 - 500.00 Ω m and thickness of between 5.00 - 10.00 m. This layer is shallow but constitutes the first aquifer at a

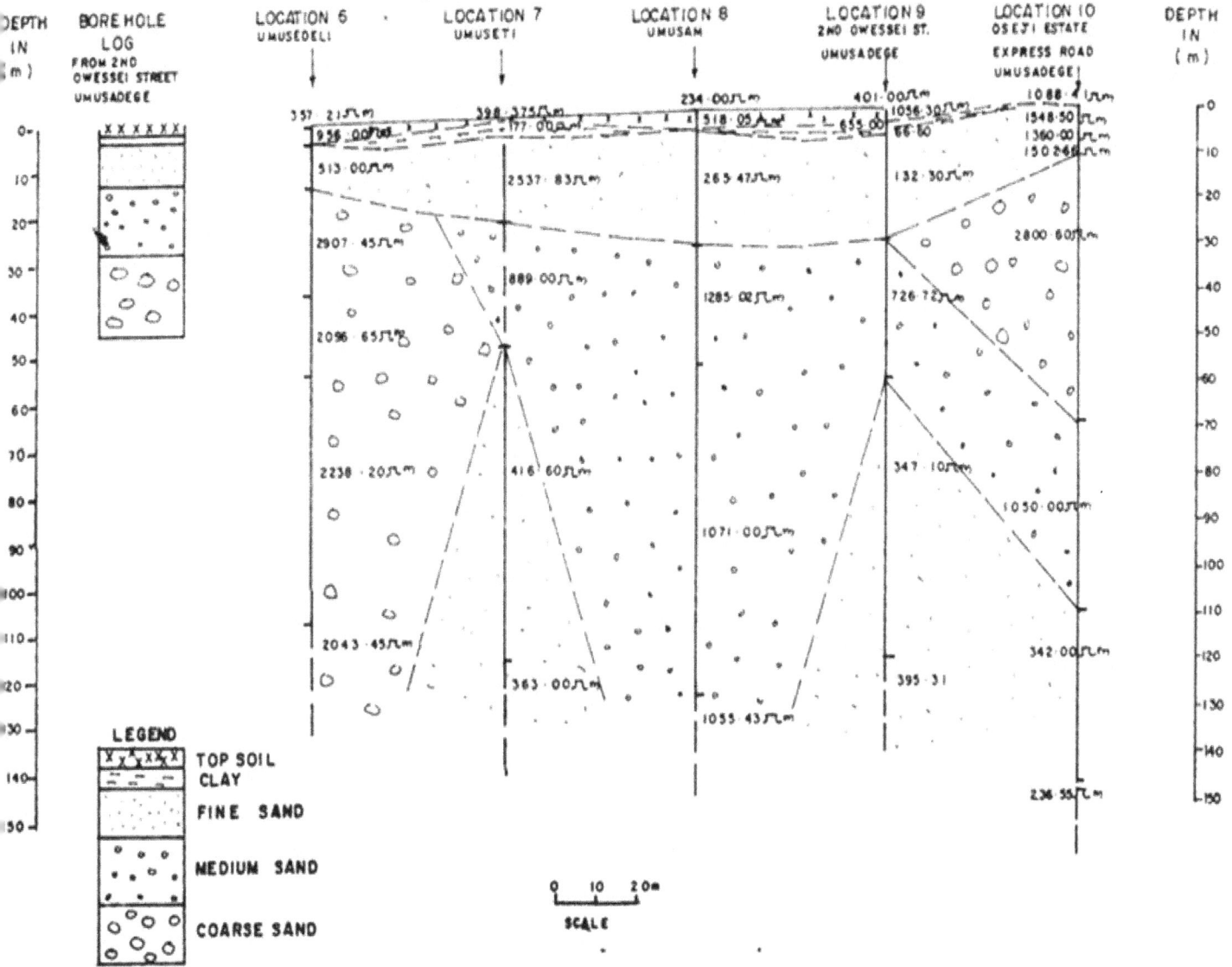

Figure 1. Geoelectric section of Utagba-Ogbe kingdom.

depth of between 10.00 - 15.00 m.

The third geoelectric layer at Umusedeli, Umusam and Oseji estate consists of medium to coarse-grained sand. This layer constitutes an aquifer of very good quantity of groundwater. The average depth to this aquifer is between 35.00 - 45.00 m with an undefined thickness since it is the last layer. However, in Umuseti, second Owessei Street and Oseji estate, there exists a fourth geoelectric layer whose resistivity ranges from 200.00 - 400.00 Ω m and consist fine grain sand. This is the third layer, whose thickness is not defined, since it is the last layer. It occurs at an average depth of between 60.00 - 100.00 m. Based on the resistivity measurements, it is clear that three near surface layers, which are not confined, have been

identified in Utagba-Ogbe and environs. In the event of pollution, groundwater may be contaminated.

The first aquifer is at an average depth of between 5.00 - 15.00 m and has a thickness of 10.00 - 20.00 m. This is a very shallow aquifer and it is not an encouraging prospect for groundwater development.

The second layer consist medium to coarse grained sand formation. This layer has good potential for groundwater development and occurs at an average depth between 35.00 - 45.00 m.

The third layer in Utagba-Ogbe is at a depth of between 60.00 - 100.00 m with an undefined thickness. It consists of fine-grained sand formation which is not an encouraging formation for groundwater development.

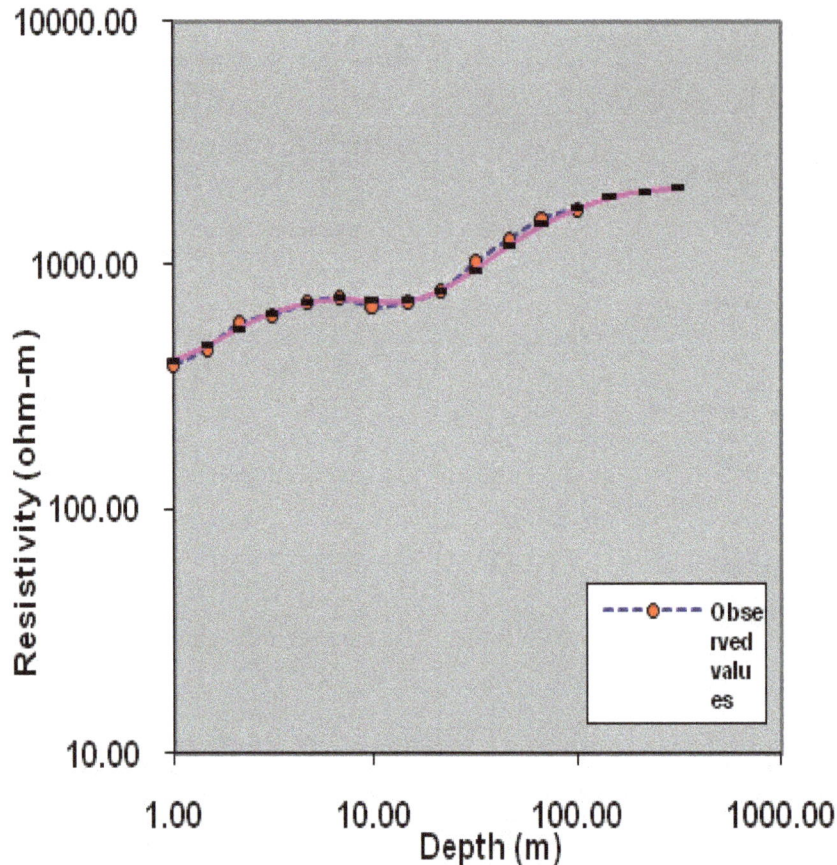

Site 1. Umusedeli-Ogbe by Agip petrol station, Kwale. VES No : 1.

The best layer in Utagba-Ogbe and environs for groundwater development is at a depth of between 35.00 - 45.00 m within the second layer and in the medium to coarse grained sand formation. Generally, materials that lack pore spaces will show high resistivity. Materials whose pore spaces lack water will show high resistivity such as dry sand or gravel. Materials whose water content is clean will show high resistivity such as clean gravel or sand even if water is saturated. Weathered rocks and clay will show medium to low resistivity. Frozen ground will show much higher resistivity than unfrozen ground.

In the sedimentary environment, high resistivity may broadly be associated with the presence of fresh groundwater in porous medium aquifer while low resistivity may be due to the presence of clay and/or brackish water (Emenike, 2000). The type of curve (Selemo et al., 1995), the modified water and sediments resistivity table by Oyedele (2001) and Zohdy and Martins (1993) as well as the knowledge of the local geology of Utagba-Ogbe Kingdom were used as guides in the interpretation of the VES data in terms of probable aquifer in this work. The results were correlated with lithologic log from a producing borehole in second Owessei Street Umusadege between VES 4 and 9, and they were found to be consistent.

Conclusion and Recommendation

The study revealed Utagba-Ogbe Kingdom as an extensive sandy unit. The best layer of the aquifer in Utagba-Ogbe kingdom for groundwater development is at a depth between 35.00 - 45.00 m within the second layer. This layer consist medium-grained sand to coarse grained sand formations, which is the best formation to obtain an appreciable quantity of water for sustainable groundwater development since it has high porosity.

The research did not only pave way for a clear picture of the hydro geological knowledge of Utagba-Ogbe Kingdom in other to create awareness on the productive and prolific aquifer for sustainable groundwater supply but act as guides to both the Government and individuals especially those involved in groundwater development on the type of near surface aquifers, the formation of the aquifer as well as the thickness of the aquifer and the depths boreholes could be drilled for sustainable water supply (Site 1 - 5).

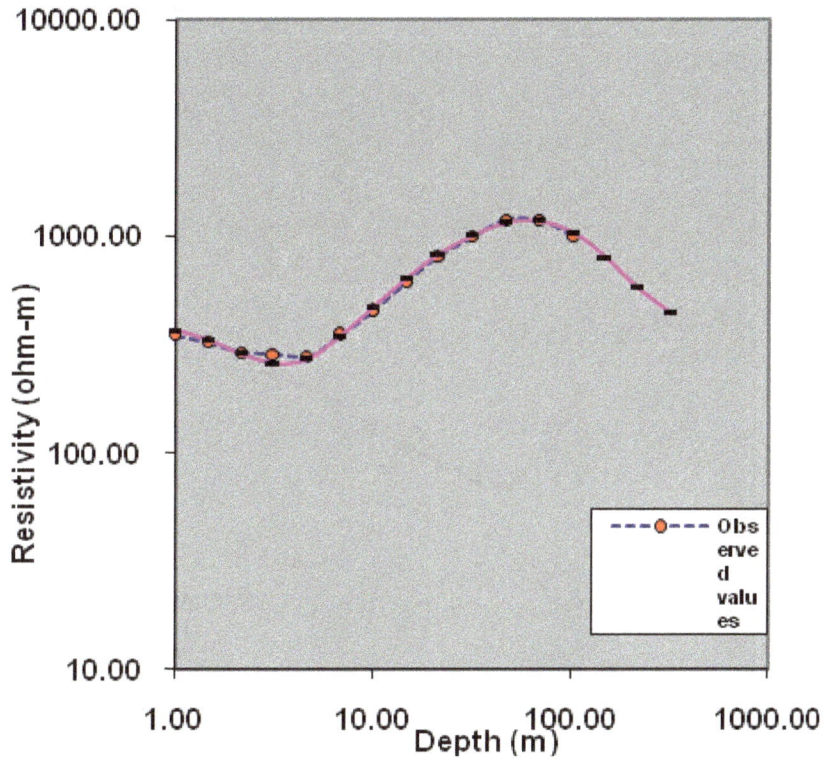

Site 2. Umuseti-Ogbe opposite chief Onefeli's Compound, Kwale. VES No: 2.

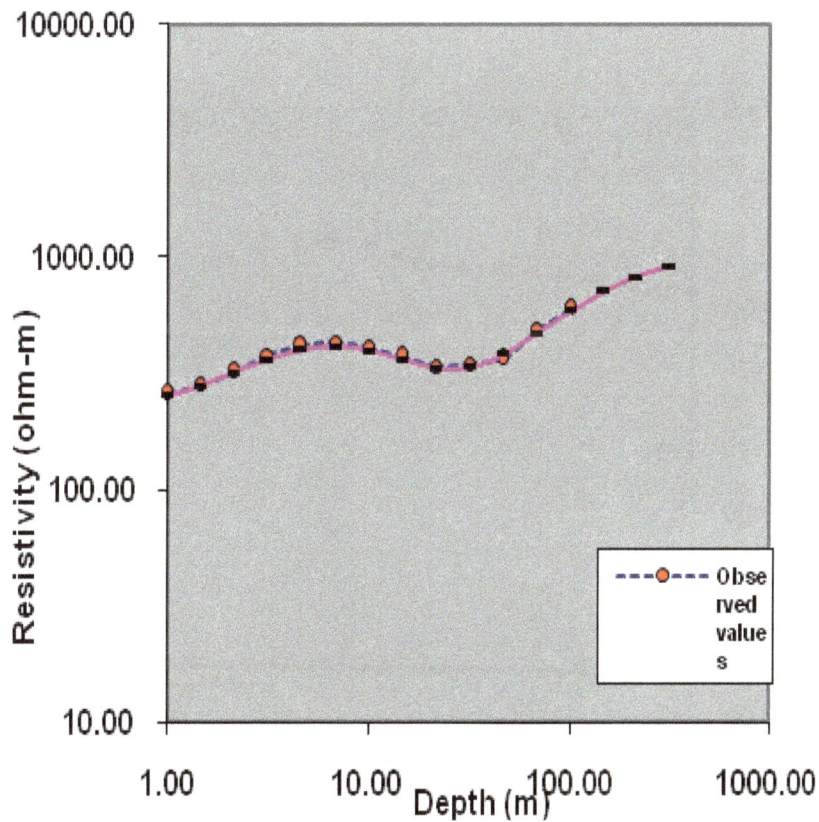

Site 3. Umusam-Ogbe by the Obiogwa, Kwale. VES No: 3

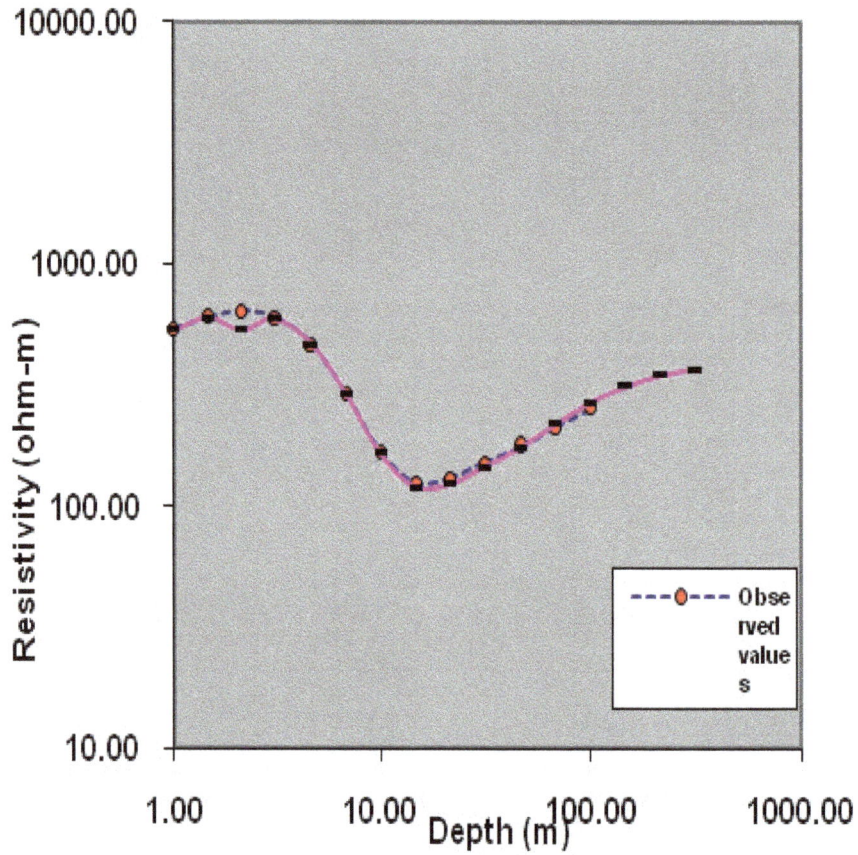

Site 4. Umusadege-Ogbe by SECOND Owessei Street, Kwale. VES No: 4.

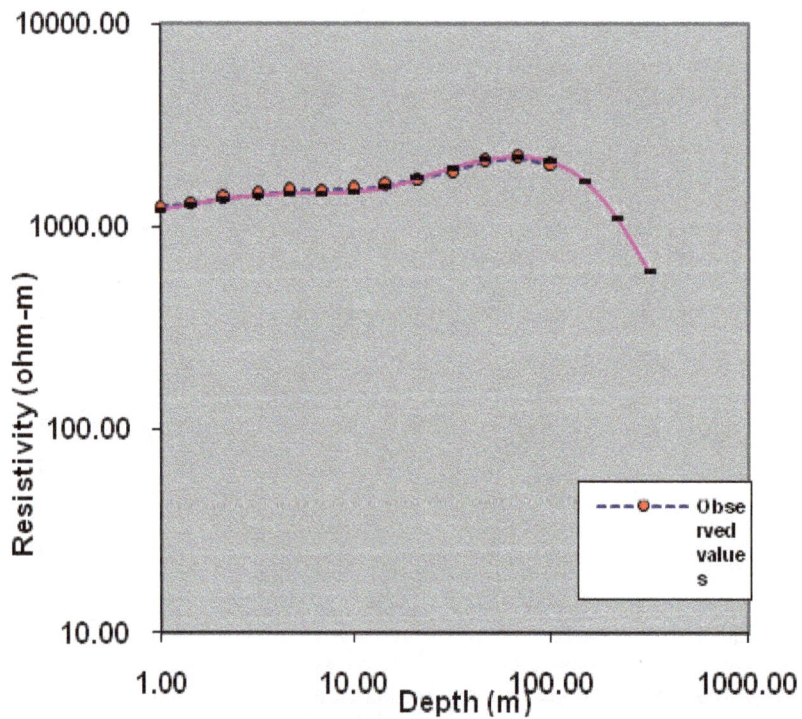

Site 5. Umusadege-Ogbe (Oseji Estate, Ozoro/Asaba Express Rd.) Kwale. VES No: 5.

REFERENCES

Ako AO, Olorunfemi MO (1989). "Geoelectric survey for groundwater in the newer Basalts of Vom Plateau State. Niger. J. Min. Geol. 25(1-2): 247-450.

Ako BD, Osunde VC (1982). "Electrical Resistivity Survey of Kerri Formation darazo". Niger. J. Afr. Earth Sci. 5: 527-534.

Akpokodje EG (1979). "The importance of engineering geological mapping" in the development of the Niger Delta Basin. Bull. 1 Assoc. Eng. Geol. 19: 102-108.

Akpokodje EG (1987): "The engineering geological characteristics and classification of the major superficial soils of the Niger Delta. Eng. Geol. 23: 193-211.

Emenike EA (2000). "Geophysical Exploration for Groundwater in a Sedimentary Environment": A case study from Nanka over Nanka formation in Anambra Basin, Southeastern Nigeria. Global J. Pure Appl. Sci. 7(1): 97-110.

Ekine AS, Osobonye GT (1996). "Surface Geoelectric Sounding for the determination of Aquifer Characteristics in parts of Bonny Local Government Area of River State". Niger. J. Phys. 85: 93-97.

Etu-Efeotor JO, Akpokodje EG (1990). "Aquifer systems of the Niger Delta". Niger. J. Min. Geol. 26(2): 279–284.

Kearey P, Brooks M (1991). "An introduction to Geophysical exploration" Second Edition. Blackwell Scientific Publication Ltd. Oxford. pp. 173-193.

Okolie EC, Osemeikhian JEEA, Oseji JO, Atakpo E (2005). "Geophysical Investigation of the source of River Ethiope" in Ukwuani Local Government area of Delta State. Niger. Inst. Phys. (17): 21-26.

Okwueze EE (1996). "Preliminary Findings of the Groundwater Resources Potentials from a regional geoelectric survey of the Obudu basement area, Niger. Global J. Appl. Pure Sci. 2: 210-211.

Oseji JO, Atakpo E, Okolie EC (2005). "Geoelectric Investigation of the Aquifer Characteristics and Groundwater Potential in Kwale, Delta State, Niger. J. Appl. Sci. Environ. Mgt. 9(1): 157 – 160. ISSN 1119 – 8362 www. Bioline. Org.br.ja.

Oseji JO, Asokhia MB, Okolie EC (2006). "Determination of Ground-water Potential in Obiaruku and Environs Using Surface Geoelectric Sounding". The Environmentalist, Springer Science + Business Media, DO1 10.10669-006-0159-x 26: 301-308, Netherlands.

Oseji JO, Ujuanbi O (2009). "Hydrogeophysical investigation of groundwater potential in Emu Kingdom, Ndokwa land of Delta State, Nigeria". Int. J. Phys. Sci. 4(5): 275 – 284 May. Available online at htt: / www.academicjournalsorg/IJPS.

Oyedele KF (2001). "Hydro geophysical and Hydro geochemical Investigation of Groundwater Quality in some parts of Lagos, Nigeria. African J. Environ. Stud. 2(1): 31-37.

Selemo AOI, Okeke PO, Nwankwor GI (1995). "An Appraisal of the usefulness of VES in Groundwater Exploration in Nigeria" Water Res. 6(1-2): 61-67.

Shanker RK (1994). "Selected Chapters in Geology". Shell Petroleum Development Company, Warri. pp.10-148.

Zohdy AA, Eaton CP, Mabey DR (1974). "Application of Surface Geophysics to Groundwater Investigation". Tech. water resources investigation, Washington, U.S Geological Survey pp. 2401-2543.

Zohdy AAR, Martin RJ (1993). "A study of sea water intrusion using Direct Current Sounding in the Southern part of the Ox ward Plain California". Open-file reports 93 – 524 U.S. Geological Survey 139p.

17

Influence of pre-oxidation with potassium permanganate on the efficiency of iron and manganese removal from surface water by coagulation-flocculation using aluminium sulphate: Case of the Okpara dam in the Republic of Benin

Zogo D.[1], Bawa L. M.[2*], Soclo H. H.[3] and Atchekpe D.[1]

[1]Société Nationale des Eaux du Bénin (SONEB), Republic of Bénin.
[2]Laboratoire de Chimie de l'Eau, Faculté Des Sciences, Université de Lomé, BP 1515, Lomé, Togo.
[3]Unité de Recherche en Ecotoxicologie et Etude de Qualité (UREEQ), Université d'Abomey-Calavi, Republic of Bénin.

Okpara dam water contained significant amounts of iron and manganese that could create problems for consumers. This study perfected a deferrization and demanganization technique preceded or followed by oxidation with potassium permanganate. The maximum iron and manganese contents varied from 30 to 50 and 1.5 to 4.5 mg/L respectively. Under the best conditions, simple coagulation-flocculation was allowed to obtain iron removal yields of 18 to 75%. Manganese was eliminated between 8 and 24%. Pre-oxidation with 2.5 mg/L potassium permanganate allowed attaining of about 99% elimination of iron and about 72% of manganese at a pH of 6.5. The treatment plant composed of a clarification stage followed by oxidation with potassium permanganate at pH 8.5 allowed a complete elimination of iron and manganese.

Key words: Surface water, water clarification, chemical coagulant, potassium permanganate, deferrization, demanganization.

INTRODUCTION

Iron, and to a lesser degree manganese, are some of the most abundant elements in the earth's crust. They are found in waters emanating from soil leaching and industrial pollution. These elements pose no danger to human health or to the environment. But they cause esthetic and organoleptic inconveniences. Iron and manganese gives water colour that can stain linen and sanitary appliances. Iron and manganese, when not eliminated, could be progressively oxidized in the distribution network giving water colour, taste, smell, turbidity and favouring the development of micro-organisms with serious consequences for users. In surface or ground waters, one finds iron and manganese

in different chemical forms (dissolved, precipitated, free or complexed) in variable concentrations (Myint and Barry, 1999; Omoregie et al., 2002; Muwanga and Barifaidjo, 2006). They may be present in concentrations of the order of 2 to 5 mg/L of iron and 0.5 to 2 mg/L of manganese (Ellis et al., 2000; Roccaro et al., 2007) or that are markedly higher and capable of attaining 20 mg/L iron and 5 mg/L manganese (Berbenni et al., 2000).

The methods of iron and manganese removal from water consist of transforming the dissolved forms (Fe^{2+} and Mn^{2+}) by oxidation, into precipitates ($Fe(OH)_3$ and MnO_2) followed by filtration. Oxidation can be carried out using powerful chemical oxidants like oxygen, chlorine, chlorine dioxide, ozone or potassium permanganate (El Araby et al., 2009; Katsoyiannisa et al., 2008) or biologically (Katsoyiannisa et al., 2008; Qin et al., 2009; Tekerlekopoulou et al., 2008; Tekerlekopoulou and Vayenas, 2007; Burgera et al., 2008). Predominantly,

---

*Corresponding author. E-mail: bawamoktar@yahoo.fr.

physical processes like clarification and adsorption on precipitates or activated carbon are evoked (Aziza and Smith, 1992; Llofd et al., 1983; Okoniewsk et al., 2007). Simple membrane techniques often preceded by oxidation are also cited among the methods of water deferrization and demanganization (Ellis et al., 2000; Teng et al., 2001; Choo et al., 2005). The plant treatment of Okpara water dam which consisted of a preoxidation phase with chlorine, a clarification phase with aluminium sulphate and a disinfection phase with chlorine did not allow a sufficient removal of iron and manganese contained in the water.

The objective of this work was to study a process for the optimum removal of iron and manganese contained in the dam water and to propose a scaled up process with procedures for dealing with the water treatment problem. The coagulant used was aluminium sulphate and the oxidizing agent, potassium permanganate.

MATERIALS AND METHODS

Study area

The dam is built at Parakou on a tributary of the Ouémé River called Okpara, in the North-East of Benin 450 km from Cotonou. The watershed of this dam is situated in the eastern part of the town and extends to the districts of Tchaourou, Pèrèrè, Nikki, N'dali, then to part of the south-east of Bembereke district. Parakou is situated at latitude 9°21′ north and longitude 2°36′ east. The dam supplies the potable water treatment plant of the town.

Methods of analysis and testing

Raw water samples were taken from the surface (0.2 m) or at depths of from 2.0 to 4.5 m in 25 L plastic containers on the day of testing. The electrical conductivity and pH of the samples were measured using an LF 340-A/SET conductimeter, and a WTW pH 340/ion SET pH-meter, respectively. Iron and manganese were measured using the orthophenanthroline at 510 nm and the acid medium potassium periodate at 525 nm methods respectively. A DR 4000U spectrophotometer model 48100 was used to measure absorbance. Alkalinity was obtained by volumetric titration. The solutions were prepared from bidistilled water. Orthophenantroline (> 99%) and potassium periodate (> 99%) were purchased respectively from Acros Organics and Pancreac Quimica SA. The methods of analysis were based on AFNOR standards (Association Française de Normalisation). Oxidation with potassium permanganate was done in beakers containing the water to be treated and into which variable amounts of the stock solution (1% $KMnO_4$) prepared from crystals of 99% $KMnO_4$ (from Merck) were introduced.

Coagulation–flocculation tests were carried out according to the jar test protocol using an Orchidis 6-post miniflocculator equipped with a time switch and a tachometer. The volumes of water in the beakers were fixed at 1 L each. The different coagulation-flocculation phases consisted of a rapid agitation phase at a speed of 150 rpm for 3 min and a slow agitation phase at a speed of 25 rpm for 15 min. The samples were then allowed to settle for at least 30 min and the settled water filtered using a 0.22 μm pore size Whatman filter paper. The addition of reagents like the coagulant made of aluminium sulphate ($Al_2(SO4)_3.18H_2O$) and lime for pH correction took place during the rapid agitation phase. Aluminium

sulphate (17 to 18% as Al_2O_3) and lime (> 92%) were purchased respectively from Société Tunisienne des Produits Alumineux, STPA and Balthazard et Cotte. The raw water tested did not undergo any pretreatment. The tests took place within 24 h of sampling.

EXPERIMENTAL RESULTS

Iron and manganese contents of the dam water

The raw water was characterized by low total salinity (60 to 120 μscm^{-1}) and low alkalinity (30 to 50 mg/L $CaCO_3$). The pH varied from 6.7 to 5.6. The seasons of the year had a notable effect on the characteristics of the water. We monitored variations in iron and manganese concentrations in time and space. Water samples were taken from the surface at about 0.2 m depth and from a depth of 4.5 m. The results showed low variation in iron and manganese contents at the water surface in all seasons. On the other hand, the concentrations of these metals in depth were practically zero between the months of June and October. One noted an increase in these concentrations from the month of November, reaching a peak between May and June. These periods correspond to the beginning of the big rainy season and the small rainy season. The maximum values for iron varied between 30 and 50 mg/L and manganese 1.5 and 4.5 mg/L.

Figure 1 shows the evolution of dissolved iron and manganese concentrations according to depth; one observed generally that the concentrations of these elements varied little for depths below 2 m. From the 3 m depth, the concentrations increased rapidly, reaching the maximum value at the bottom of the dam. The study showed moreover that, the variation in oxygen concentration was reversed compared to those of iron and manganese. Iron and manganese concentrations were more significant in the zones of low oxygen content. In the superficial more aerated part of the dam, dissolved oxygen could reach 4.5 to 5.0 mg/L. At the 4.5 m depth oxygen content was low or practically zero.

Coagulation-flocculation

Tests were carried out for doses of aluminium sulphate between 10 and 60 mg/L in the water. The residual concentrations of iron and manganese were measured on decanted and filtered water. The results without pH adjustment in Figure 2 shows that, clarification allowed eliminating part of the iron and manganese. The residual concentration of iron was in the order of 1.5 to 2.0 mg/L for an initial concentration of 9.1 mg/L and a coagulant dose between 50 and 60 mg/L. For the same doses of coagulant, the manganese concentration was brought down from 2.9 to about 2.3 mg/L. The same experiments were carried out at pH 6.5. The results (Figure 3) show

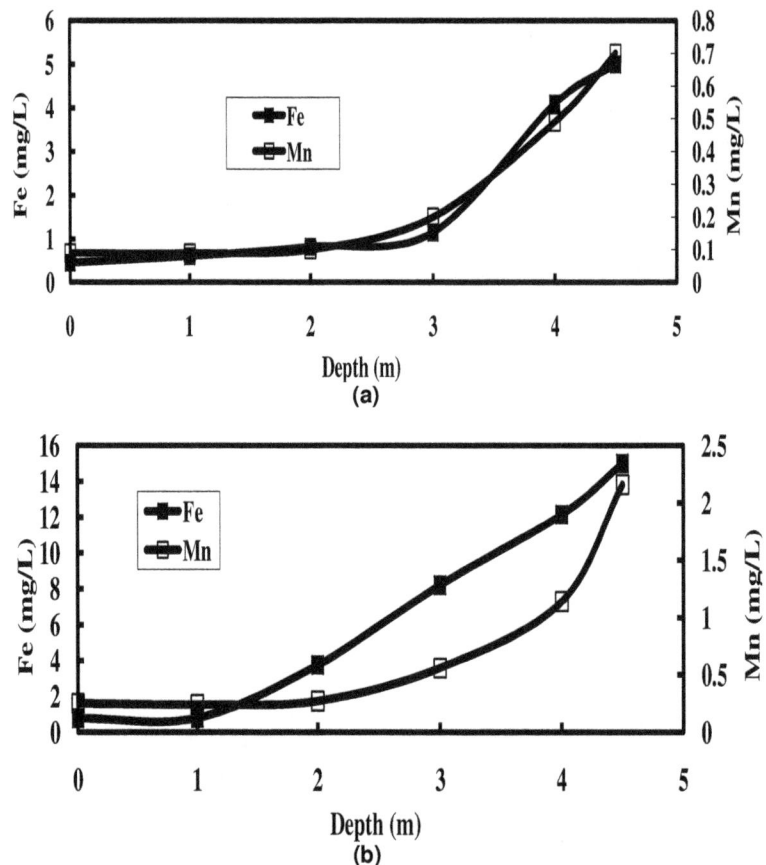

Figure 1. Evolution of iron and manganese concentrations as a function of depth in July (a) and June 2006 (b).

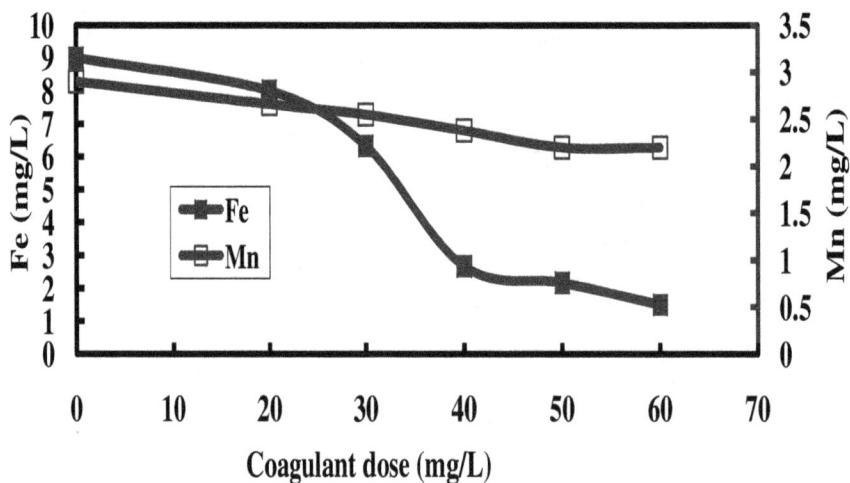

Figure 2. Removal of iron and manganese by coagulation without pH adjustment.

that at this pH, the elimination yields for iron and manganese improved significantly. Using between 40 and 60 mg/L of coagulant, the residual concentrations of iron and manganese were lower than 0.5 and 2.0 mg/L respectively (initial concentrations of 9.1 mg/L of iron and 2.9 mg/L of manganese).

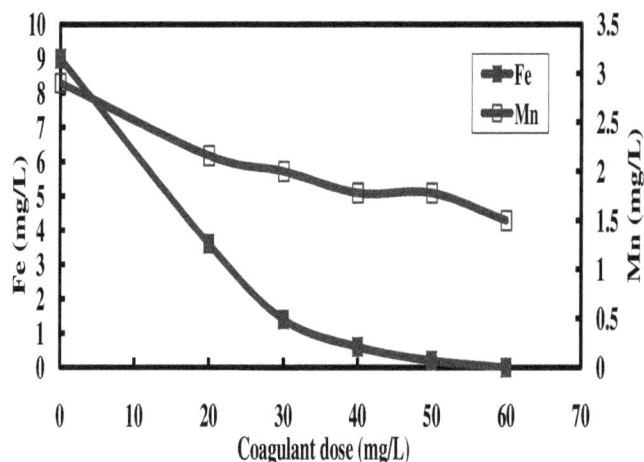

Figure 3. Removal of iron and manganese by coagulation at pH 6.5.

Table 1. Removal yields (%) of iron and manganese by coagulation (20 to 40 mg/L of coagulant) without pH adjustment and at pH 6.5.

Treatment	% Fe removal	% Mn removal
Coagulation without pH adjustment	10 - 73	2 - 24
Coagulation pH = 6.5	18 - 75	8 - 20

Table 1 shows orders of magnitude of the percentages of iron and manganese removed without pH adjustment or pH 6.5 for several values of initial concentrations of iron and manganese in raw water sampled at different depths of the dam. The results showed that, for initial iron concentrations between 8 and 35 mg/L and for a coagulant dose from 20 to 40 mg/L, iron removal yields were between 10 and 73% during treatment without pH adjustment. At pH 6.5, the yields could reach 75%. Manganese was eliminated with a maximum yield of 24% (initial concentrations of manganese varied from 0.9 to 4.1 mg/L).

Influence of pre-oxidation

The potassium permanganate concentrations applied in pre-oxidation varied between 1 and 6 mg/L. The coagulant dose was fixed at 40 mg/L. The tests were carried out without pH adjustment and at pH 6.5. Figures 4 and 5 show that removal yield increased with potassium permanganate dose. In particular, for a $KMnO_4$ concentration of 4 mg/L and an initial iron concentration of 8.7 mg/L, the final concentration was in the order of 2.2 mg/L without pH adjustment and 2.0 at pH 6.5. Under these conditions, the residual manganese concentrations

Table 2. Removal yields (%) of iron and manganese, using 40 mg/L of coagulant without pH adjustment, pH 6.5 and a $KMnO_4$ dose of 2.5 mg/L.

Fe (mg/L)	Mn (mg/L)	without pH adjustment		pH = 6.5	
		% Fe	% Mn	% Fe	% Mn
8.0	1.1	98	66	99	72
17.0	0.9	74	51	95	61
24.0	2.0	76	28	85	35
30.0	3.5	75	26	80	29

were 0.4 and 0.3 mg/L without pH adjustment and pH 6.5 respectively (initial concentration of 1.8 mg/L).

The results showed moreover that, with larger doses (5 to 6 mg/L) of $KMnO_4$, one could completely remove iron and manganese from water. Table 2 shows the removal yields for a dose of 40 mg/L of coagulant and a pre-oxidation dose of 2.5 mg/L $KMnO_4$. At pH 6.5, the yields varied from 80 to 99% for iron concentration from 8 to 30 mg/L. For manganese, the highest yields (60 to 72%) were obtained when the initial concentration was between 0.9 and 1.1 mg/L. For larger (4 mg/L of $KMnO_4$) oxidant doses and for raw water concentrations of from 8 to 17 mg/L of iron and from 0.9 to 1.1 mg/L of manganese, the removal yields were greater than 99% for iron and between 94 and 96% for manganese.

Clarification and oxidation at pH = 8.5

To improve removal yields of the elements (manganese in particular) we implemented the following treatment scheme, based on optimal coagulation-flocculation conditions:

1. coagulation at pH 6.5 with 40 mg/L of aluminium sulphate
2. oxidation of coagulated and decanted water with potassium permanganate at pH 8.5
3. filtration and measurement of iron and manganese concentrations.

The results show a clear improvement in removal yields (Figures 6 and 7). Manganese, even at high concentration (3.5 mg/L) was completely removed from the water with oxidant doses of between 3 and 4 mg/L. The sequence/ clarification then oxidation and filtration (two filters)/ was more efficient than the sequence/ oxidation and clarification (only one filter). Table 3 shows the evolution of removal yields. One observed that manganese was removed at between 65 and 95% for concentrations varying between 0.9 and 3.5 mg/L. Iron was completely removed from the water for concentrations

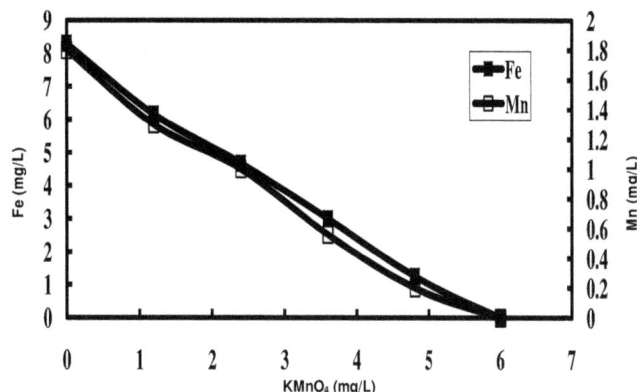

Figure 4. Influence of pre-oxidation on iron and manganese removal by coagulation without pH adjustment.

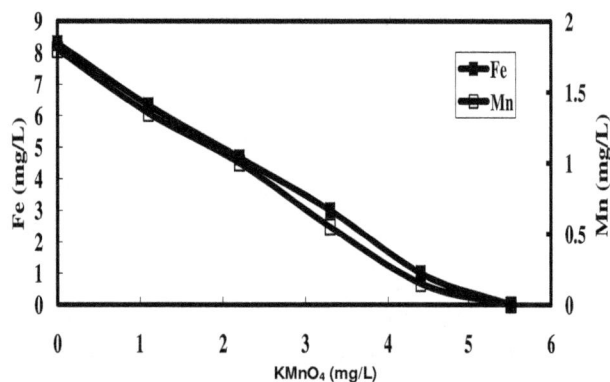

Figure 5. Influence of pre-oxidation on iron and manganese removal by coagulation at pH 6.5

Figure 6. Removal of manganese with $KMnO_4$. Initial water contents: 0.9 mg Mn/L, 17 mg Fe/L (TI : oxidation by $KMnO_4$ of coagulated water with 40 mg/L at pH 6.5 ; T2 : coagulated water with 40 mg/L at pH 6.5, decanted, filtered and oxidized with KMnO4 at pH 6.5 and filtered).

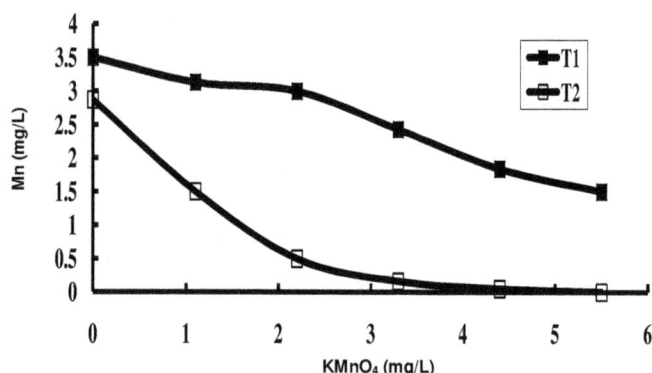

Figure 7. Removal of manganese with $KMnO_4$. Initial water contents: 3.5 mg Mn/L, 30 mg Fe/L (TI : oxidation by $KMnO_4$ of coagulated water with 40 mg/L at pH 6.5 ; T2 : coagulated water with 40 mg/L at pH 6.5, decanted, filtered and oxidized with $KMnO_4$ at pH 6.5 and filtered).

of potassium permanganate of about 2.5 mg/L; higher concentrations than 2.5 mg/L (3 to 4 mg/L) are required to completely eliminate manganese.

DISCUSSION

Mechanisms of iron and manganese removal

Without pH adjustment or at pH 6.5 we determined iron and manganese removal yields by coagulation. The influence of pH was not systematically studied because several studies show that the range of pH, favourable for coagulation with aluminium sulphate lies between 5.0 and 7.0 (Bawa et al., 1998a; Van Benschoten and Edzwald, 1990; Gregor et al., 1997; Zhong et al., 2010). The study showed that coagulation allowed the removal of a significant proportion of iron and manganese. The eliminated fractions took different forms (hydroxide, oxide) sometimes associated with solid particles like clays. We observed that iron removal declined steadily as

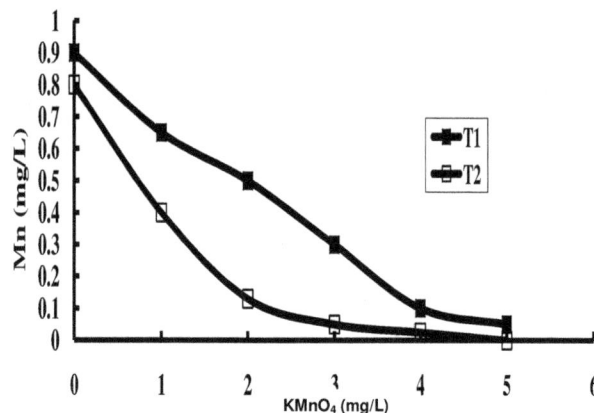

the dose of coagulant increased and the elimination of manganese tended to stabilize from 30 to 40 mg/L of coagulant applied. Higher doses of up to 60 mg/L did not improve manganese elimination, neither did it cause a restabilization of flocs. Knocke et al. (1992) obtain a 63% total iron removal during the coagulation of raw river water using aluminium sulphate (coagulant dose of from 40 to 50 mg/L, pH between 6.3 and 6.5, initial iron concentration from 0.6 to 2.4 mg/L) and Montiel and Welte, (1990) obtained a manganese elimination yield of 25 to 35% (initial concentration of 0.5 to 1.0 mg/L) during coagulation using ferric chloride in the presence of alginate at pH between 7.6 and 7.8.

Many mechanisms of particle elimination by coagulation such as complexation between the hydrolyzed soluble forms of metals and organic or colloidal

Table 3. Removal yields (%) of iron and manganese for 40 mg/L of coagulant at pH 6.5 and a $KMnO_4$ dose of 2.5 mg/L at pH 8.5.

Fe (mg/L)	Mn (mg/L)	% Fe	% Mn
8.0	1.1	> 99	97
12.2	1.1	> 99	95
17.0	0.9	> 99	75
24.0	2.0	> 99	79
30.0	3.5	> 99	69

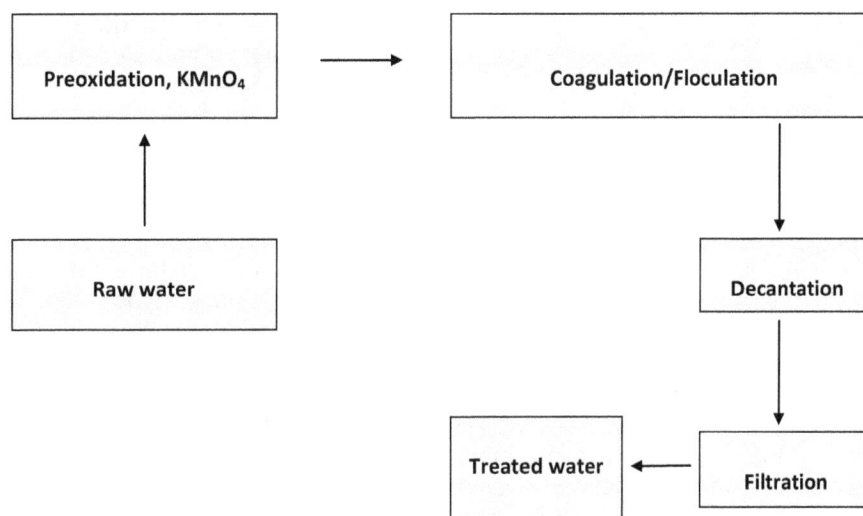

Figure 8. Structure of the treatment plant (plant a).

matter, double layer reduction, adsorption on flocs of metallic hydroxides, charge neutralization, particle trapping in a mesh and co- precipitation are generally cited in coagulation-flocculation studies. The conditions of treatment, particularly the pH, may favour some approaches. Thus, Jian-Jun et al. (2006) and Zhong et al. (2010) showed that, between pH 4.5 and 5.2, the contribution of the charge neutralization mechanism is more important.

At pH greater than 5.5, the contribution of coagulation by neutralization declined and the mechanism of coagulation by adsorption combined with trapping and co-precipitation dominated. It is therefore probable that in this study, the mechanism of particle neutralization was dominant during coagulation without pH adjustment (the pH decreased rapidly after coagulant addition from 5.4 to 4.5). The mechanism by adsorption was dominant at pH 6.5 (the pH having varied between 6.5 and 6.0). It was recently shown that at pH between 4.5 and 9.0, aluminium in the form of sulphate in water hydrolyzes forming 8 products of hydrolysis (Al^{3+}, $Al(OH)^{2+}$, $Al(OH)_2^+$, $Al(OH)_3$, $Al(OH)_4^-$, $Al_2(OH)_2^{4+}$, $Al_3(OH)_4^{5+}$ and AlO_4 $Al_{12}(OH)_{24}(H_2O)^{7+}_{12}$). The solid form of aluminium hydroxide $Al(OH)_3$ is the dominant type of the total

aluminium added to water and the cation containing 13 atoms of aluminium (AlO_4 $Al_{12}(OH)_{24}(H_2O)^{7+}_{12}$) is most important for the neutralization and destabilization of colloids which are generally negatively charged (Xiao et al., 2008). Franceshi et al. (2002) also showed that between pH 6 and 8, the mechanism of clay elimination, responsible for water turbidity, is adsorption and neutralization. With regard to the coagulation-flocculation carried out at pH 6.5, pre-oxidation allowed the increase of removal yield from 50 to 95%. Iron was practically eliminated. Manganese was still present in high residual concentrations. It was shown that potassium permanganate was a good chemical reagent for the oxidation of manganese in water. But pH is a limiting factor in this treatment. It is necessary to operate at a pH greater than 7.0 (Doré, 1989; EPA, 1999; Roccaro et al., 2007).

In these conditions, potassium permanganate is more stable in water and reacts with its highest oxidation state. Roccaro et al. (2007) obtained a 95% removal yield for manganese in underground water, after oxidation with potassium permanganate at pH 8.5 and an initial manganese concentration of about 2 mg/L. Moreover, studies show that the oxidized forms of iron ($Fe(OH)_3$)

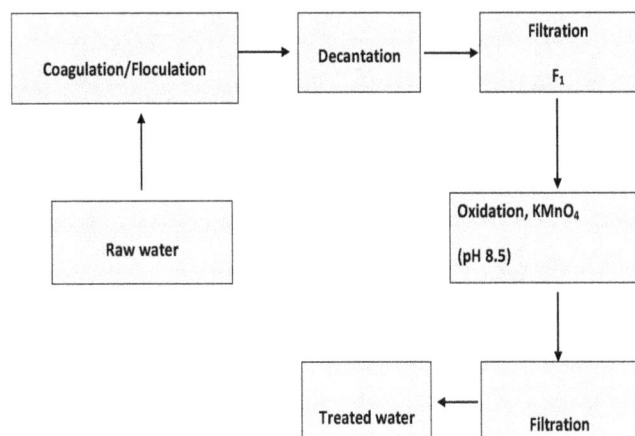

Figure 9. Structure of the treatment plant (plant b).

and manganese (MnO_2), like aluminium hydroxide ($Al(OH)_3$), are good adsorbents (Mazet et al., 1990; Berbenni et al., 2000). It is therefore probable that oxidized forms of iron and manganese, in playing an adsorptive role, might have contributed a lot to improving the efficiency of coagulation (coagulation by adsorption, trapping and co-precipitation).

Next, we treated the water, associating physicochemical treatments with oxidation, using potassium permanganate ($KMnO_4$) at pH 8.5. The treatment plant included first, a clarification stage (coagulation with aluminium sulphate at pH 6.5, decantation and filtration) and second, an oxidation stage with potassium permanganate at pH 8.5 followed finally by filtration. This plant allowed a complete removal of iron and manganese for coagulant doses of 40 mg/L and potassium permanganate concentrations greater than 2.5 mg/L. The first module of the plant allowed clarifying water and reducing high molecular weight organic compounds like humic substances (Bawa et al., 1998b; Van Benschoten and Edzwald, 1990; Jian-Jun et al., 2006), the second module allowed, under the best conditions, the oxidation of iron and particularly manganese.

Proposed treatment plants

We tested treatment plants including, firstly, physicochemical (coagulation/flocculation/decantation/ filtration) and secondly, chemical oxidation procedure using potassium permanganate (Figures 8 and 9). With plant a), the results showed that it was possible to get a complete removal of iron and manganese, if their initial concentrations were low (< 8 mg/L iron and < 1 mg/L manganese). In the dam, the lowest iron and manganese concentrations were encountered in the superficial parts (depths < 2 m). We therefore concluded that when the

dam water was sampled at shallow depth, plant a) could allow attaining the objectives of the treatment.

Plant b) represented in Figure 9 included a clarification phase with coagulation at pH 6.5 and an oxidation phase with potassium permanganate ($KMnO_4$) at pH 8.5. Potassium permanganate oxidation was preceded by one filtration, followed by a second filtration stage. The first stage eliminated the residual flocs resulting from the physicochemical treatment and the second allowed to retain the metallic oxides and hydroxides of iron and manganese (mainly $Fe(OH)_3$ and MnO_2). The results showed that this plant could also be used but it was indispensable that the decantation and filtration phases after coagulation-flocculation be well executed, to ensure a complete removal of flocs, for if they escaped to the clarification module, they could reach the $KMnO_4$ oxidation post at pH 8.5. At this pH, the residual flocs, rich in aluminium salts, are restabilized, water is recoloured and soluble aluminium increases in the treated water. Actually at pH 8.5, the dominant form of aluminium in the water is no longer the solid $Al(OH)_3$ but the very soluble type, the aluminate AlO_2^{2-}, (Van Benshoten and Edzwald,1990, Chow et al., 2009). In fact, the sampling and implementation conditions, if respected, one could supply water that conforms to the norms of portability with regard to iron particularly to manganese (< 0.4 mg/L) (WHO, 2006).

Conclusion

The water studied contained significant amounts of iron and manganese. For a better appreciation of its quality by consumers, it was necessary to propose a method of deferrization and demanganization. In this study, we tested two treatment plants including an oxidation stage with potassium permanganate. The first plant was adapted for waters with low iron and manganese contents (depths < 2 m) and the second was more suited for waters sampled in depth. In both cases the treated water conformed to the norms of potability relative to iron and manganese.

REFERENCES

Aziza HA, Smith PG (1992).The influence of pH and coarse media on manganese precipitation from water. Water Res., 26(6): 853-855

Bawa LM, Djanéyé-Boundjou G, Boukari Y, Agbozo AK (1998a). Optimisation de la clarification d'une eau de surface. Univ. Bénin,Togo, J. Rech. Sci., 2(1): 88-95.

Bawa LM, Boukari Y, Djanéyé-Boundjou G, Passem A (1998b).Coagulation d'une eau de surface et de solutions d'un acide fulvique par le sulfate d'aluminium: Influence du charbon actif en poudre". Environ. Technol., 19: 717-724.

Berbenni P, Pollice A, Canziani R, Stabile L, Nobili F (2000). Removal of iron and manganese from hydrocarbon-contaminated groundwaters. Bioresour. Technol., 74: 109-114.

Burgera MS, Mercera SS, Shupe GD, Gagnon GA (2008). Manganese removal during bench-scale biofiltration. Water Res., 42(19): 4733-4742.

Choo KH, Lee H, Choi SJ (2005).Iron and manganese removal and membrane fouling during UF in conjunction with prechlorination for drinking water treatment. J. Memb. Sci., 267: 18-26

Chow CWK, van Leeuwena JA, Fabris R, Drikas M (2009).Optimised coagulation using aluminium sulfate for the removal of dissolved organic carbon, Desalination, 245(1-3): 120-134.

Doré M (1989). Chemistry of the oxidants and water treatment, Tech and Doc, Paris, Lavoisier, pp. 396-406.

El Araby R, Hawasha S, El Diwania G (2009).Treatment of iron and manganese in simulated groundwater via ozone technology. Desalination, 249(3): 1345-1349.

Ellis D, Bouchard C, Lantagne G (2000).Removal of iron and manganese from groundwater by oxidation and microfiltration. Desalination, 130: 255-264.

EPA (1999). Alternative disinfectants and oxidants. EPA guidance manual., April, 5-1 to 5-15.

Franceschi M, Girou A, Carro-Diaz, AM, Maurette MT, Puech-Costes E (2002). Optimisation of the coagulation-flocculation process of raw water by optimal design method. Water Res., 36: 3561-3572.

Gregor JE, Nokes CJ, Fenton E (1997).Optimising natural organic matter removal from low turbidity waters by controlled pH adjustment of aluminium coagulation . Water Res., 31(12): 2949-2958

Jian-Jun Q, Maung HO, Kiran AK, Frans K, Peter M (2006).Impact of coagulation pH on enhanced removal of natural organic matter in treatment of reservoir water . Sep. Purif. Technol., 49(3): 295-298.

Katsoyiannisa IA, Zikoudib A, Huga JS (2008).Arsenic removal from groundwaters containing iron, ammonium, manganese and phosphate: A case study from a treatment unit in northern Greece. Desalination, 224(1-3): 330-339.

Knocke WR, Conley L, Van Benschoten JE (1992).Impact of dissolved organic carbon on the removal of iron during water treatement. Wat. Res., 26(1): 515-1522.

Mazet M, Angbo L, Serpaud B (1990). Adsorption of humic acids onto preformed aluminium hydroxide flocs. Wat. Res., 24(12): 1509-1518.

Montiel A, Welte B (1990). Manganese in water: Manganese removal by biological treatment.Rev. Sci. Eau., 3 : 469-481.

Muwanga A, Barifaidjo E (2006). Impact of industrial activities on heavy metal loading and their physico-chemical effects on wetlands of lake Victoria basin (Uganda). Afr. J. Sci. Tech., 7(1): 51-63.

Myint Z, Barry C (1999). Iron and manganese dynamics in lake water Water Res., 33(8): 1900-1910

Llofd A, Grzeskowiak R, Mendham J (1983). The removal of manganese in water treatment clarification processes. Water Res., 17(11): 1517-1523.

Okoniewsk E, Lacha J, Kacprzaka, Neczaja E (2007).The removal of manganese, iron and ammonium nitrogen on impregnated activated carbon. Desalination, 206(1-3): 251-258

Omoregie E, Okoronkwo MO, Eziashi AC, Zoakah AI (2002). Metal concentrations in water column; Benthic macro invertebrates and tilapia from Delimi river, Nigeria. J. Aquat. Sci., 17(1): 55-59.

Qin S, Mab F, Huang P, Yang J (2009).Fell and Mnll removal from drilled well water: A case study from a biological treatment unit in Harbin. Desalination, 245(1-3): 183-193.

Roccaro P, Barone C, Mancini G, Vagliasindi FGA (2007).Removal of manganese from water supplies intended for human consumption: a case study. Desalination, 210: 205-214.

Tekerlekopoulou AG, Vasiliadoua IA, Vayenas DV (2008).Biological manganese removal from potable water using trickling filters. Biochem. Eng. J., 38(3): 292-301.

Tekerlekopoulou AG, Vayenas DV (2007).Ammonia, iron and manganese removal from potable water using trickling filters. Desalination, 210(1-3): 225-235.

Teng Z, Huang JY, Fujita K, Satoshi T (2001).Manganese removal by hollow fiber micro-filter. Membrane separation from drinking water. Desalination, 139: 411-418.

Van Benschoten JE, Edzwald JK (1990). Chemical aspects of coagulation using aluminum salts—II. Coagulation of fulvic acid using alom and polyaluminum chloride. Water Res., 24(12): 1527-1535.

WHO (World Health Organisation) (2006) Guidelines for drinking water quality, Vol 1, Recommendation,Geneva, Switzerland 3rd ed., pp. 296-460.

Xiao F, Zhang B, Lee C (2008).Effects of low temperature on aluminum III) hydrolysis: Theoretical and experimental studies. J. Environ. Sci., 20: 907-914.

Zhong LY, Bao-Yu G, Qin-Yan Y, Yan W (2010). Effect of pH on the coagulation performance of Al-based coagulants and residual aluminium spéciation during the treatment of humic acid–kaolin synthetic water. J. Hazard. Mater., 178(1-3): 596-603.

18

Monitoring of basic and acid radicals load in main canal, Giza governorate: A risk to health of consumers

Abd El-Moneim M. R. Afify, Sayed A. Fayed and Emad A. Shalaby*

Biochemistry Department, Faculty of Agriculture, Cairo University, Giza, Egypt, 12613.

The main canal (Masraf El-Moheet) is one of the largest canals in Giza governorate (Egypt). Due to the extended domestic activities and urbanization as well as the continuous industrial and agricultural growth of the region, water quality is potentially changing and causes fish floating and death. This study was conducted to measure toxic metals concentrations in water samples along the main channel using ICP Spectrometer (iCAP 6000 Series; Thermo Scientific). The most pronounced feature is the highest concentration of Fe, NH$_4$, Al, Pb and Cr in main canal (35.28, 11.55, 9.318, 0.386 and 0.748 mg L^{-1}, respectively) when compared with World Health Organization, United States Environmental Protection Agency and Egyptian Organization for Standardization and Quality Control. Analysis of variance (ANOVA) of metals levels showed significant difference among the regions. The present results showed that the collected water samples based on the higher levels of metal accumulation could be the first reason for fish floating and death in these regions, in addition to unsafe use for human consumption.

Key words: Water pollution, heavy metals, anions radicals, fish floating, Giza canals.

INTRODUCTION

The poor quality of available water supplies is a major environmental concern and it is one that afflicts the world, including Egypt. Water contamination by point sources of pollution from industrial, domestic and/or agricultural discharge occurs in many parts of Egypt, particularly in the rural areas. Many industries discharge inadequately treated-wastes into waterways network which include irrigation canals and drains. Water degradation from non point sources of pollution, due to the agrochemicals improper management by the farmers, mainly fertilizers and pesticides, is already a serious problem in the Nile Delta Region (Watts and El-Katsha, 1995). Water pollution, if allowed to grow uncontrolled, is likely to cause substantial economic and health ramifications to the country. Within this context, it is estimated that the proportion of urban population in Egypt having access to safe drinking water dropped from 93% in 1982-85 to 82% in 1990-96, and the proportion of the rural population from 61 to 50% while sanitation services were available

for only 23% of the urban population and 6% of the rural population (World Bank, 2000). Polluted water related diseases are very frequent in Egypt, accounting for 90,000 deaths each year. Outbreaks of diseases such as cholera, typhoid, and infectious hepatitis have occurred in several provinces because of drinking water pollution. For instance, 10% of the rural population in 1995 was affected by diseases as a result of contamination of surface water with human waste. The contamination of those surface and ground waters could also affect the food chain causing deadly intestinal diseases or poisoning (Watts and El-Katsha, 1995).

Chemicals derived from agricultural activities (pesticides, fertilizer and herbicides) and industrial effluents, such as metals, ultimately find their way into a variety of different water bodies and can produce a range of toxic effects in aquatic organisms, ranging from alterations to a single cell, up to changes in whole populations (Bernet et al., 1999; Marcovecchio et al., 2007; Al-Kahtani, 2009). The accumulation of toxic metals to hazardous levels in aquatic biota has become a problem of increasing concern. Excessive pollution of surface waters could lead to health hazards in man, either

---

*Corresponding author. E-mail: dremad2009@yahoo.com.

through drinking of water and/or consumption of fish (Mathis and Cummings, 1973). It seems without sense the increasing importance of fish as a source of protein and the interest in understanding the accumulation of heavy metals at the tropic levels of the food chain, extend the focus towards finfish (Obasohan, 2007). Pollution enters fish through five main routes: Via food or non-food particles, gills, oral consumption of water and the skin. On absorption, the pollutant is carried in blood stream to either a storage point or to the liver for transformation and/or storage. Pollutants transformed in the liver may be stored there or excreted in bile or transported to other excretory organs such as gills or kidneys for elimination or stored in fat, which is an extra hepatic tissue (Nussev et al., 2000). The concentration of any pollutant in given tissue therefore depends on its rate of absorption and the dynamic processes associated with its elimination by the fish. Stocks of both freshwater and marine fish within Saudi Arabia are increasingly threatened by aquatic pollution, but no data are available on the extent pollution impacts. Fathi and Al-Kahtani (2009) showed that main canal water had an obvious increase in electrical conductivity, Chemical Oxygen Demand (COD), total alkalinity, nitrates, phosphorus, chloride and potassium. These features indicated pollution with organic wastes, increased salinity and deteriorated oxygenated state.

The effect of exposure of organisms to any potentially toxic substance depends on its concentration as well as exposure time. One of the effects frequently observed by environmental agents is the chemical alteration of the DNA affecting vital processes like DNA duplication and transcription, gene regulation and cell division, and leading cells to pathologic processes and/or cell death (Barberio, 2009).

Potentially toxic metals, resulting from some agriculture and industrial activities are one of the most common environmental contaminants and several have been shown to be mutagenic and/or carcinogenic in both human and animal studies (Egito et al., 2007)

Heavy metal can cause serious health effects with varied symptoms depending on the nature, quantity and exposure time of the metal ingested (Adepoju-Bello and Alabi, 2005; Adepoju-Bello et al., 2009). They produce their toxicity by forming complexes with proteins, in which carboxylic acid (–COOH), amine (–NH2), and thiol (–SH) groups are involved. These modified biological molecules lose their ability to function properly and result in the malfunction or death of the cells. When metals bind to these groups, they inactivate important enzyme systems or affect protein structure, which is linked to the catalytic properties of enzymes. This type of toxin may also cause the formation of radicals which are dangerous chemicals that cause the oxidation of biological molecules.

The most common heavy metals that humans are exposed to are Aluminium, Arsenic, Cadmium, Lead and Mercury. Aluminium has been associated with Alzheimer and Parkinson's disease, senility and presenile dementia.

Arsenic exposure can cause among other illness or symptoms, cancer, abdominal pain and skin lesions.

Cadmium exposure produces kidney damage and hypertension. Lead is a commutative poison and a possible human carcinogen (Bakare-Odunola, 2005) while for Mercury, toxicity results in mental disturbance and impairment of speech, hearing, vision and movement (Hammer and Hammer, 2004). In addition, Lead and Mercury may cause the development of autoimmunity in which a person's immune system attacks its own cells. This can lead to joint diseases and ailment of the kidneys, circulatory system and neurons. At higher concentrations, Lead and Mercury can cause irreversible brain damage.

In the present work, heavy metals and other metals accumulation levels as well as acidic radical in water samples were determined from Giza canals to identify the first reason for fish floating or death in this area.

MATERIALS AND METHODS

Sample collection and location

Water samples were randomly collected from 12 sampling sites (from Masraf El Moheete) in 7 different areas: 1-Sugar Company (El-hawamdya, 3 samples), 2- Sewage treatment region (3 samples), 3- Sakara canal (2 samples), 4- Abou Cir canal, 5- Shapramant canal, 6- El maryotia canal, 7- El harania canal. The samples were collected during the month of February 2010. The collected water samples individually were placed in clean bags and immediately taken to the laboratory where they were kept at 4°C in refrigerator until used for analysis.

Sample digestion

For the analysis of total ions (dissolved and suspended), water (200 mL) samples were digested with 5 mL of di-acid mixture (HNO3:HClO4 : 9: 4 ratio) on a hot plate and filtered by Whatman No. 42 filter paper and made up the volume to 50 mL by double distilled water for analysis of 19 ions (APHA, 1995).

Electric conductivity (EC) and pH values of water samples

The pH of water samples was determined using pH-meter with electronic glass electrode (LI 127 of Elico, India) and conductivity (EC) was measured by conductivity-meter (Systronics 304). (Kar et al., 2008).

Analysis of basic radicals

The digested water samples were analyzed for the presence of aluminum, cadmium, chromium, iron, magnesium, manganese, strontium, zinc and lead using the ICP Spectrometer (iCAP 6000 Series; Thermo Scientific). The calibration plot method was used for the analysis (APHA, 1995).

Analysis of acid radicals

The digested water samples were analyzed for the presence of

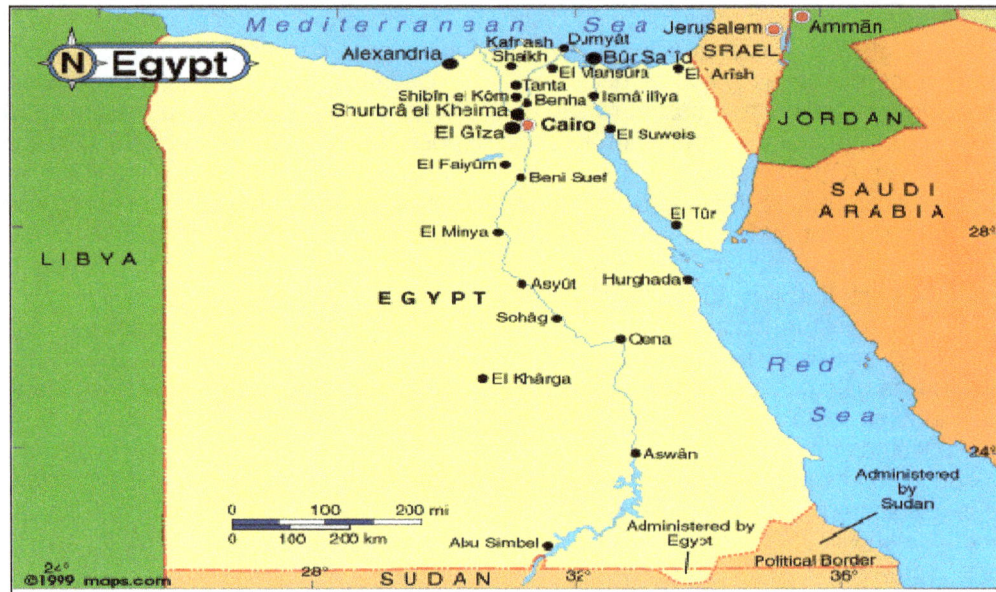

Figure 1. Map of the study area the arrow (↑) pointes the place of the study.

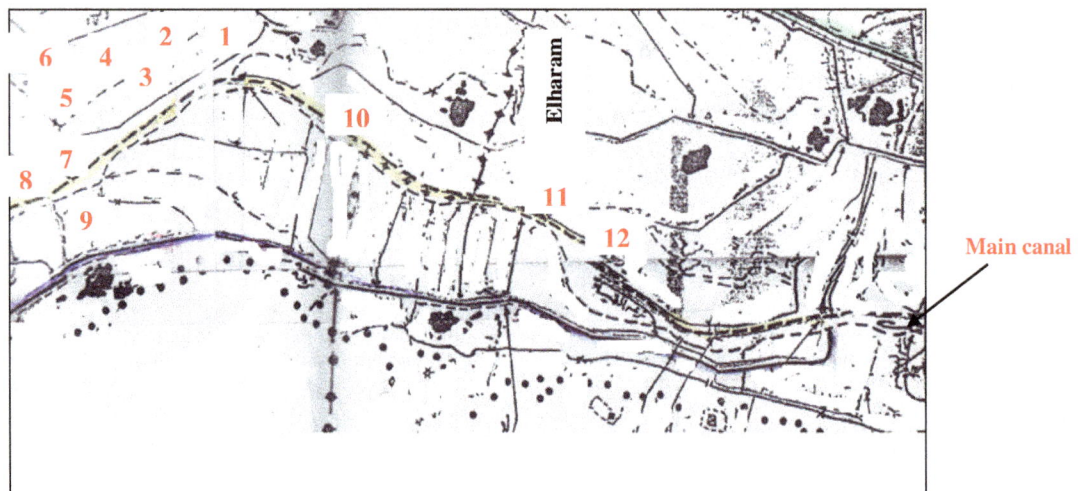

Figure 2. Map of the main canal, Giza governerate. 1- Sugar company (El Hawamdya) after treatment; 2- Sugar company (El Hawamdya) before treatmen; 3- outside the wall of sugar company; 4- Sewage treatment station before treatment; 5- Sewage treatment station after treatment; 6- Sample from canal (Sewage treatment station); 7- Sakara canal; 8- Sakara canal station 2; 9- Abou Cir canal.; 10- Shapramant canal; 11- El maryotia canal; 12- El harania.

chloride (calorimetrically and titer metrically) according to Clarke (1950); Phosphorus (colorimeterically) according to El-Merzabani et al. (1977); sulphate was determined accordingto Furniss et al. (1989); nitrate was determined according to Cataldo et al. (1975) and carbonate and bicarbonate were determined according to Jackson (1967).

Statistical analysis

Data were subjected to an analysis of variance, and the means were compared using the Least Significant Difference (LSD) test at the 0.05 level, as recommended by Snedecor and Cochran (1982).

RESULTS AND DISCUSSION

Electric conductivity as mmos/cm and pH values of water samples collected from different regions (Figures 1 and 2) on main canal is presented in Table 1. The data show that there is difference in pH values in all regions (ranged

Table 1. The physical properties of water samples collected from main canal at different regions.

Sample no.	Area classification	EC mmos/cm	pH	Color
1	Industrial area	4.186^c	b7.79^b	Brown
2	Industrial area	4.65^a	4.35 ^{h}	Brown
3	Industrial area	4.30^b	7.47 ^{de}	Brown
4	Agricultural area	2.47^e	7.50 ^{d}	Light brown
5	Agricultural area	1.93^h	6.88 ^{g}	Light brown
6	Agricultural area	1.95^h	7.50^d	Light brown
7	Residential area	0.94^i	7.437^e	Brown
8	Residential area	2.56^d	8.10^a	Light brown
9	Agricultural area	2.25^g	7.52^{cd}	Yellow
10	Agricultural area	0.867^j	7.32 ^{f}	Black
11	Residential area	1.946^h	6.83 ^{g}	Black
12	Agricultural area	2.33^f	7.57 ^{c}	Black
LSD at 0.05		0.05329	0.05329	

Each value is presented as mean of triplet treatments. Values with different superscript letters within the same column are significantly different ($P < 0.05$).

Figure 3. Disaster of the fish floating or death in the main canal, Giza governorate.

from 4.35 and 8.10) but the EC value ranged between 0.867 at region 10 to 4.65 mmos/cm at region 2. The most pronounced feature is the highest values of EC in the first three regions (El-Hawmdya Sugar Company). In addition, the color of the collected samples were ranged from yellow at region 9 to black at region 10, 11 and 12 and these variables in water physical parameter may be due to the amount and kind of contamination from different factories in this area (Figure 3).

The mean concentration of the heavy metals and other ions in main canal in Giza governorate compared with

World Health Organization (WHO, 1985), United States Environmental Protection Agency (USEPA, 1986) and Egyptian Organization for Standardization and Quality Control (EOSQC, 1991) allowable limits are presented in Tables 2 and 3. The data in Table 3, show that the concentration of cadmium in water ranged between 0.002 mg L^{-1} at region 5, 6, 7, 8, 10 and 12 to 0.012 mg L^{-1} at region 3, but the concentration of copper in water ranged between 1.218 mg L^{-1} at region 10 to 9.22 mg L^{-1} at region 3. However, aluminum, lead, iron and magnesium concentrations at region 1 were found to be much higher

Table 2. The acid radical content (as µg/ml) of water samples collected from main canal at different regions.

Sample no.	S^{--}	Cl^-	SO_4^{--}	NO_3^-	$CO3^{--}$	HCO_3^-	PO_4^{---}	CN^-
1	534[e]	456.41[cd]	1602.2[c]	11.431[d]	ND	1611.35[a]	52.4[b]	ND
2	469.2[d]	521.6[a]	1407.5[d]	26.96[a]	ND	0.0[j]	57.8[a]	ND
3	251.4[h]	476.6[b]	754.3[f]	13.27[c]	ND	1475.25[b]	51.8[b]	ND
4	236.96[j]	245.06[h]	710.85[f]	7.10[g]	ND	1068.3[c]	25.6[d]	ND
5	235.44[k]	269.9[g]	706.28[f]	3.84[i]	ND	577.13[e]	53.2[b]	ND
6	293.52[e]	159.6[i]	880.6[e]	4.64[h]	ND	779.06[d]	53.2[b]	ND
7	275.56[f]	89.9[j]	826.7[e]	3.67[i]	ND	347.26[g]	30.5[cd]	ND
8	239.24[i]	472.1[bc]	717.74[f]	4.64[h]	ND	347.53[g]	36.2[c]	ND
9	788.4[b]	409.19[e]	2365.4[b]	9.55[e]	ND	344.43[h]	28.03[d]	ND
10	858[a]	101.2[j]	2574.2[a]	3.29[j]	ND	339.56[h]	24.2[d]	ND
11	260.6[g]	308.01[f]	782[f]	22.10[b]	ND	201.62[i]	32.3[c]	ND
12	232.68[l]	447.4[d]	698.04[fg]	7.63[g]	ND	388.1[f]	32.7[c]	ND
WHO	50	100	250	50	35	80	0.4	0.0
USEPA	40	100	250	50	50	80	0.4	0.0
EOSQC	40	400	300	40	50	77	0.4	0.0
LSD at 0.05	0.255	18.61	50.869	0.436	NS	45.635	11.4352	NS

Each value is presented as mean of triplet treatments, means within each row with different letters (a-l) differ significantly at P # 0.05 according to Duncan's multiple range test, WHO: World health organization; USEPA: United States environmental protection agency; EOSQC: Egyptian Organization for standardization and quality control.

(9.318, 0.386, 35.28 and 105.3 mg L^{-1} respectively) than other eleven regions. On the other hand, cadmium, chromium and copper were found to be much higher concentration on region 3 (0.012, 0.748 and 9.22 mg L^{-1}, respectively). But strontium and zinc were found to be much higher on region 4 (4.676 and 6.18 mg L^{-1}, respectively).

The most pronounced feature is the highest concentration of iron, ammonia, aluminum, lead and chromium in main canal (35.28, 11.55, 9.318, 0.386 and 0.748 mg L^{-1}, respectively) in comparison to other metals. Analysis of variance (ANOVA) of metals levels showed significant difference among the regions.

The levels of heavy metals recorded in water in this study were generally high, when compared to World Health Organization (WHO, 1985), United States Environmental Protection Agency (USEPA, 1986), and Egyptian Organization for Standardization and Quality Control (EOSQC, 1991) recommended levels in water. The high level of these elements at the mentioned regions could be attributed to discharge of either treated sewage water or re-use drainage water on the main canal In Glza governorate. The data in Table 3 showed that most determined heavy metals had the highest value in the first three regions (industrial area) and these results indicated El-Hawmdya Sugar Company to be one of the main reasons responsible for realizing of heavy metals in the main canals and may lead to the decreasing of available oxygen in these regions and cause fish floating and death.

The crucial roles of zinc and copper in several enzymatic processes are classified as highly toxic metals by Hellawell (1986) and are bio-accumulated in aquatic organisms. On the other hand, organs of aquatic animals may accumulate copper when exposed to toxic concentrations (Mazon et al., 2002), which can lead to redox reactions generating free radicals and, therefore, may cause biochemical and morphological alterations (Monteiro et al., 2005). In addition, Pb++ residues could result in hematological, gastrointestinal and neurological dysfunction in animals. Severe or prolonged exposure to Pb++ may also cause chronic nephropathy, hypertension and reproductive impairment. Pb++ inhibits enzymes, alters cellular calcium metabolism and slows nerve conduction (Lockitch, 1993). In parallel, cadmium is a widespread environmental pollutant that is highly toxic and is considered to have no biological function (Hallenbeck, 1984). It causes severe membrane integrity damage with a consequent loss of membrane-bound enzyme activity which can result in cell death (Younes and Siegers, 1984). This and other nonessential metals have been reported to cause anemia both in mammals (Kostic et al., 1993) and in fish (Gwozdzinski et al. 1992; Al-Kahtani, 2009).

The results of acid radicals analysis of polluted water are shown in Table 2. The data reported that, S-- and SO4-- concentration on region 10 were found to be much higher (858 and 2574.2 mg L^{-1}, respectively) than other region but Cl- and HCO3- concentration were found to be much higher (521, 6 and 1611.35 mg L^{-1} at regions 1 and 2 respectively). On the other hand, CO3-- and CN- are non detectable (ND) in samples analysis. The data

Table 3. The basic radical content (as µg/ml) of water samples collected from main canal at different stations.

Sample no.	Al+++	Cd++	Cr+++	Cu++	Pb++	Fe++	Mg++	Mn++	Sr+++	Zn++	NH4+
1	9.318 [a]	0.01 [b]	0.498 [b]	2.638 [c]	0.386 [a]	35.28 [a]	105.3 [a]	1.42 d[e]	1.666 [d]	2.234 [e]	4.9 [d]
2	3.16 b[cd]	0.004 [b]	0.408 [b]	3.17 [b]	0.132 [cde]	12.383 [bcd]	92.7 [c]	0.338 [i]	0.984 [g]	1.568 [gh]	11.5 [a]
3	2.455 b[cd]	0.012 [a]	0.748 [a]	9.22 [a]	0.378 [b]	30.9 [a]	95.34 [b]	1.306 [e]	1.486 [e]	2.46 [cd]	5.68 [c]
4	2.044 [cd]	0.004 [b]	0.28 [c]	1.434 [ef]	0.142 [cde]	14.554 [bc]	65.24 [d]	4.292 [a]	4.676 [a]	6.18 [a]	3.04 [g]
5	2.278 [bcd]	0.002 [b]	0.234 [c]	1.462 [def]	0.13 [cde]	9.214 [bcd]	38.54 [f]	2.694 [b]	2.998 [b]	3.764 [b]	1.65 [l]
6	4.584 [b]	0.002 [b]	0.25 [c]	1.274 [gh]	0.174 [c]	10.218 [bcd]	35.46 [h]	1.942 [c]	2.1 [c]	2.606 [c]	1.98 [h]
7	1.472 [cd]	0.002 [b]	0.222 [c]	1.448 [de]	0.122 [de]	5.96 [bcd]	26.08 [k]	1.334 [e]	1.59 [d]	2.544 [cd]	1.57 [j]
8	0.964 [bc]	0.002 [b]	0.236 [c]	1.342 [fg]	0.168 [cd]	4.09 [cd]	37.06 [g]	0.9 [f]	1.518 [e]	1.352 [h]	1.98 [h]
9	1.328 [d]	0.004 [b]	0.182 [cd]	1.358 [efg]	0.144 [cde]	3.436 [d]	32.26 [i]	0.652 [g]	1.232 [f]	1.616 [g]	4.09 [e]
10	1.796 [cd]	0.002 [b]	0.252 [c]	1.218 [h]	0.106 [e]	3.272 [d]	21.24 [l]	0.624 [h]	0.912 [g]	1.694 [fg]	1.41 [k]
11	2.456 [bcd]	0.006 [b]	0.24 [c]	1.534 [d]	0.172 [cd]	11.396 [bcd]	57.5 [e2]	1.484 [d]	1.634 [d]	2.396 [de]	9.5 [b]
12	2.232 [cd]	0.002 [b]	0.108 [d]	1.442 [dfg]	0.082 [f]	5.93 [bcd]	26.32 [j]	0.344 [i]	0.806 [h]	1.89 [f]	3.27 [f]
WHO	0.05	0.05	0.05	1.0	0.05	1.0	50	0.01	0.001	5.0	0.1 [m]
USEPA	0.05	0.01	0.05	1.0	0.05	1.0	50	0.05	0.002	1.0	0.1 [m]
EOSQC	0.05	0.005	0.05	1.0	0.1	0.3	100	1.0	0.001	10	0.5 [l]
LSD at 0.05	2.443	0.0533	0.1066	0.1192	0.0532	11.24	0.1685	0.141	0.0923	0.1994	0.0111

Each value is presented as mean of triplet treatments, means within each row with different letters (a-l) differ significantly at P # 0.05 according to Duncan's multiple range test, WHO: World Health organization; USEPA: United states environmental protection agency; EOSQC: Egyptian organization for standardization and quality control.

Table 4. Summary statistics of acid radical's analysis.

	S-	Cl-	SO4-	NO3-	CO3-	HCO3-	PO4-	CN-
Number of samples	12	12	12	12	12	12	12	12
Number with elements detected	12	12	12	12	0.0	11	12	0.0
% detected	100	100	100	100	0.0	91.66	100	0.0
Minimum concentration detected (mg/l)	232.68	89.9	710.85	3.29	-	0.0	24.2	-
Maximum concentration detected (mg/l)	858	521.6	2574.2	26.96	-	1611.35	57.8	-
WHO Maximum Contaminant Level (MCL)	50	100	250	50	35	80	0.4	0.0
Number above MCL	12	11	12	0.0	0.0	12	12	-
% above MCL	100	91.66	100	0.0	0.0	100	100	-

in Table 2 confirmed that the general trend of anions concentration was highly in the first three regions when compared with most 12 regions and these results are correlated with the results obtained with cations concentrations. The summary of statistical analysis of acid and basic radicals are shown in Tables 4 and 5 and the results illustrated that, most of collected samples had ion concentration above WHO maximum contaminant level (MCL) and this causes dangerous state for using this water in all consumption fields (industrial, agriculture) and its

Table 5. Summary statistics of basic radical analysis.

	Al+++	Cd++	Cr+++	Cu++	Pb++	Fe++	Mg++	Mn++	Sr+++	Zn++	NH4+
Number of samples	12	12	12	12	12	12	12	12	12	12	12
Number with elements detected	12	12	12	12	12	12	12	12	12	12	12
% detected	100	100	100	100	100	100	100	100	100	100	100
Minimum concentration detected (mg/l)	1.328	0.002	0.108	1.218	0.082	3.272	21.24	0.338	0.806	1.568	1.41
Maximum concentration detected (mg/l)	9.318	0.012	0.748	9.22	0.386	35.28	105.3	4.292	4.676	6.18	11.5
WHO M aximum Contaminant Level (MCL)	0.05	0.05	0.05	1.0	0.05	1.0	50	0.01	0.001	5.0	0.1
Number above MCL	12	0.0	12	12	12	12	5.0	12	12	1.0	12
% above MCL	100	0.0	100	100	100	100	41.66	100	100	8.33	100

living organisms (e.g. fish). These results are in harmony with the results obtained by Momodu and Anyakora (2010) who reported that high concentration of heavy metals (above WHO) suggests a significant risk to this population, given the toxicity of these metals and the fact that for many, hand dug wells and bore holes are the only sources of their water supply in this environment.

Overall, all samples contained detectable amounts of all the heavy metals studied and in some cases the levels were above WHO specified Maximum Contaminant Level with focused on samples 1, 2 and 3. Eight samples contain high levels of these metals.

Conclusion

The present results showed that, the collected water samples especially from the first three regions (El-Hawmdya Sugar Company) based on the higher levels of metal accumulation could be the first reason for fish death or floating in these regions.

REFERENCES

Adepoju-Bello AA, Alabi OM (2005). Heavy metals: A review. The Nig J Pharm., 37: 41-45

Adepoju-Bello AA, Ojomolade OO, Ayoola GA, Coker HAB (2009). Quantitative analysis of some toxic metals in domestic water obtained from Lagos metropolis. Nig. J. Pharm., 42 (1): 57-60.

Al-Kahtani MA (2009). Accumulation of Heavy Metals in Tilapia Fish (Oreochromis niloticus) from Al-Khadoud Spring, Al-Hassa, Saudi Arabia. Am. J. Appl. Sci., 6 (12): 2024-2029.

APHA (1995). Standard methods for the examination of water and waste water, 19th. Ed, American Public Health Association, American

Water Works Association and Water Environment Federation, Washington, DC.

Bakare-Odunola MT (2005). Determination of some metallic impurities present in soft drinks marketed in Nigeria. Nig. J. Pharm., 4(1): 51-54.

Barberio A (2009). Cytotoxic and genotoxic effects in root meristem of Allium cepa exposed to water of the river Paraíba do Sul - São Paulo state - Tremembé and regions. PhD thesis, University of Campinas - Unicamp, Campinas - São Paulo, Brazil.pp.54-62

Bernet D, Schmidt H, Meier W, Burkhardt- Hol P, Wahli T (1999). Histopathology in fish: Proposal for a protocol to assess DOI: 10.1046/j.1365- 2761.1999.00134.x aquatic pollution. J. Fish Dis., 22: 25-34.

Cataldo DA, Haroon M, Schrader LE, Youngs VL (1975). Rapid calometric determination of nitrate in plant tissues by nitration of salicylic acid. Common Soil Sci. Plant Ana., 6: 71-80.

Clarke FE (1950). Determination of chloride in water improved colorimeteric and titrimetric methods. Anal. Chem., 22(4): 553-555.

Egito LCM, Medeiros MG, Medeiros SRB, Agnez-Lima LF (2007). Cytotoxic and genotoxic potential of surface water from the Pitimbú river, northeastern/RN Brazil. Genet. Mol. Biol., 30: 435-441.

Egyptian Organization for standardization and quality control (1991). Frozen fish, Cairo, Egypt.

El-Merzabani MM, El-Aaser AA, Zakhary NI (1977). Determination of inorganic phosphorus in serum. J. Clin. Chem. Clin. Biochemist., 15: 715-718.

Fathi AA, Al-Kahtani M (2009). Water quality and planktonic communities in Al-khadoud spring, Al-Hassa, Saudi Arabia. Am. J. Environ. Sci., 5: 434-443.

Furniss BS, Hannaford AJ, Rogers V, Smith PWG, Tatchell AR (1989). Vogels textbook of practical Organic Chemistry, pp. 831-832.

Gwozdzinski K, Roche H, Peres G (1992). The comparison of the effects of heavy metal ions on the antioxidant enzyme activities in human and fish Dicentrarchus labrax erythrocytes. Comput. PMID: 1358529 Biochem. Physiol. C., 102: 57-60.

Hellawell JM (1986). Biological Indicators of Freshwater Pollution and Environmental Management. Elsevier Applied Science Publishers Ltd., London and New York, ISBN: 10- 1851660011, p. 546.

Jackson ML (1967). Soil chemical analysis; Hall of India Pvt. Ltd., New Delhi, p. 498.

Kar D, Sur P, Mandal SK, Saha T, Kole RK (2008). Assessment of heavy metal pollution in surface water. Int. J. Environ. Sci. Tech., 5

(1): 119-124.

Kostic MM, Ognjanovic B, Dimitrijevic S, Zikic, RV, Stajn A (1993). Cadmium-induced changes of antioxidant and metabolic status in red cells of rats: In vivo effects. Eur. J. Haematol., 51: 86-92.

Lockitch G (1993). Perspectives on lead toxicity. Clin. Biochem., 26: 371-381.

Hallenbeck WH (1984). Human health effects of exposure to cadmium. Cell. Mol. Life Sci., 40: 136-142. DOI: 10.1007/BF01963576.

Hammer MJ, Hammer Jr MJ (2004). Water Quality. In: Water and W aste W ater Technology. 5th Edn. New Jersey: Prentice-Hall, p. 139-159.

Marcovecchio JE, Botte, SE, Freije RH (2007). Heavy Metals, Major Metals, Trace Elements. In: Handbook of Water Analysis. L.M. Nollet, (Ed.). 2nd Edn. London: CRC Press, pp.275-311.

Mathis BJ, Cummings TF (1973). Selected metals in sediments, water and biota of the Illinois River. J. Water Poll. Cont. Trop., 45: 1573-1583. PMID: 4720140.

Mazon AF, Cerqueira CCC, Fernandes MN (2002). Gill cellular changes induced by copper exposure in the South American tropical freshwater fish Prochilodus scrofa. Environ. Res., 88: 52-63.

Momodu MA, Anyakora CA (2010). Heavy Metal Contamination of Ground Water: The Surulere Case Study. Res. J. Environ. Earth Sci., 2(1): 39-43.

Monteiro SM, Mancera JM, Fonta Dnhas- Fernandes A, Sousa M (2005). Copper induced alterations of biochemical parameters in the gill and plasma of Oreochromis niloticus. Comput. Biochem. Physiol. C., 141: 375-383. DOI: 10.1016/j.cbpc.2005.08.002.

Nussev G, Van Vuren, JHJ, Du Preez HH (2000). Bioaccumulation of chromium, manganese, nickel and lead in the tissues of the moggel, *Labeo umbratus* (Cyprinidae), from Witbank Dam, Mpumalanga. Water SA, 26: 269-284.

Obasohan EE (2007). Heavy metals concentrations in the offal, gill, muscle and liver of a freshwater mudfish (Parachanna obscura) from Ogba River, Benin city, Nigeria. Afr. J. Biotechnol., 6: 2620-2627. http://www.bioline.org.br/pdf?jb07468.

Snedecor GW, Cochran WG. (1982). Statistical Methods. The Iowa State Univ. Press., Ames., Iowa, USA, p.507.

United States Environmental Protection Agency (USEPA), 1986. Quality Criteria for Water. EPA- 440/5-86-001, Office of Water Regulations and Standards, Washington DC., USA.

Watts Sj, El-Katsha S (1995) Schistosomiasis control through rural health units. World health forum, 16: 252-254.

World Bank (2000). World development report. Investing in health. New york, Oxford University press, pp. 25-27.

World Health Organization (WHO) (1985). Guidelines for Drinking Water Quality (Recommendations). WHO, Geneva, ISBN: 92-4- 154696-4: 130.

Younes M, Siegers CP (1984). Interrelation between lipid peroxidation and other hepatotoxic events. DOI: 10.1016/0006-2952(84)90564-1. Biochem. Pharmacol., 33: 2001-2003.

19

Trace element concentrations of soils of Ife-Ijesa area Southwestern Nigeria

Aderonke A. Okoya[1]*, Olabode I. Asubiojo[2] and Adeagbo A. Amusan[3]

[1]Institute of Ecology and Environmental Studies, Obafemi Awolowo University, Ile-Ife, Nigeria.
[2]Department of Chemistry, Obafemi Awolowo University, Ile-Ife, Nigeria.
[3]Department of Soil Science, Obafemi Awolowo University, Ile - Ife, Nigeria.

Accumulation of selected trace elements on the surface and sub-soil of the Ife-Ijesa area of southwestern Nigeria was investigated. This was with a view to ascertaining the levels of these elements on some agricultural land that have been receiving chemical fertilizers and pesticides for high crop productivity in the study area. From the nine selected soil types, composite soil samples were taken to the depths 0 to 15 and 15 to 30 cm using simple random technique method. The soil samples were air-dried, crushed, sieved and digested using standard methods. The concentrations of the trace elements were thereafter read on atomic absorption spectrophotometer. Trace elements concentrations were higher in the surface than sub-surface soils. This may be attributed to the application of some fungicides and fertilizers for agricultural purposes in the area. High trace elements concentrations in Itagunmodi area when compared with other areas may be attributed to high clay mineral deposit in the area.

Key words: Trace elements, soil contamination, soil series, enrichment factor, Nigeria.

INTRODUCTION

Metals occur naturally in soil in minute amounts, and life on earth has evolved to cope with only small exposure to these elements. Many industrial processes concentrate metals like copper, cadmium, lead and zinc. These can then end up in the earth (Hendershot, 2005). A large number of trace metals are transported to the oceans from natural sources. However, these natural sources are supplemented by releases from anthropogenic processes which, for some metals, can exceed natural in-puts (Manahan, 1991). Trace metals are found in the soil, water, biota and sediment compartments of the environment, but potentially the most hazardous environmental effects to human health arise when they enter the food chain (Ayodele and Oluyomi, 2011). Trace metals studies have been an area of active investigation over the years (Fatoki et al., 2002). They are important in many fields of human endeavor such as human and animal nutrition,

human health and disease, geochemist-try (Davies, 1992) and environmental pollution (Nriagu, 1986). All trace elements are toxic to living organisms at excessive concentrations but some are essential for normal healthy growth and reproduction by either plants and / or animals at low but critical concentrations (Alloway, 1995; Anyakora et al., 2011).

Nigeria is not left behind in this wind of trace element consciousness blowing across the globe, but most trace element studies in Nigeria have been on the water systems (Asubiojo et al., 1997; Mombeshora et al., 1983; Ndiokwere and Cumie, 1983; Nriagu, 1986; Nriagu and Pacyna, 1988). In recent times however, there have been some studies of trace (and major/minor) element concentrations of Nigerian soils (Akanle et al., 1994; Ogunsola et al., 1994; Onianwa, 2001; Oyedele et al., 1995). In all of the soil studies, trace element data of soils from other parts of the world have been used as references. However, for better interpretation of these and other subsequent trace element, data of Nigerian soils that have been receiving various agro-inputs for enhanced crop

*Corresponding author. E-mail: ronkeokoya@yahoo.com

productivity, the levels of trace elements in these soils are important. Hence, this paper presents the trace (and some major/minor) element concentrations of the different soil types of Ife-Ijesa area of Southwestern Nigeria, an area of considerable geological and environmental interest.

Trace element data have found tremendous use in geochemical and environmental pollution studies (Ajayi and SUH[+], 1999). In geochemistry, elemental concentration may be used to detect anomalous concentrations of elements that constitute an expression of mineralization. Also, in view of the high degree of variation in the metal contents of rocks (Alloway and Ayres, 1993; Krauskopf, 1967; Rose et al., 1979), there is a possibility that the soils and stream sediments in a locality suspected of being polluted may have developed from rocks with anomalously high concentrations of certain heavy metals and that pollution, in the strict sense of the definition, has not occurred. Nevertheless, the natural enrichment of metals in the soils may still give rise to harmful effects in living organisms. In environmental pollution studies, trace elements have been found adequate in describing the environmental exposure to which the organism has been subjected (Ojo et al., 1994). The natural members of the ecosystem in an area of geochemical enrichment will have evolved tolerance to the elevated concentrations of metals, but newly introduced plant and animal species may be adversely affected. It is therefore important to determine the local background concentrations of heavy metals in order to determine whether the concentrations in the soils and sediments under investigation are significantly higher than those of the area.

MATERIALS AND METHODS

Study sites

The study was carried out within Ife-Ijesa area of Southwestern Nigeria. The major soil types in the region were identified using the semi-detailed soil map of Central Western Nigeria produced by Smyth and Montgomery (1962). A total of nine soil series which differ widely in parent material, texture, drainage, topographic position and chemical composition were sampled, as presented in Table 1.

Sample collection and preparation

The identified soil types (Table 1) were sampled using Dutch soil auger to collect core samples at 0 to 15 and 15 to 30 cm soil depths. For each soil series, ten core samples, randomly taken were homogenized and a composite sample was taken. The homogenized composite samples were air-dried, crushed and sieved using 2 mm sieve. The less-than-2 mm fraction of each sample was kept in a polythene bag and labeled.

Sample analyses

All reagents used were of analytical grade and from which standard solutions were prepared. Glassware were thoroughly washed with

detergent and rinsed with distilled water. The digestion method previously described by Francek et al. (1994) was adopted for the extraction of trace metals in this study. One gram each of air-dried soil sample was crushed to fine powder in an agate mortar and digested in 10 ml of 1:1 concentrated HNO_3. The mixture was evaporated to near dryness on a hot plate and then cooled. This procedure was repeated with a 15 ml solution of 1:1 concentrated HCl. The extracts were filtered with No. 40 Whatman filter paper and then made up to 100 ml volume with 2% HNO_3. Solutions of the sample and blanks were run using Atomic Absorption Spectrometer (AAS) (200A Model), at the Centre for Energy Research and Development of Obafemi Awolowo University, Ile-Ife, Nigeria.

Data analyses

The standard deviation of concentrations of replicate measurements was determined for each element. For the subsequent general evaluation of the data, the mean values and calculated enrichment factors of the elements were used. The enrichment factor (EF) was calculated with aluminium as the reference element and using the formula:

$$EF = (CEs / CAs) \times (CAr / CEr)$$

Where CEs = concentration of element in sample; CEr = concentration of element in crustal rock; CAs = concentration of aluminium in sample, and CAr = Concentration of Aluminium in crustal rock.

The new Duncan multiple range test was also carried out to separate mean concentrations of elements that are significantly different. The Pearson correlation was used to test the relationship between the soil parameters.

RESULTS AND DISCUSSION

The results of the total elemental concentrations of the surface and sub-surface soil samples are presented in Tables 2 and 3, respectively. Generally, the results show that for most of the soils, the concentration of Zn, Cu, Ni and Mn is higher at the surface than at the sub- surface. This could be due to applications of copper fungicides and Mn and Zn-containing fertilizers to the soils which lead to higher concentrations of the residues of these elements on the surface. However, there is no particular trend in the changes in concentration of Pb and Cr with increasing depth. This is because, Pb and Cr are not components of the fertilizers and pesticides applied as for those elements which vary substantially with soil depth. The unusually high concentrations of Cu in Ondo and Itagunmodi surface soils are due to the effects of copper fungicides being applied to cocoa plantations in the area, cocoa being a major cash crop of the areas (Akinnifesi et al., 2006). This is also evident from the relatively high soil pH values of the study area in Table 1 (Okoya et al., 2010).

In the study area, the trace element mean concentrations at the top soil (Table 2) are generally lower than (except Itagunmodi soil series) the average given by Alloway (1995) for industrialized countries such as Netherland (Zn 72.5, Cu 18.6, Ni 15.6 Cr 25.4 and Pb 60.2 mg kg[-1]); Canada (Zn 74.0, Cu 22.0, Ni 20.0 Cr 43.0

Table 1. Characteristics of soil.

Soil series	Sampling site	Major land use in sampling location	Brief description of the soil	*pH (0 to 15 cm) depth	*pH (15 to 30 cm) depth
Iwo	O. A. U. Teaching and Research farm, Ile-Ife	Yam and pepper plot	Well drained, coarse textured soils, overlying weathered rock material, derived from coarse-grained granitic rocks and gneisses	6.53[d]	5.90[e]
Ondo	Near OWENA town	Cocoa and Kolanut plantations	Well drained, medium to fine, textured soil, overlying orange brown, yellow brown and white mottled clay, mainly derived from medium ground granitic rocks and gneisses	6.53[d]	6.17[c]
Egbeda	O.A.U. Teaching and Research Farm, Ile-Ife	Cocoa, kolanut, orange, plantain/ banana plantations	Well drained fine textured soil, overlying red brown, yellow brown and white mottled clay, mainly derived from fine –ground biotite gneiss and schist.	7.36[a]	6.10[d]
Itagun-Modi	Near Itagunmodi village	Cocoa farm	Well drained, very fine textured soils of uniform brownish red or dark chocolate brown colour to depth, derived from amphibolite and related basic rocks.	6.28[c]	5.64[f]
Jago	O.A.U. Teaching and Research farm, Ile-Ife	Sugar cane, bamboo trees banana trees	Soils of various textures in low topographical sites, with drainage affected by seasonally high water table derived from alluvium and local colluvium.	6.18[f]	4.85[h]
Oba	O.A.U. Teaching and Research farm, Ile-Ife	Maize and yam plot	Well drained, coarse textured soils, overlying weathered rock material, derived form coarse-grained granitic rocks and gneisses.	6.83[b]	6.24[a]
Gambari	O.A.U. Teaching and Research farm ,Ile-Ife	Citrus plantation	Sandy and clayey soils either containing very large quantities of ferruginous concretions and fragments of iron stones or overlying massive iron stone pan	5.88[h]	4.81[i]
Apomu	O.A.U. Teaching and Research farm, Ile-Ife	Maize farm	Well drained drift soils, pale brown to reddish brown, sandy to fairly clayey soils, with no gravel concretions and quartz grains.	6.09[g]	5.28[g]
Efon	Ipetu –Ijesa road	Maize, pepper, yam farm	Well drained drift soils, sandy or fairly clayey, gravel soils merging to rotten rock.	6.61[c]	6.22[h]

*: Means with the same letter are not significantly different by new Duncan's multiple range test at p < 0.05.

and Pb 20.0 mg kg^{-1}) and England and Wales (Zn 78.2, Cu 15.6, Ni 22.1 Cr 44.0 and Pb 48.7 mg kg^{-1}).

Table 3 showed values lower than those in Table 2 which agreed with Alloway (1995) that observed

higher values of minerals in the topsoil than the subsoil as a result of cycling through vegetation,

Table 2. Total elemental concentration (mg/kg) of the soil's surface (0 to15 cm) (n = 9).

Soil series	Zn	Cu	Ni	Pb	Fe	Cr	Na	Mn	Mg	Al
Apomu	9.60^h± 0.0	2.30^g±0.01	10.00^g±0.08	9.00^i±0.06	7990^i±0.1	2.00^h±0.00	248^i±0.03	272^i±0.01	441^b±0.00	4670^g±3.0
Efon	31.60^c±0.01	8.9^f±0.01	16.00^c±0.02	10.00^f±0.06	30200^b±4.1	34.00^b±0.01	352^e±0.12	687^d±0.06	213^g±0.00	6370^f±1.2
Egbeda	36.30^b±0.00	9.4^c±0.00	19.00^b±0.00	30.00^b±0.09	16000^c±2.3	22.00^c±0.00	350^f±0.10	795^b±0.05	365^c±0.06	9940^d±1.8
Gambari	15.90^f±0.01	9.20^d±0.00	11.00^e±0.00	36.00^a±0.07	14040^e±1.03	12.00^g±0.00	298^h±0.02	367^g±0.01	180^i±0.02	11310^b±3.1
Itagunmodi	59.80^a±0.01	50.7^a±0.00	36.00^a±0.01	29.00^c±0.00	111400^a±19.4	75.00^a±0.02	426^c±0.04	860^a±0.15	167^a±0.01	21240^a±2.7
Iwo	3.60^i±0.00	2.10^h±0.01	3.00^i±0.08	10.00^g±0.15	8940^i±2.1	9.0^i±0.01	408^d±0.03	207^f±0.02	87^h±0.00	3850^i±3.1
Jago	16.70^e±0.00	1.5^i±0.01	14.00^d±0.01	9.00^h±0.00	15630^d±0.1	17.00^f±0.00	321^g±0.06	398^e±0.06	200^f±0.01	11260^c±1.4
Oba	12.40^g±0.00	9.1^e±0.00	10.00^f±0.01	15.00^d±0.04	12250^f±1.3	18.00^d±0.02	454^a±0.17	791^c±0.06	199^g±0.01	8500^d±6.0
Ondo	27.10^d±0.00	30.00^b±0.00	7.00^h±0.00	10.00^e±0.05	9450^h±1.4	18.00^e±0.02	430^b±0.11	298^h±0.03	221^d±0.00	4080^h±4.1

Means with the same letter are not significantly different by new Duncan's multiple range test at $p < 0.05$.

Table 3. Total elemental concentration (mg/kg) of the soil's sub- surface at 15 to 30 cm depth (n = 9).

Soil series	Zn	Cu	Ni	Pb	Fe	Cr	Na	Mn	Mg	Al
Apomu	9.1^g±0.00	2.0^g±0.00	5.00^g±0.05	9^i±0.04	7270^g±1.2	11.00^h±0.01	342^g±0.06	131^g±0.00	404^a±0.00	4670^g±3.7
Efon	32.8^b±0.00	8.6^d±0.00	10^e±0.03	10^f±0.04	37700^b±3.0	30.00^b±0.01	311^h±0.09	501^d±0.08	198^e±0.01	8130^b±0.04
Egbeda	27.80^d±0.01	12.0^b±0.01	18.00^b±0.00	34^a±0.12	2780^h±0.2	28^c±0.00	613^a±0.27	665^c±0.09	307^c±0.00	7970^c±11.5
Gambari	3.9^i±0.00	7.80^e±0.01	6.0^f±0.03	21^d±0.07	20110^c±0.5	16^f±0.00	230^i±0.10	232^f±0.03	63^i±0.01	9510^a±7.6
Itagunm-odi	45.30^a±0.00	36.00^a±0.00	36^a±0.01	24^c±0.02	112620^a±4.2	72^a±0.01	555^c±0.13	1075^a±0.18	351^b±0.07	1379^i±15.60
Iwo	7.00^f±0.02	1.7^a±0.00	5.00^d±0.04	10.0^h±0.27	13650^d±2.6	18^e±0.00	388^d±0.03	305^e±0.08	165^f±0.00	6790^e±4.4
Jago	28.30^c±0.01	3.6^f±0.01	1^i±0.03	15^e±0.09	1580^i±0.4	1.0^i±0.00	358^f±0.03	120^i±0.02	216^d±0.01	2590^g±4.4
Oba	12.5^e±0.00	9.4^c±0.00	11^d±0.04	30^b±0.03	17310^d±0.01	16^g±0.01	380^e±0.16	791^b±0.01	87^g±0.01	7700^d±9.2
Ondo	9.60^h±0.00	1.50^h±0.00	12^c±0.03	10^g±0.01	8520^f±0.4	22^d±0.00	582^b±0.16	129^h±0.01	65^h±0.00	1730^h±0.08

Means with the same letter are not significantly different by new Duncan's multiple range test at $p < 0.05$.

atmospheric deposition and adsorption by the soil organic matter. The trace element levels in Itagunmodi soil series compare well with the European/American soils; however this is an area where mining activity takes place. The distribution of trace metals among soil textural classes (e.g. sandy, silty, loamy soils) are differentiated more strongly than among taxonomic units.

Hence, a comparison of the trace elements in the loamy soils of the study area determined with the loamy soils of Poland fall within the same range. However, it is higher than the sandy soils in Poland (Kabata-Pendias et al., 1992) as expected since trace metal contents in sands are significantly lower than in loams and clays.

The results of the new Duncan multiple range test (DMRT) is indicated on the values in Tables 2 and 3 showed that the mean elemental concentrations were statistically different from one soil series to another. This variability of in concentrations of trace elements in these soils may be

Table 4. Enrichment factors for the elements at 0 to 15 cm soil depth (n = 9).

Soil series	Zn	Cu	Ni	Fe	Pb	Cr	Na	Mn	Mg
Apomu	2.93	1.76	3.82	3.21	13.74	0.15	0.60	4.89	1.08
Efon	7.07	4.98	4.48	8.90	11.19	1.90	0.63	9.05	0.38
Egbeda	5.21	3.37	3.41	3.02	21.52	0.79	0.40	6.71	0.42
Gambari	2.00	2.90	1.73	2.33	22.69	0.38	0.30	2.72	0.18
Itagunmodi	4.01	8.5	3.02	9.84	9.73	1.26	0.22	3.40	0.36
Iwo	1.33	1.94	1.39	4.63	18.52	0.08	1.20	5.86	0.55
Jago	2.11	0.47	2.22	2.60	5.70	0.54	0.32	2.96	0.20
Oba	2.08	3.82	2.10	2.70	12.58	0.75	0.60	7.81	0.26
Ondo	9.47	26.21	3.06	4.35	17.48	1.57	1.19	6.13	0.61

Table 5. Enrichment factors for the elements at 15 to 30 cm soil depth (n = 9).

Soil series	Zn	Cu	Ni	Fe	Pb	Cr	Na	Mn	Mg
Apomu	2.78	1.53	2.92	13.74	13.74	0.84	0.83	2.35	0.98
Efon	5.75	5.98	8.70	13.90	13.90	2.08	0.43	5.17	0.28
Egbeda	4.97	5.37	0.65	30.42	30.42	1.12	0.87	7.0	0.44
Gambari	0.58	2.92	3.97	15.74	15.74	0.60	0.27	2.05	0.07
Itagunmodi	46.84	93.07	153.23	124.09	124.09	18.61	4.55	65.39	2.88
Iwo	1.47	0.89	1.31	10.50	10.50	0.95	0.65	3.77	0.28
Jago	15.58	4.96	0.69	41.29	41.29	0.14	1.56	3.89	0.94
Oba	22.94	4.35	2.55	27.78	27.78	0.74	5.53	85.39	1.27
Ondo	7.91	3.09	12.36	9.24	41.21	4.53	3.81	6.25	0.43

caused by the variety of underlying rocks of different ages and lithology (Davies, 1985; McGrath, 1987; Bini et al., 1990). It may also be due to local phenomena of contamination (Angelone, 1991). The enrichment factors (EF) for the elements in the soil samples were calculated with aluminium as the reference element and the results are presented in Tables 4 and 5. The fact that there is no enrichment for most of the elements in the various locations imply that there is not much pollution in these areas. The enrichment factors of Zn in Efon, Egbeda, Itagunmodi and Ondo soil series are greater than 3. This implies that these soils are enriched with Zn from anthropogenic source(s). High Zn concentrations in these soils could be due to the use of agrochemicals such as fertilizers and pesticides (Fatoki et al., 2002). The enrichment factor of Cu in Itagunmodi and Ondo soil series are also high (EF >3). This may be due to the effects of copper fungicides being applied to the soils. Incidentally, cocoa is the major crop grown on these soils and is usually sprayed with copper fungicides, whose residues directly increase the pH and the organic matter contents of the soils (Akinnifesi et al., 2006). The resultant effects bring about variation in the concentration and distribution of some major and trace elements in the soils.

The high enrichment of Pb in all the soils is attributed to a substantial contribution from motor vehicle exhaust; Nigerian gasoline being leaded (Mombeshora et al., 1983). It may also be due to the use of agrochemicals such as Pb- containing pesticides (Fatoki et al., 2002). Zn enrichment could be due to tyre wear.

Another element with high enrichment factor in these soils is manganese. Soils derive virtually all their Mn content from the parent materials, and the concentrations found in mineral soils reflect the composition of these parent materials. Apart from these natural mineralogical sources, the only other significant source of Mn in soils, that explains its very high enrichment factors in all the soils considered, is its application to soils in the form of fertilizers. This is normally in the form of $MnSO_4$, MnO, or as an addition to macronutrient fertilizer (Fatoki et al., 2002). Nigeria has become home to various forms of fertilizers for some time now, hence some Mn containing fertilizers could be responsible for the high enrichment (EF>3) observed for Mn in almost all the Ife-Ijesa soils. Itagunmodi series is highly enriched with virtually all the elements in the sub-surface. This may be associated with the fact that Itagunmodi soil is very rich in clay. Clayey soils usually have strong affinity for binding metal ions.

A possible relationship between total metal concentrations and selected soil properties for both top soil and sub-soil was also made as presented in Tables 6 and 7.

Table 6. Correlation between total metal and the selected soil properties at depth 0 to 15 cm.

Soil properties	Zn	Cu	Ni	Pb	Fe	Cr	Na	Mn	Mg	Al
pH	0.2047 (ns)	-0.03844 (ns)	0.04146 (ns)	0.00804 (ns)	-0.12033 (ns)	0.04920 (ns)	0.44724 (ns)	0.58532(ns)	-0.03717 (ns)	-0.15826 (ns)
OM	0.55696(ns)	0.33584 (ns)	0.52845 (ns)	0.34145 (ns)	0.33774 (ns)	0.48418 (ns)	-0.31236 (ns)	0.14180 (ns)	0.19390 (ns)	0.48714 (ns)
Sand	-0.76948*	*-0.73296 (ns)	**-0.85402 (ns)	-0.37844 (ns)	***-0.96574	0.90276***	-0.26557 (ns)	-0.50358 (ns)	-0.65948 (ns)	-0.83970*
Silt	0.25201(ns)	0.42037 (ns)	0.38993 (ns)	-0.27372 (ns)	0.63891(ns)	0.47825 (ns)	0.22621 (ns)	0.00579 (ns)	0.41676 (ns)	0.42990 (ns)
Clay	0.86346**	*0.74279	***0.90962	0.58717 (ns)	***0.94111	***0.93233	0.23647 (ns)	0.63543 (ns)	0.65129 (ns)	*0.87381

Table 7. Correlation between total metal and the selected soil properties at depth 15 to 30 cm.

Soil properties	Zn	Cu	Ni	Pb	Fe	Cr	Na	Mn	Mg	Al
pH	0.12868 (ns)	0.06458 (ns)	0.32261 (ns)	0.15122 (ns)	0.06190 (ns)	0.29604 (ns)	0.51044 (ns)	0.43370 (ns)	-0.12972 (ns)	0.11056 (ns)
OM	0.53333 (ns)	0.59114 (ns)	0.37478 (ns)	0.07585 (ns)	0.63727 (ns)	0.48458 (ns)	-026643 (ns)	0.27608 (ns)	0.09543 (ns)	-0.01046 (ns)
Sand	-0.80741**	**-0.87399	*0.69821	-0.21427 (ns)	**-0.88540	-0.77503*	-0.10058 (ns)	-0.60922 (ns)	-0.37190 (ns)	0.28969 (ns)
Silt	0.40740 (ns)	0.32154 (ns)	0.10352 (ns)	-0.30712 (ns)	0.49553 (ns)	0.19590 (ns)	-0.22958 (ns)	0.13527 (ns)	0.22782 (ns)	0.54585 (ns)
Clay	*0.79028	***0.89816	*0.76053	0.33847 (ns)	**0.85150	**0.81910	0.18531 (ns)	0.64967 (ns)	0.35156 (ns)	0.15963 (ns)

The selected soil properties are pH, organic matter (OM), sand, silt and clay.

Significant negative correlations were observed between the total metals and the sand content of the soil at both depths, while the correlations between clay and total metal concentrations (Zn, Cu, Ni, Fe, Cr, Al) at both depths are positive and significant at the different probability levels indicated in the tables. This follows the trend observed in the correlation between the extractable metals and physico-chemical properties of some southwestern Nigerian soils (Okoya et al., 2010) in which at both depths, none of the metals correlated significantly with silt while Ni, Cr, Al, Fe, and Mg are correlated significantly with clay.

This is in line with the general principle that trace metal contents in sands are significantly lower than in loams and clays (Davies, 1985).

Clayey soils usually have strong affinity for binding metal ions. The heavy metals bound in clayey soils are therefore protected from leaching by water and made available in soil as plant nutrient (Nriagu and Pacyna, 1988). As the sand content is increasing, silicon content is expected to increase. Increase in silicon content (being a major element in the earth crust) implies a decrease in trace and minor elements content. However there is no significant correlation between the total metal and OM for both surface soils (0 to 15 cm) and sub-surface soils (15 to 30 cm).

concentrations of most of the soils of the area varied from one soil type to another. Higher trace element levels in the surface than sub-surface soils may be attributed to applications of some fungicides and fertilizers for agricultural purposes in the area. On the other hand, the anomalously high trace element concentrations found between the trace element concentrations and physico-chemical characteristics of the soils of the area may be attributed to the effects of the anthropogenic activities on the soils.

Conclusions

The study showed that trace element

REFERENCES

Ajayi TR, SUH[+] SE (1999). "Partially extractable metals in the Amphibolites of Ife- Ijesa Area". J. Minn. Geol., 33(2): 103-116.

Akanle OA, Ogunsola OJ, Oluwole AF, Asubiojo OI, Olaniyi

HB, Akeredolu FA, Spyrou NM, Ward NI, Ruck W (1994). Traffic pollution: Preliminary elemental characterization of roadside dust in Lagos, Nigeria. Sci. Tot. Environ., 146/147: 175-184.

Akinnifesi TA, Asubiojo OI, Amusan AA (2006). Effects of fungicide residues on the physico-chemical characteristics of soils of a major cocoa-producing area of Nigeria. Sci. Total Environ., 366: 876-879.

Alloway BJ (1995). Heavy metals in soils. Blackie and Son Ltd, Glasgow, London, p. 339.

Alloway BJ, Ayres DC (1993). Chemical principles of environmental pollution. Blackie Academic and Professional, Glasgow, London, p. 291.

Angelone M, Vaselli O, Bini C, Coradossi N, Pancani MG (1991). Total and EDTA- extractable element contents in ophiolitic soils from Tuscany (Italy). Z. Pflanzenernaehr, Bodenkd, 154: 217-223.

Asubiojo OI, Nkono NA, Ogunsola AO, Oluwole AF, Ward NI, Akanle OA, Spyrou NM (1997). Trace elements in drinking and ground water samples in Southern Nigeria. Sci. Total Environ., 208: 1-8.

Anyakora C, Nwaeze K, Awodele O, Chinwe Nwadike C, Arbabi M, Herbert CH (2011). Concentrations of heavy metals in some pharmaceutical effluents in Lagos, Nigeria. J. Environ. Chem. Ecotoxicol., 3(2): 25-31

Ayodele JT, Oluyomi CD (2011). Grass contamination by trace metals from road traffic. J. Environ. Chem. Ecotoxicol., 3(3): 60-67.

Bini C, Coradossi N, Vaselli O, Pancani MG, Angelone M (1990). Weathering and soil mineral evolution from Mafic rocks in temperate climate. Proc. 14th Int. Congress Soil Sci. Kyoto, 7: 54.

Davies BE (1992). Trace metals in the environment: Retrospect and prospect In: Adriano DC (eds.). Biogeochemistry of trace metals. Lewis Publishers, Boca Raton Ann Arbor London, Tokyo, pp. 1-17.

Davies BE (1985). Baseline survey of metals in Welsh soils. In: Environmental Geochemistry and Health, I. Thornton ed., London, pp. 87-93

Fatoki OS, Lujiza N, Ogunfowokan AO (2002). Trace metal pollution in Umtata river. Water SA, 28: 183-189.

Francek MA, Makimaa B, Pan V, Hanko JH (1994). Small town Lead levels: A case study from the homes of Pre – Schooler in MT – Pleasant, Michigan. Environ Pollut. 0269 –7491/94/307.00 © Elsevier Science Limited, England, pp. 159-167.

Hendershot W (2005). Digging up the dirt. The Reporter, Mc. Grill, 27(9): 37

Kabata-Pendias A, Dudka S, Chlopecka A, Gawinowka T (1992). In: Andriano DC (eds). Biogeochemistry of trace metals. Lewis Publishers, Boca Raton Ann Arbor London, Tokyo. pp 61-84

Krauskopf KB (1967). Introduction to geochemistry. McGraw – Hill, New York, p. 721.

Manahan SE (1991). Environmental chemistry. 5th edn. Lewis Publishers, pp. 10-565.

McGrath SP (1987). Computerized quality control, statistics and regional mapping of the concentrations of trace and major elements in the soil of England and Wales. Soil Use Manage., 1(3): 31-38.

Mombeshora C, Osibanjo O, Ajayi SO (1983). Pollution studies on Nigerian rivers: The onset of lead pollution ofsurface waters in Ibadan. Environ. Int., 9: 81-84 .

Ndiokwere CL, Cumie VP (1983). Determination of some toxic trace metals in Nigeria Rivers and harbour water samples by INAA. J. Radioanal. Nucl. Chem., 79: 147-151.

Nriagu JO (1986). Chemistry of the river Niger, I: major ions, II: trace metals. Sci. Tot. Environ., 58: 81-92 .

Nriagu JO, Pacyna JM (1988). Quantitative assessment of world wide contamination of air, water and soils by trace metals. Nature, 333: 134-139.

Ogunsola OJ, Oluwole AF, Asubiojo OI, Olaniyi HB, Akeredolu FA, Akanle OA, Spyrou NM, Ward NI, Ruck W (1994). Traffic pollution: Preliminary elemental characterization of road side dust in Lagos, Nigeria. Sci. Total Environ. 146 / 147: 175-184.

Ojo OJ, Oluwole AF, Asubiojo OI, Durosinmi MA, Ogunsola OJ (1994). Correlations between Trace Element levels in Head Hair and Blood components of Nigerian Subjects, Biol. Trace Elem. Res., 13: 123-129.

Okoya AA, Asubiojo OI, Amusan AA (2010). Extractable metals and physicochemical properties of some Southwestern Nigerian Soils. In Biotechnology Development and threat of Climate Change in Africa: The case of Nigeria, Cuvillier Verlag Gottingen German, 2: 166-176.

Onianwa PC (2001). Roadside topsoil concentrations of Lead and other heavy metals in Ibadan. Nigeria. Soil Sediment Contam., 10: 577-591.

Oyedele DJ, Obioh IB, Adejumo JA, Oluwole AF, Aina PO, Asubiojo OI (1995). Lead contamination of soils and vegetation in the premises of a smelting factory in Nigeria. Sci. Total Environ., 172: 189-195.

Rose AW, Hawkes HE, Webb JS (1979). Geochemistry in mineral exploration. 2nd edn. Academic press London, p. 657

Smyth AJ, Montgomery RF (1962). Soils and land use in Central Western Nigeria. The government printer, Ibadan, Western Nigeria, p. 265

20

A study on removal characteristics of para-nitrophenol from aqueous solution by fly ash

Alinnor I. J.[1]* and Nwachukwu M. A.[2]

[1]Department of Pure and Industrial Chemistry, Federal University of Technology, P. M. B. 1526, Owerri, Imo State, Nigeria.
[2]Department of Earth and Environmental Studies, Montclair State University, New Jersey, USA.

The removal characteristics of para-nitrophenol from aqueous solution by fly ash were investigated under various conditions of contact time, pH and temperature. The influence of pH on the para-nitrophenol uptake by the fly ash was carried out between pH4 and pH 10. The level of uptake of para-nitrophenol by fly ash increased at higher pH values. The effect of temperature on the uptake of para-nitrophenol was investigated between 30 and 60°C; the adsorption increased at lower temperature. Rate constants were evaluated in terms of first-order kinetics. The rate constants k for uptake of different concentrations of para-nitrophenol was 1.10×10^{-2} s^{-1} and 1.14×10^{-2} s^{-1}, respectively. The experimental results underlined the potential of coal fly ash for recovery of para-nitrophenol from waste water. The main mechanisms involved in the removal of para-nitrophenol from solution by fly ash were electron – withdrawing effect of NO_2 group of benzene ring and adsorption at the surface of the fly ash.

Key words: Fly ash, para-nitrophenol, adsorption, kinetics, mechanism.

INTRODUCTION

Fly ash is a waste product obtained from burning of coal. Given the large amount of fly ash generated in coal-burn power plants and large dumping sites required for the safe disposal, any means of reuse, recycling and recovery of fly ash will be a welcome development. Reports have shown that fly ash has been used as a binding reagent for the fixation of heavy metal and nutrients contained in hazardous wastes and organic wastes (Lin and Hsin, 1996; Vincini et al., 1994; Shende et al., 1994; Parsa et al., 1996). It has been reported that many researchers have reused fly ashes as adsorbents for waste water or air pollutants control (Rivatti et al., 1988; Sell et al., 1994). Alinnor (2007) reported the use of fly ash for the removal of heavy metal ions from aqueous solution.

The removal of toxic solvent from polluted environment has received much attention in recent years, especially in Nigeria. Removal of toxic solvent from polluted

environment can be done by several techniques but the adsorption technique is widely used due to its high rate, high uptake capacity, effective treatment in dilute solution, low cost and regeneration (Saleem et al., 1993). Adsorption is a reliable technique that achieves rapid results. Adsorption of organic compounds on the surface of carbon has been studied extensively (Uranowski, 1998; Tanju and James, 1999; Lin and Liu, 2000; Yu and Chou, 2000; Wiessner et al., 1998).

The role of granular activated carbon surface chemistry in the adsorption of trichloroethylene and trichlorobenzene has been reported (Tanju and James, 1999). In fact, a number of studies were conducted to show the effectiveness of fly ash in the removal of organic materials from aqueous solutions (Nollet el al., 2003; Wang et al., 2005; Kumar et al., 2005; Gupta et al., 1990). Khalid et al. (2004) have reported the use of zeolites for the removal of para-nitrophenol from waste water by adsorption technique. Iqbal et al. (2005) have studied adsorption of phenol on activated charcoal from aqueous solution. The most important characteristics of fly ash are calcium content that provides alkalinity in the

---
*Corresponding author. E-mail: alijuiyke@yahoo.com.

Table 1. Chemical composition of the fly ash.

Constituent	Wt. %
SiO_2	57.25
Al_2O_3	22.03
Fe_2O_3	8.36
CaO	2.97
MgO	0.97
SO_3	0.76
TiO_2	0.68
K_2O	0.52
Others	6.49

system raising pH to strongly alkaline value (~ 12) and the (SiO_2 + Al_2O_3 +Fe_2O_3) content (Komnitas et al., 2004).

Research efforts on fly ash to date have been focused on the study of mechanisms involved in contaminant uptake. A lot of research works have been carried out on the removal of toxic solvents from aqueous solution by activated charcoal and other adsorbents. But work on the removal characteristics of para-nitrophenol by fly ash is very scanty. In view of this, para-nitrophenol was chosen. The aim of present study was to investigate the use of fly ash as a low – cost adsorbent for the removal of toxic solvent such as para-nitrophenol from aqueous solutions. The kinetics and mechanism of para-nitrophenol uptake by the fly ash was investigated.

MATERIALS AND METHODS

The fly ash used as the adsorbent in this present investigation was obtained from Nigeria Coal Corporation, Enugu. The fly ash samples were dried at 105 ± 1°C for 2 h before tests. The fly ash samples were ground and sieved to a particle size of 250 μm before use (Alinnor, 2007). Table 1 shows the chemical composition of the fly ash samples used in this study, SiO_2 and Al_2O_3 contents make up about 79% of the fly ash, while Fe_2O_3 and CaO compose about 11%

About lg portions of fly ash were taken in different Erlenmeyer flasks. 100cm^3 of para-nitrophenol solution was added to each flask having different concentrations (2.0×10^{-3} – 5×10^{-3} m). The content of the flasks was stirred in a water bath for different durations of time at 30°C using magnetic stirrer. The pH was noted before and after stirring with pH meter. The slurries were then filtered through ordinary filter paper. The clear filtrate was then analyzed for para –nitrophenol (PNP) content by Spectrophotometer (Spectronic 21D) at wavelength 400 nm.

Blank determinations were performed under similar experimental conditions. The amount of para-nitrophenol adsorbed by the fly ash was then calculated from the difference in concentration of blank and sample. The reported values of PNP adsorbed by fly ash in each test were the average of at least three measurements (Iqbal et al., 2005).

The effects of pH on the uptake of para–nitrophenol by fly ash were determined by adjusting the pH of the slurry in the range of pH4 to pH10. At the end of agitation period, the PNP uptake by fly ash was determined as earlier described. The influence of

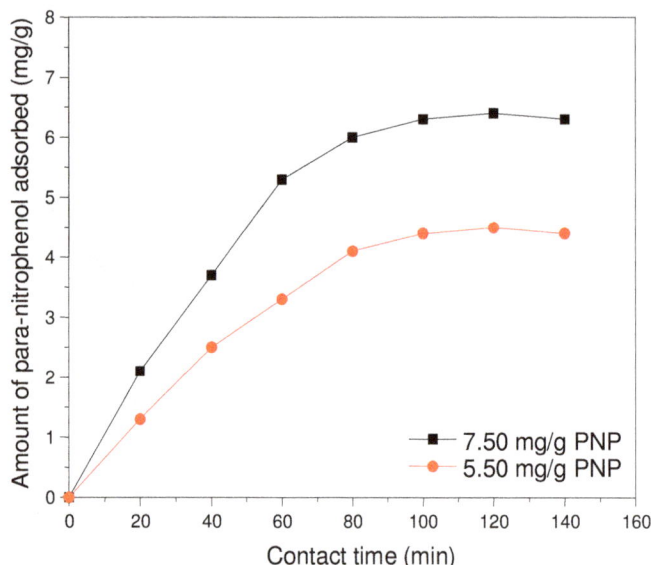

Figure 1. Effect of contact time on the adsorption of para-nitrophenol on fly ash at pH 6.8.

temperature on the adsorption of PNP on fly ash was carried out between 30 and 60°C using thermostated water bath. Also, at the end of agitation period, the para – nitrophenol adsorbed by fly ash was determined as described above. All the chemicals used were of analytical reagent grade.

RESULTS AND DISCUSSION

Reaction kinetics

The effect of contact time on adsorption of 5.50 and 7.50 mg/g para-nitrophenol is shown in Figure 1. The rate of uptake of PNP by fly ash increases with time initially. The PNP removal in the first 20 min was 1.3 and 2.1 mg/g, respectively for the two concentrations studied. Figure 1 indicates that equilibrium was established within 2 h in both concentrations of PNP. At equilibrium, 4.1 mg/g or 74.55% and 6.1 mg/g or 81.33% were removed from the initial concentrations of PNP by the fly ash. The results indicated that level of removal of PNP by fly ash depends on the initial concentration of para-nitrolphenol.

The rate constants for adsorption of PNP on fly ash were determined using first-order kinetics (Eligwe and Okolue, 1994; Alinnor, 2007):

$$\ln (C_o/C_t) = kt \qquad ....(1)$$

Where C_o is the initial PNP solution concentration, C_t is the concentration at time t, and k is the rate constant.

Figure 2 shows that the initial rate of PNP uptake conforms to first-order kinetics as shown in Equation (1). A plot of in C_o/C_t versus t should yield a straight line

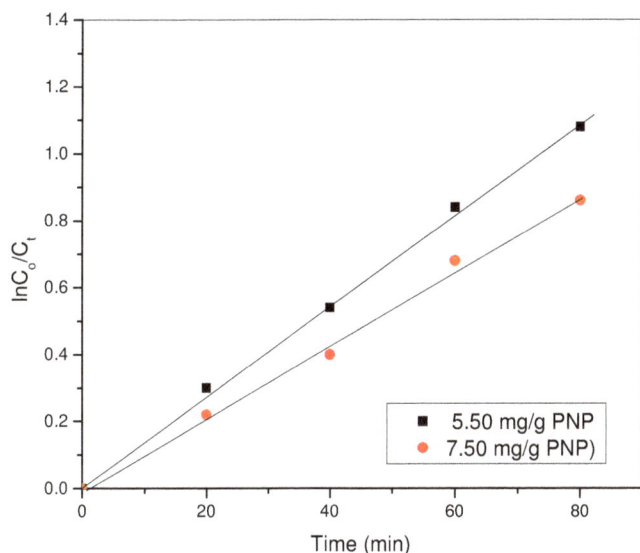

Figure 2. First- order kinetic plot for para-nitrophenol adsorption on fly ash.

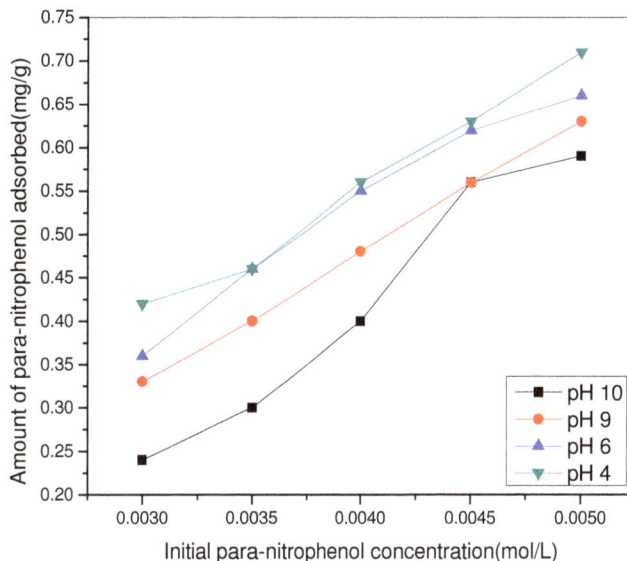

Figure 3. Effect of pH on adsorption of para-nitrophenol on fly ash.

from the slope of which the rate constant k was calculated to be 1.10×10^{-2} s^{-1} and 1.14×10^{-2} s^{-1} for PNP concentrations, 7.50 and 5.50 mg/g, respectively.

Effect of pH

The adsorption isotherms of para-nitrophenol at different pH ranges are shown in Figure 3. The range of concentration varied from 3.0×10^{-3} to 5.0×10^{-3} M and the pH ranges were pH 4 to 10. Figure 3 shows that the amount of para – nitrophenol adsorbed from aqueous solution is significantly high at pH9 when compared to pH10, pH6 and pH4 respectively. At low pH4 the adsorption of para-nitrophenol on fly ash increases rapidly. This may be attributed to substitution of nitro (NO_2) group, an electron-withdrawing group in the benzene ring. This NO_2 group substitution will enable the ring to withdraw more electrons from the oxygen atom thereby increasing the acid strength of para-nitrophenol. In view of the increase in acidic strength of PNP, the PNP will rapidly adsorb on the negatively charged surface of fly ash. Bhatgava and Sheldarkar (1993) working on adsorption of para-nitrophenol on clay reported an increase in adsorption of para-nitrophenol on clay due to presence of nitro group in the benzene ring. At intermediate pH6 there was a gradual increase on the removal of PNP from aqueous solution onto fly ash. This reduction of adsorption of PNP onto fly ash at intermediate pH6 may be attributed to the reduction in acidic strength of PNP at pH6 when compared to pH4.

Figure 3 shows that the removal of PNP from aqueous solution was more at pH9 when compared to pH 10. At pH 9 the surfaces of the adsorbent were negative and

there was an increase on the uptake of PNP by the fly ash. The increase in PNP uptake by fly ash may be explained in terms of electrostatic interaction. This can be seen by the sharp increase on the amount of PNP uptake at pH9. The increase in PNP uptake by fly ash at higher pH values may also be attributed to calcium content and (SiO_2 + Al_2O_3 + Fe_2O_3) content of fly ash that provides alkalinity in the system raising the pH to strongly alkaline values, thereby facilitating the adsorption of PNP by the fly ash.

However, the decrease in PNP uptake by fly ash at pH10 compared to pH9 may be attributed to dissociation of PNP into phenolate anions. At pH 10 fly ash surface acquires more negative charge due to presence of OH^- ions on the surface of fly ash. The more negatively charged fly ash surface would have repulsed the resultant PNP phenolate anion with each other, which would have resulted in lower adsorption of para-nitrophenol onto fly ash at pH10 when compared to pH9

A schematic representation of the electrostatic repulsion between the PNP phenolate anion and the negatively charged fly ash surface is depicted in Figure 4.

Moreover, PNP phenolate ions have more affinity for an aqueous solution than neutral PNP and thus provide another contributing factor to low adsorption rates. Solubility is another factor, which can affect the adsorption of PNP. Solubility of PNP changes with changes in the pH of the medium and thus would have its own contributing role in adsorption.

Effect of temperature

The removal of para–nitrophenol from aqueous solution

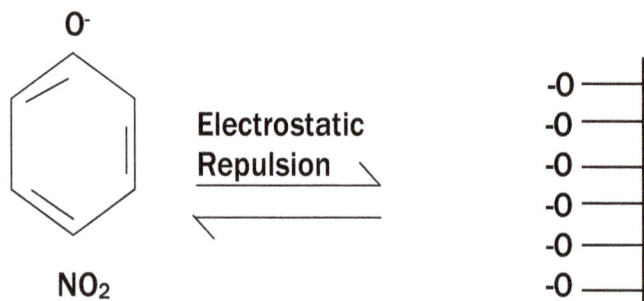

Figure 4. Electrostatic repulsion mechanism between phenolate ion and charged fly ash surface.

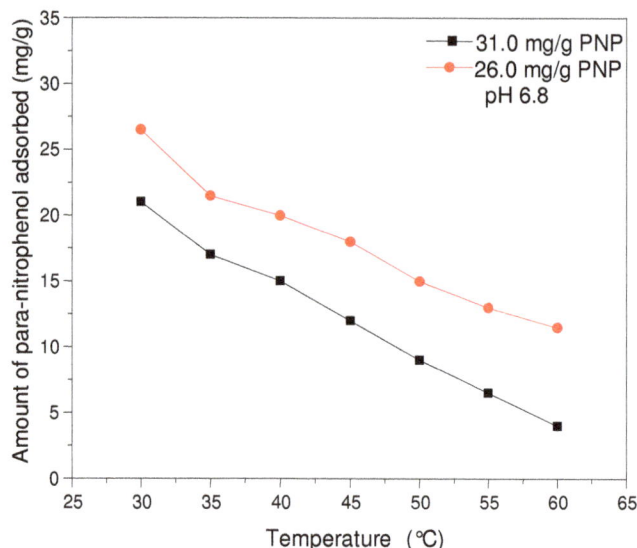

Figure 5. Effect of temperature on adsorption of para-nitrophenol on fly ash.

is temperature dependent. Figure 5 shows the amount of PNP removed from aqueous solution as a function of temperature at PNP concentrations of 26.0 and 31.0 mg/g, respectively. There was a gradual decrease on the uptake of PNP by fly ash from 30 to 60°C. At initial concentration 26.0 mg/g of PNP at 30°C the amount of PNP adsorbed was 21.0 mg/g or 80.72%. As the temperature increased, the amount of PNP adsorbed decreased. At 40°C the amount of PNP uptake by fly ash was 15.0 mg/g or 57.69%. As temperature increased to 60°C, the amount of PNP removed from aqueous solution was 4.5 mg/g or 17.31%

At initial concentration 31.0 mg/g of PNP, at 30°C, the amount of PNP removed from aqueous solution was 26.5 mg/g or 85.48%. At 40°C the amount of PNP removed by fly ash decreased to 19.8 mg/g or 63.87%. From 40 to 60°C there was a progressive decrease in the amount of PNP removed by fly ash. At 60°C the amount of PNP removed was 11.5 mg/g or 37.10%.

It can be seen that the PNP adsorption follows a similar pattern in the two concentrations studied, but the amount adsorbed at a particular temperature differs. These results indicate that the uptake of PNP increases at lower temperature. The decrease on the uptake of PNP with the increase in temperature may be explained as a result of the increase in the average kinetic energy of the PNP phenolate anions; thus increasing the repulsive forces between the phenolate anions and negatively charged fly ash surface. This could lead to desorption or cause the PNP to bounce off the surface of the fly ash instead of colliding and combining with it. Therefore, the increase in temperature may be associated with the decrease in the stability of PNP phenolate anion-adsorbent complex. Jain et al. (2004) reported a decrease in adsorption of phenol onto fly ash as temperature increases from 30 to 50°C.

Conclusion

This study revealed the application of coal fly ash as an adsorbent for removal of toxic solvents from waste water. The kinetic study shows that PNP was adsorbed onto the fly ash very rapidly within the first 20 min, while equilibrium was attained within 2 h for both concentrations of PNP studied. This investigation revealed that PNP uptake was high at pH9 due to the increase of negatively charged fly ash surface, which enhances adsorption. Also the substitution of NO2 group in benzene ring enhances adsorption. This study revealed that the increase in temperature decreases uptake of PNP by the fly ash.

RECOMMENDATION

Fly ash could be used by industries for the treatment of waste water to remove heavy metal ions and organic materials. Using of fly ash for waste water treatment will reduce cost since it is a waste product from coal. Also treatment of soil with fly ash will reduce the population of plant parasites. Reduction of plant parasite will increase agricultural yield. Fly ash could also be used to neutralize soil acidity, thereby increasing agricultural yield. It could be used in the manufacture of cement. This will reduce the cost of reduction of cement since it is a waste product from coal. Reduction in price of cement will enhance development of any country.

ACKNOWLEDGEMENTS

The authors are grateful to Miss Njunbemere Nkajima for technical assistance in performing some measurements. The authors are also grateful to Nichben Pharmaceutical Industry, Awo-omama, Imo State, Nigeria for making use of the facilities in their laboratory.

REFERENCES

Alinnor IJ (2007). Adsorption of heavy metal ions from aqueous solution by fly ash. J. Fuel, 86: 853-857.

Bhatgava DS, Sheldarkar SB (1993). Adsorption of Para–nitrophenol on clay. Water Res., 27: 313.

Eligwe CA, Okolue BN (1994). Adsorption of iron (II) by a Nigerian brown coal. J. Fuel, 73(4): 569-572.

Environmental Protection Agency (1984). Methods 604, Phenols in Federal Register,October 26, part VII,40,CFR,USA, p. 58.

Gupta GS, Prasad G, Singh VN (1990). Removal of chrome dye from aqueous solution by mixed adsorbents. Water Res., 24: 45-50.

Iqbal U, Khan MA, Ihsanullah NA (2005). Effect of selected parameters on the adsorption of phenol on activated charcoal. Int. J. Environ. Stud., 62(1): 47-57.

Jain Ak, Gupta VK, Jain S, Suhas S (2004). Adsorption of phenol on fly ash. Environ. Sci. Technol., 38: 1195.

Khalid M, Joly G, Renand A, Magnoux P (2004). Removal of para-nitrophenol from waste water by adsorption using zeolites. Ind. Eng. Chem. Res., 43: 5275-5280.

Komnitsas K, Bartzas G, Paspaliaris I (2004). Clean up of acidic leachates using fly ash barriers:Laboratory column studies. Global Nest. Int. J., 6(1): 81-89.

Kumar KV, Ramamurthi V, Sivanesan S (2005). Modeling the mechanism involved during the sorption of methylene blue onto fly ash. J. Colloid Interface Sci., 284: 14-21.

Lin CC, Liu HS (2000). Adsorption in a centrifugal field of basic dye adsorption by activated carbon. Indus. Eng. Chem. Res., 39(1): 161-167.

Lin CF, Hsin HC (1996). Resource recoveries of waste fly–ash–synthesis of zeolite–like materials. Environ. Sci. Tech., 29 (4): 1109-1117.

Nollet H, Roels M, Lutgen P, Meeren P, Verstrate W (2003). Removal of PCBs from Wastewater using fly ash. Chemosphere, 56: 655-665.

Parsa J, Stuart H, Munson M, Robert S (1996). Stabilization/solidification of hazardous wastes using fly ash. J. Environ. Eng., 935-939.

Rovatti P, Peloso MA, Ferraiolo G (1988). Susceptibility to regeneration of fly ash as an adsorbent material. Resour. Conser. Recycl., 1:137-143.

Saleem M, Afzal M, Qadeer R, Hanif J (1993). Effect of temperature on the adsorption of zirconium ions on activated charcoal from aqueous solution. Proc. 5th Nat. Chem. Conf., Islamabad, Pakistan, 319-324.

Sell NJ, Norman JC, Vandembasch MB (1994). Removing color and chlorinated ouhanics from pulp mill bleach plant effluent by use of fly ash. Resour. Conser. Recycl., 10: 279-299.

Shende A, Juwarkar AS, Dara SS (1994). Use of fly ash in reducing heavy metal toxicity to plants. Resour. Conser. Recycl., 12: 221-228.

Tanju K, James Ek (1999). Role of granular activated carbon surface chemistry on the adsorption of organic compounds. Environ. Sci. Tech., 33(18): 3217-3224.

Uranowski JL (1998). The effect of surface metal oxides on activated carbon adsorption of phenolics. Water Res., 32 (6): 1841-1851.

Vincini M, Carini F, Silva S (1994). Use of alkaline fly ash as an amendmentor for swine manure. Biosour. Tech., 49: 213-222.

Wang S, Boyjon Y, Choueih A, Zhu ZH (2005). Removal of dyes from aqueous solution using fly ash and red mud. Water Res., 39: 129-138.

Wiessner A, Remmler M, Kuschk P, Stottmeister U (1998). The treatment of a deposited lignite pyrolysis waste water by adsorption using activated carbon and activated coke.Colloids and surfaces, 139(1): 91-97.

Yu JJ, Chou SJ (2000). Contaminated site remedial investigation and feasibility removal of chlorinated volatile carbon fibre adsorption. Chemosphere, 41(3): 371-378.

21

Chlorpyrifos induces hypertension in rats

Alvarez A. Anthon* and Campaña-Salcido A. D.

Faculty of Medicine, University A. of Sinaloa, Mexico.

This study using a non-invasive measurement with a working range from 0–300 mm Hg and a silent built-in air pump that ensured automated cuff inflation/deflation at a constant rate blood pressure (BP) demonstrated an initially elevated BP in all treatment groups for several hours after exposure. This parameter was monitored for 3 and 14 days. BP was elevated at 3- and 14 –day exposure, but not heart rate (HR), except at the second day at 500 bpm only with the 25 mg kg^{-1} dose. The systolic pressure (SBP) increased with the 25 mg kg^{-1} dose at acute treatment P=<0.001; not only but also the media resulted increased with 5 and 25 mg kg^{-1}, P < 0.007, increased too at 5 an 25 mg kg^{-1} and at 14 days of treatment P < 0.001. And the diastolic pressure increased with the 25 mg kg^{-1} dose at 3-day exposure; P< 0.024, but there was an elevation with 5 and 25 mg kg^{-1} doses at 14-day exposure; P< 0.001. The media resulted increased with 2.5, 5, and 25 mg kg^{-1}; P< 0.001. Data suggest that the alteration of the regulation of blood pressure may be involved in the effects of CHP on rats.

Key words: Chlorpyrifos, hypertension, toxicity, oral exposure.

INTRODUCTION

Chlorpyrifos is an organophosphate (OP) pesticide was commonly used in residences and agro industry until the U.S. Environmental Protection Agency (EPA) banned it from domestic use in 2001 due to its alleged but controversial impact on organic functions in exposed individuals. Chlorpyrifos competitively inhibit pseudocholinesterase and acetylcholinesterase, preventing hydrolysis and inactivation of acetylcholine. Acetylcholine accumulates at nerve junctions, causing malfunction of the sympathetic, parasympathetic, and peripheral nervous systems and some of the central nervous system. The substance Chlorpyrifos: (LD50 135 mg/kg/ VO rat Wistar female, 163 male) and bioactivated in liver to chlorpyrifos oxon (Drevenkar et al., 1993).

In patients with mild to moderate poisoning, the nicotinic effects may include tachycardia, hypertension, mydriasis, and muscle cramps. Chlorpyrifos exposure is generally thought to target cholinesterase but chlorpyrifos may also act on cellular intermediates (Song et al., 1997; Pope, 1999). Individually living in poverty and amid dilapidated and crowded conditions, are more likely to be exposed to these environmental toxins.

There are also several studies on the effects these kinds of pesticides on the circulatory system (Davies et al., 2008; Richardson et al., 1975; Smith et al., 2001; Berberian and Enan, 1987; Saldana et al., 2009). In addition, a number of studies reported that acute and chronic toxicity of OP insecticides may lead to degeneration of collagenous and elastin fibers of vascular wall (Yavuz et al., 2005). However, at present there are no studies investigating the effects of Op compounds on the mechanical properties of arteries.

MATERIALS AND METHODS

The Wistar adult rats (4 months old) used for this project, weighting between 250 and 300 g, quickly adapted to the cage and the appliances, and even though some of them were initially disturbed, after spending some time in the cage they became calm. We administered the pesticide chlorpyrifos with corn oil (Table 1).

1. Non-invasive equipment for blood pressure (LE 5160/R Pulse transducer and pressure cuff, for tails between 5 and 10 mm diameter; for rats from 100 to 500 gr). Characteristics: 5001 (basic unity) LE model, "LETICA" for determination of blood pressure in animals.
2. Heater container LETICA (LSI) Mod. LE5610 Ser. No: 7137/03 AC – 110V120W 50/60 Hz. range: 0 – 300 mm Hg. Silent pump inflating/deflating manually to constant proportion. Digital expression of systolic, diastolic, media pressure and heart rate.

---

*Corresponding author. E-mail: Chiaaa_1999@yahoo.com.

Table 1. Scheme of exposure.

Groups *	Dosage p.o.	Days of exposure	
		Acute	Subchronic
Control	Management – Vehicle**	3	14
Chlorpyrifos	2.5, 5, y 25 mg/kg	3	14

*n = 5; N = 60 Wistar male rats; ** corn oil.

Non-invasive measurement: The arterial tension of the rats was registered as follows: A controlled external pressure is applied on the artery at the same time as the blood pulse is detected. The pulse signal detection is carried out by means of a transducer based on a silicon semiconductor full Wheatstone bridge mounted on the base of a clamp to be located around the artery. This technique allows the accurate pulse detection and a high sensitivity regardless of the transducer's position around the tail. The pulse signal level is permanently monitored on the LCD display of the instruments, making it possible to ensure the carrying out of the measurement with a proper pulse level. The mean blood pressure (MP) is computed as MP = DP + 0.33 (SP) – DP). The heart rate was obtained from the pulse signal and evaluated permanently.

The Heater, Model: LE 5610 heater for single holder provides an easy way to increase the measurement capability of all the indirect blood pressure meters.

The animals warming-up facilitates blood pressure measurement by producing a peripheral vasodilatation (including the tails vessels) and inducing the animals rest.

Nevertheless, the heaters offer the ideal environment for an easy performance of the preparation of rats for a reliable pressure measurement, because, besides heating, the Heater isolates the animals from external noise and light (up to reasonable limits), to create a quiet environment to help the animals to relax. Its use greatly increases the number of animals able to be measurement, shortening the interval between successful successive determinations (rodent restrainers helpful in keeping the animals while measuring pressure). MODEL: LE 5024 for Rats from 300 to 440 g.

Statistical analysis

From the data of diastolic blood pressure (DP) and systolic (SP) mean arterial pressure (MAP) = (DP + [SP - PD] 1 / 3) was calculated. From the experimental results of each of the variables under study and for each group of animals under the same combination of treatments (taxon and time), calculating the mean and standard error of the mean. The data were analyzed with analysis of variance test for repeated measures two-factor (RM ANOVA) followed by the test of Student-Newman-Keuls for comparison of means. In all cases it is considered a significance level of P <0.05 in two-way test (Armitage et al., 2002). We used the statistical program Sigma Stat version 2.03 and Sigma Plot version 8.0.

RESULTS

The experimental subjects exposed to the pesticide chlorpyrifos had altered cardiovascular parameters, heart rate, and blood pressure; effects which lasted for 24 h with each of the daily doses in both acute exposure to three days and sub-chronic exposure to fourteen days. A search for 24 h. in the control experiments showed circadian changes of variables without importance for the study (± 5 mmHg). The elevation of blood pressure during the night was higher than that presented in the morning.

This difference in the effects of the pesticide with respect to dose and schedule of administration and also controls the stress of handling (5 mmHg) at doses of 5 and 25 mg/kg was presented with blood pressure in the dark phase. It also presented a significant elevation of systolic blood pressure in groups of 5 and 25 mg/kg during the second night of exposure. Heart rate was not significantly changed; i.e., the changes were within the minimum values, average and maximum according to this species with changing heart rate ranging from 350 to above 500 beats per minute, only highlights the increase of heart rate on the second day of acute exposure dose of 25 mg/kg decreased in the first half of the night with 25 mg, however, with the dose of 5 mg/kg there were no changes but increased although not significantly in the second with no variability with respect to baseline (350-500 beats per minute). The acute and sub-chronic oral exposure to chlorpyrifos at doses of 2.5, 5, and 25 mg/kg for 3 and 14 days resulted in a prolonged hypertensive response in experimental subjects, but not in the control groups (P <0.05), (Figures 1 to 4).

The CHP-induced hypertension persisted for 24 h after each dose. Blood pressure was recorded within 6 h in the morning and 14 h pm to groups of rats in each respective dose in both the periods of exposure.

DISCUSSION

According to LD-50 (82-155 mg/kg) p.o. for Wistar male rat (Kousba et al., 2004), literature survey showed that different doses have been used to observe the chronic effects of chlorpyrifos (2.5-38 mg/kg/day) in rats (Raheja and Gill, 2007; Verma et al., 2009; Kobayashi et al., 1980; Kaur et al., 2007).

Chlorpyrifos we used mainly influenced the pressure regions. We think that the elevated blood pressure of the rats could be the response of the exposure to this pesticide and due the activity of its metabolite over pressor areas in the body. In addition, this pesticide decreased the rate constant of power function, according to other authors (Anand et al., 1990; Gordon and Padnos, 2000; Smith et al., 2001; Smith and Gordon, 2005) or even other components as the content of dichlorvos in chlorpyrifos (Gao et al., 2010). Results show the mean

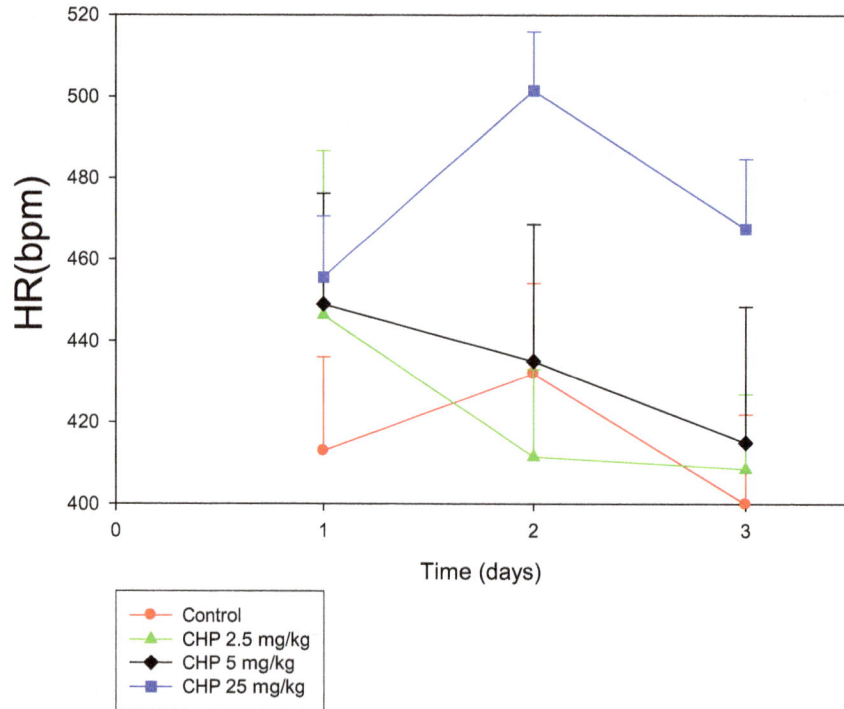

Figure 1. Heart rate was with variations in physiological parameters. Only exposure to one dose of 25 mg/kg increased beats to 500/min the second day, but returned to normal mean values on the third day of exposure to pesticide.

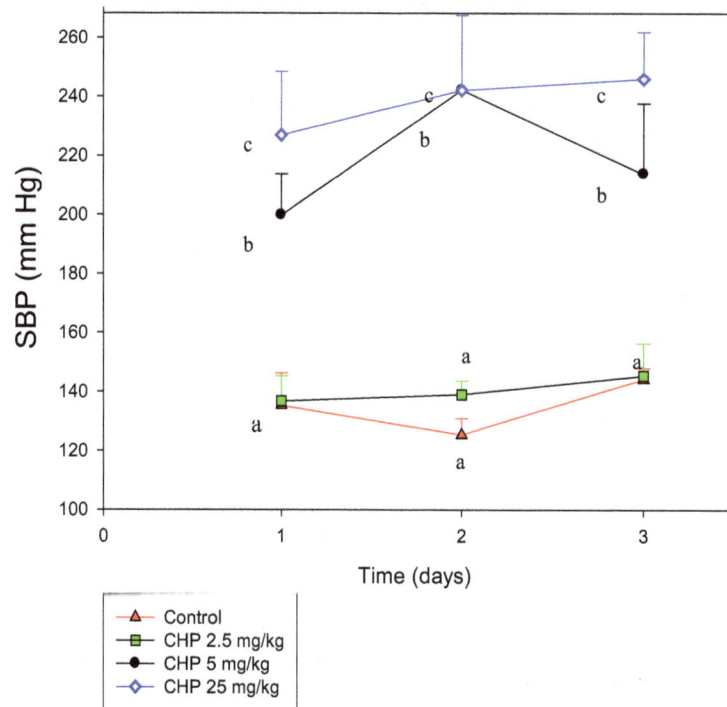

Figure 2. The systolic pressure of rats exposed to chlorpyrifos three days, increased at 25 mg/kg dose; it was statistically significant for all doses, and with respect to the control. With the dose of 2.5 mg/kg no statistical significance resulted compared with the control or the dose of 5 mg/kg. (P = <0.001).

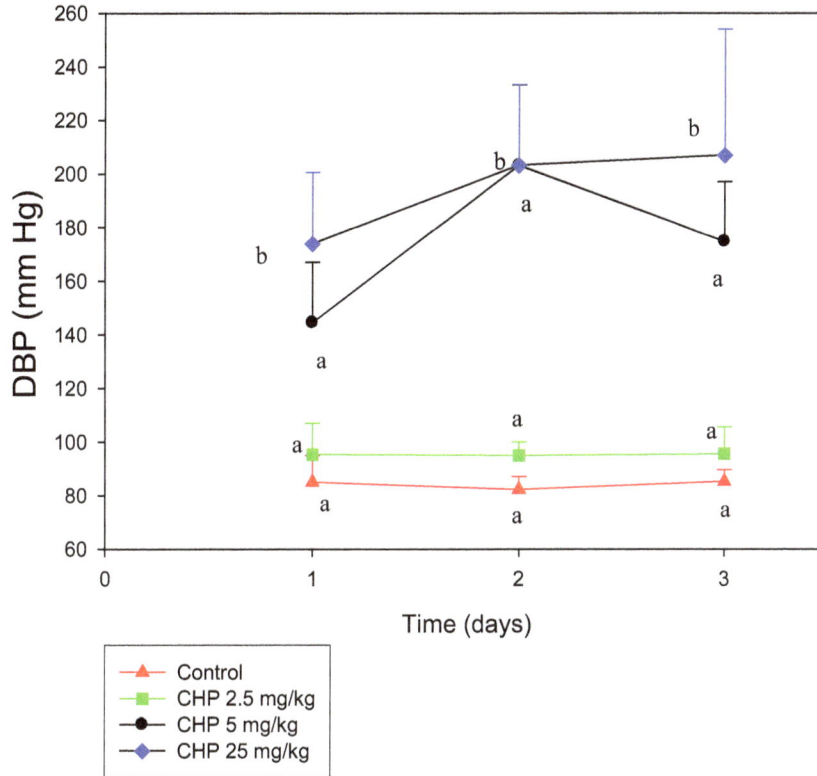

Figure 3. Three days of exposure to pesticide diastolic pressure observed with statistically significant increase in the dose of 25 mg/kg compared to control group (P = 0.024).

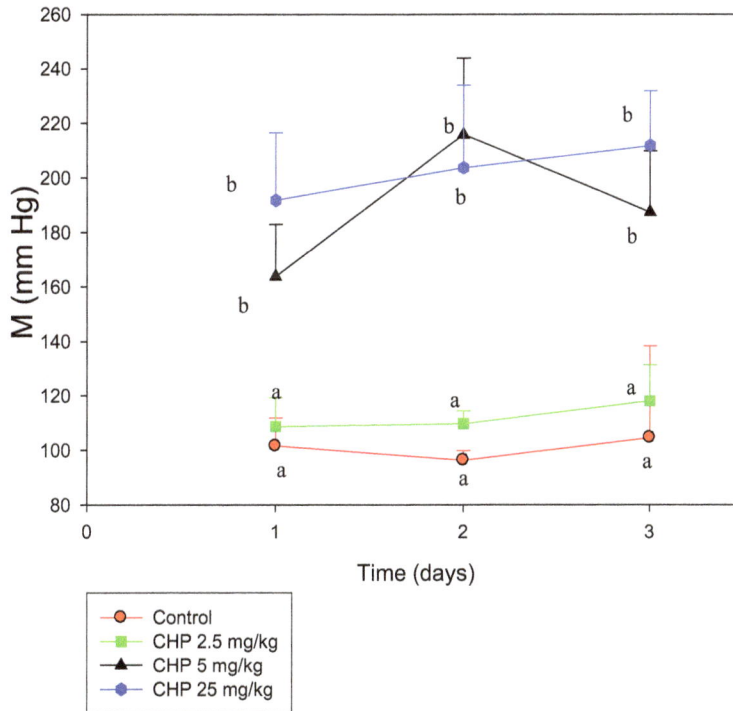

Figure 4. The media pressure of the acute exposure period was high at the doses of 25 and 5 mg/kg compared with the control group. Not being so with the dose of 2.5mg/kg compared to control group (P = 0.007).

values and standard error of the mean obtained for the 4 indicated parameters: SBP=113±13.7mmHg (range 92.6 to 148 mmHg), HR=352±50.8 beats/min (range 222 to 447 beats/min DBP=91±13.6 mmHg (range 76 to 127 mmHg), and the Media Blood Pressure = 85 105.63 ± 19.81 mmHg.

Conclusions

We obtained registers from 22 rats weighing between 200 and 350 g. Exposure to pesticides such as CHP may increase the risk of hypertensive disorders. Laboratory research may provide insights into relationships between pesticides exposure and hypertensive diseases. Comparison of pressure arterial values in the rats: Parameter Evaluated Observed Systolic pressure, mm / Hg 116 113.12 ± 13.73 Diastolic pressure, mmHg 90 91.22 ± 13.56 Heart rate/min. 300-500 351.31 ± 50.78 beats/min. Mean pressure, mm /Hg 85 105.63 ± 19.81. It is according with others authors that implicate this pesticide in a number of studies that workers exposed chronically to pesticides or living in the neighborhood of the exposed area had high blood pressure (Berberian and Enan, 1987; Saldana et al., 2009; Guvenc et al., 2011).

ACKNOWLEDGMENTS

We thank CONACyT (Nos. 174311 and 126555), by support this work, and the University of Zacatecas. We also acknowledge Elaine Bell for her review of the manuscript.

REFERENCES

Anand M, Gulati A, Gopal K, Gupta GS, Khanna RN, Ray PK (1990). Hypertension and myocarditis in rabbits exposed to hexachlorocyclohexane and endosulfan. Vet. Hum. Toxicol., 1990, 32(6):521–523.

Armitage P, Berry G, Matthews JNS (2002). Statistical Methods in Medical Research. In Analysing variances, counts and other measures (4th edition). Oxford: Blackwell Sci., pp. 147-233.

Berberian IG, Enan EE (1987). Neurotoxic studies in humans occupationally exposed to pesticides. J. Soc. Occup. Med., 37: 126-127.

Davies J, Roberts D, Eyer P, Buckley N, Eddleston M (2008). Hypotension in severe dimethoate self-poisoning. Clin. Toxicol., 46: 880–884.

Drevenkar V, Vasilic Z, Stengl B (1993). Chlorpyrifos metabolites in serum and urine of poisoned persons. Chem Biol. Interact, 87(1-3):315-322.

Gao X, Wang XY, Wang D, Hao XH, Min SG (2010). Rapid detection of dichlorvos in chlorpyrifos by mid-infrared and near-infrared spectroscopy. Guang Pu Xue Yu Guang Pu Fen Xi. 30(11):2962-6.

Gordon CJ, Padnos BK (2000). Prolonged elevation in blood pressure in the unrestrained rat exposed to chlorpyrifos. Toxicol., 146(1):1–13.

Guvenc TB, Ozturk N, Comelekoglu UB, Yilmaz C (2011). Effects of Organophosphate Insecticides on Mechanical Properties of Rat Aorta. Physiol. Res., 60: 39-46.

Kaur P, Radotrab, Minz RW, Gill KD (2007). Impaired mitochondrial energy metabolism and neuronal apoptotic cell death after chronic dichlorvos (OP) exposure in rat brain. Neurotoxicolody, 28: 1208-1219.

Kobayashi H, Yuyama A, Imajo S, Matsusaka N (1980). Effects of acute and chronic administration of DDVP (dichlorvos) on distribution of brain acetylcholine in rats. Toxicol. Sci., 5: 311-319.

Kousba AA. Sultatos LG. Poet TS (2004). Timchalk S. Comparison of Chlorpyrifos oxon and Paraoxon Acethylcholinesterase Inhibition Dynamics: Potential Role of a Peripheral Binding Site Toxicol. Sci., 80: 239-248.

Pope CN (1999). Organophosphorus pesticides: Do they all have the same mechanism of action? J. Toxicol. Environ. Health B 2: 161-181.

Raheja G, GILL KD (2007). Altered cholinergic metabolism and muscarinic receptor linked second messenger pathways after chronic exposure to dichlorvos in rat brain. Toxicol. Ind. Health 23: 25-37

Richardson JA, Keil JE, Sandifer, SH (1975). Catecholamine metabolism in humans exposed to pesticides. Environ. Res., 9: 290-294.

Saldana TM, Basso O, Baird DD, Hoppin JA, Weinberg CR, Blair A, Michael CR, Alavanja y Sandler DP (2009). Pesticide Exposure and Hypertensive Disorders During Pregnancy. Environ. Health Perspect 117:1393-1396.

Smith EC, Padnos B, Cordon CJ (2001). Peripheral versus central muscarinic effects on blood pressure, cardiac contractility, heart rate, and body temperature in the rat monitored by radiotelemetry. Pharmacol. Toxicol., 89(1):35-42.

Smith EG, Gordon CJ (2005). The effects of chlorpyrifos on blood pressure and temperature regulation in spontaneously hypertensive rats. Basic Clin. Pharmacolog. Toxicol., 96(6):503–511.

Song X, Seidler FJ, Saleh JL, Zhang J, Padilla S, Slotkin TA (1997). Cellular mechanisms for developmental toxicity of chlorpyrifos: targeting the adenylyl cyclase signaling cascade.Toxicol Appl Pharmacol., Jul; 145(1):158-74.

Verma SK, Raheja G, Gill KD (2009). Role of muscarinic signal transduction and CREB phosphorylation in dichlorvos-induced memory deficits in rats: an acetylcholine independent mechanism. Toxicol., 256: 175-82.

Yavuz T, Delibas N, Yildirim B, Altuntas I, Candiro, Cora A, Karahan N, Ibrisime K (2005). Vascular wall damage in rats induced by organophosphorus insecticide methidathion. Toxicol. Lett., 155(1):59-64.

22

Kinetic and equilibrium studies of the adsorption of lead (II) ions from aqueous solution onto two Cameroon clays: Kaolinite and smectite

Joseph KETCHA MBADCAM[1]*, Solomon Gabche ANAGHO[2], Julius NDI NSAMI[1] and Adélaïde Maguie KAMMEGNE[1]

[1]Physical and Theoretical Chemistry Laboratory, Faculty of Science, University of Yaoundé 1, Yaoundé – Cameroon.
[2]Department of Chemistry, University of Dschang, Dschang – Cameroon.

The efficiency of the adsorbents; kaolinite (MY22s) and smectite (Sa01) for the removal of lead (II) ions from aqueous solution was investigated. Parameters such as contact time (t) and initial concentration (C_o) with a particle size of 80 µm were studied by using a batch scale adsorption technique. This is with the intension of optimising appropriate conditions to be utilised on a commercial scale to decontaminate industrial effluents. Adsorption equilibrium was reached within 60 min for the 50, 70, 90, 110 and 130 ppm initial concentration (C_o), of lead (II) ions with a 500 mg weight adsorbent. The adsorption capacity (Q_t) of lead ion removal increases with increasing initial concentration, C_o of lead (II) ions in solution. The adsorption efficiency or percentage of removal reached 92% for 130 ppm initial concentration, C_o, of lead (II) ions with Sa01. Kinetic modelling analysis using the linear correlation coefficient (R^2) values showed that the adsorption mechanism follows the pseudo-second order model for the adsorption of lead (II) ions on Kaolinite (MY22s) and smectite (Sa01). The adsorption data were also modelled by using both the Langmuir and Freundlich classical adsorption isotherms. These experimental data fitted the Langmuir isotherm for kaolinite (MY22s) and Freundlich isotherm for smectite (Sa01).

Key words: Adsorption, smectite (Sa01), kaolinite (MY22s), lead (II) ions, kinetics, equilibrium adsorption, wastewater treatment.

INTRODUCTION

The progressive increase of industrial technology results in the continuous increase in environmental pollution. Industries discharge different types of heavy metals waste into the environment at an unprecedented and at a constant increasing rate. These heavy metals waste may be discharged into streams, rivers and lakes and the continuous enrichment of these waters with these metals waste beyond the healthy level may cause poisoning, leading to various sicknesses.

Consequently the treatment of polluted industrial wastewater before they are released into the environment remains a topic of global concern. Moreover, contamination of ground water is today a major concern in the management of water resources (Waid and Hossam, 2007; Ishizaki and Marti, 1983). The majority of organic pollutants are susceptible to biological degradation while heavy metal ions can not be degraded into harmless end-product (Riaz and Sohail, 2005). The toxicity of metals depends especially on their chemical forms rather than on their total elemental contents and therefore, speciation studies increasingly gain importance (Orumwense, 1996; Norotry et al., 2000). Lead is one of the heavy metals, which is highly toxic to human, plant and animal. The

---
*Corresponding author. E-mail: jketcha@yahoo.com.

ABBREVIATIONS: MY22s, Kaolinite from Mayouom; West Region of Cameroon; Sa01, Smectite from Sabga; North West Region of Cameroon; CEC, Cationic Exchange Capacity; PSD, Pore Size Distribution; PZC, Point of Zero Charge.

metal is of special concern because of its environmental importance related to its well known toxicity and intensive use in industries (Pratik and Choksi, 2008).

The implications of lead pollution in animal and human beings are their interference with the activities of intracellular enzymes (Nagarethinam and Ananthakrishnam, 2002), causing for example anaemia, kidney failure, impairment of central nervous system and damage to DNA and RNA. According to the World Health Organization (WHO, 1971), the accepted range of lead (II) ions in water is 0.01 ppm (Waid and Hossam, 2007).

Some of the major sources of lead release into the environment are metallurgical industry, electroplating, metal finishing industries, paint manufacture, storage battery manufacture, petroleum refining, fuel combustion, photo graphic materials and drainage from ore and mines (Zamzow et al., 1990).

Developing countries suffer from water pollution; the expensive method of treatment is the main problems in these countries. This is greatly reduced by the utilization of natural material and processes (Adesola et al., 2006). Science provides many practical solutions to minimizing the present level at which these metals are introduced into the environment and for remediating (cleaning up) past problems; therefore avoiding their dangerous effect (Adesola et al., 2006; Reed and Matsumoto, 1993).

Some of the methods that are available to reduce heavy metal concentrations from wastewater are chemical precipitation, filtration, ion exchange, reverse osmosis, ultra-filtration, electrochemical deposition, coagulation and adsorption. However, the above mentioned methods are not economically feasible for small and medium size industries. It is therefore necessary to search for low cost techniques that may be effective, less environmentally degrading and economical (Orumwense, 1996).

For low concentration of metal ions in wastewater, the adsorption process is recommended for their removal because they are cheap, simple, sludge free and involve small initial cost and land investment (Norotry et al., 2000).

The process of adsorption implies the presence of a solid adsorbent that binds molecules by physical attractive forces, ion exchange, and chemical binding; it is advisable that the adsorbent is available in large quantities, easily regenerable and cheap.

Activated carbon is a potential adsorbent for the removal of several organic and inorganic pollutants but due to it high cost and (10-15%) loss during regeneration; alternative low cost adsorbents have attracted the attention of several investigators to provide an alternate for the high cost activated carbon (Emmanuel and Olalekan, 2008).

In the present study, kaolinite (MY22s) and smectite (Sa01) clays which were obtained from Mayoum and Sabga, both from the West and North West regions of Cameroon respectively were investigated as a potential

and low cost adsorbent for lead (II) ions removal from aqueous solutions.

MATERIALS AND METHODS

Two local clay materials; kaolinite (MY22s) and smectite (Sa01), lead nitrate salt $Pb(NO_3)_2$ (AR sample, S.D. Fine-Chem Ltd. India) were used in this study. Double distilled (DD) water was used for the preparation of all the reagents.

Adsorbent

These local materials were sun dried, converted into fine powder by the use of a mortar and screened through a sieve (Retsch) to get a geometrical size of 80 μm, the used beaker was kept into an oven (HEREAUS) at 110°C for a period of 24 h (Hanafiah and Ngah, 2006) in order to remove water content and volatile impurities, removed and cooled in a desiccator containing $CaCl_2$ (drying agent) for 30-60 min. These adsorbents were removed from the desiccator and the required mass 0.5 g (that was used all through), weighed and stored in an airtight plastic container for the experiment. The characterization of these local materials is tabulated in Tables 1 and 2 (Tonle et al., 2003; Njoya et al., 2007).

Adsorbate

1000 mg/L of lead (II) ions was prepared by dissolving 1.599 g of hydrous lead nitrate $(Pb(NO_3)_2)$ in a double-distilled water and made up to 1000 ml. The stock solution was diluted to obtain required standard solutions.

Batch adsorption experiments

The batch kinetic experiments of the adsorption studies were conducted at room temperature (25°C) in a 250 ml screw-cap conical flask. For each run, 0.5 g of the adsorbent was weighed and placed in the flask containing 20 ml solution of lead (II) ions of a desired concentration (ranging from 50 to 130 ppm) at pH - 4.9. The suspension was stirred for interval of time between 30-300 min, using a magnetic stirrer. After agitation, the suspensions were centrifuged at 5000 rpm for 10 min to separate the solid and liquid phases. The supernatant was capped into test tubes, labelled, and was followed by complex formation.

Formation of Lead dithizonate complex

Since lead compounds can form coloured complexes with dithizone solution together with the aid of a suitable buffer, the concentration of lead in solution could therefore be determine by colorimetric analysis (Diaper and Kuksis, 1957).

The concentration of the lead dithizonate is sensitive to and readily affected by the concentration of the lead ion, pH, the character and concentration of the buffer (Diaper and Kuksis, 1957). 2 ml of the solution containing lead (II) ions was added to 2 ml buffer solution of ammonium chloride/ammonia and 2 ml solution of dithizone in ethanol.

The chemical reaction of the complex formation and structure is as shown in Figure 1:

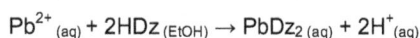

$$Pb^{2+}{}_{(aq)} + 2HDz_{(EtOH)} \rightarrow PbDz_{2\,(aq)} + 2H^+{}_{(aq)}$$

Table 1. Chemical Composition of the clays (kaolonite and smectite).

Mineral composition	SiO_2	Al_2O_3	Fe_2O_3	MnO	MgO	CaO	Na_2O	K_2O	TiO_2	P_2O_5	l.i	Total
(%) MY22s	48.4	33.0	2.3	< d.1	< d.1	<d.1	< d.1	0.1	3.1	0.3	11	98.2
(%) Sa01	63.3	14.5	4.2	< d.1	0.2	0.7	0.4	2.3	0.2	<d.1	13	98.8

d.l = detection limit; l.i = lost at ignition.

Table 2. Physical characteristics of kaolinite (MY22s) and smectite (Sa01).

Sample	Parameter	
kaolinite(MY22s)	Colour	Gray
	Mineral composition (%)	Kaolinite (75.8); quartz (9.4); illite (7.7); florencite (0.4); rutile (3.7); goethite (2.5)
	External specific area	15 m^2/g
	CEC	78.2 × 10^2 (m eq/100 g)
smectite(Sa01)	Colour	Dirty white
	External specific area	50 m^2/g
	Micropores Volume	0.01 m^3/g
	Mineral composition (%)	Montmorillonite (79.2); quartz (10) ; feldspath (5); hematite (2.53); limonite (0.1)
	CEC	86 × 10^2 (m eq/100 g)

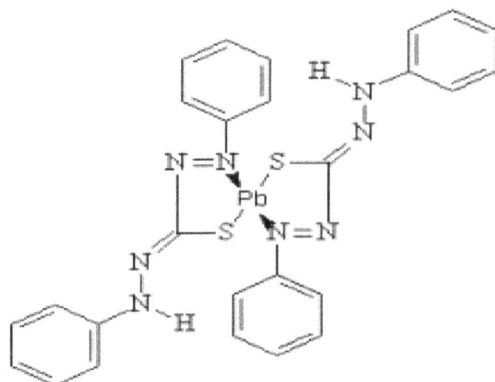

Figure 1. Structure of lead dithizonate complex.

The orange unstable complex formed was allowed for 24 h for complete formation of a stable light yellow complex.

Each experiment was duplicated under identical conditions. Lead (II) ions concentration was determined spectrophotometrically using a UV-Visible spectrophotometer set at 490 nm (Shanmugavalli, et al., 2006).

The quantity adsorbed by a unit mass of an adsorbent (Q_e) and the adsorption percentage (%R) at an instant was calculated from the differences between the concentrations of lead (II) before and after adsorption.

Adsorption kinetics

The study of adsorption kinetics describes the resistance to solute transfer from the solution up to the boundary layer at the solid-liquid interface to the pore water and then to the solid.

It is well known that adsorption kinetics is mainly controlled by the following steps;

1. Solute molecules transfer from the solution to the boundary film;
2. Solute molecules transfer from the film to the surface of the sorbent (external diffusion);
3. Diffusion from the surface to intra-particle sites and
4. Interaction of solute molecules with the available sites on the internal surface (Demirbas et al., 2007).

The evolution of the adsorption process can be followed by measuring the number of particles adsorbed per unit time. Many kinetic models have been proposed for the adsorption of solutes on solids. They include amongst others the Lagergren's pseudo-first order kinetic model, pseudo-second order model, the Elovich model, mass transfer and intraparticle diffusion models.

Pseudo-first order kinetic model

The pseudo first order equation was suggested by Lagergren, (Demirbas et al., 2007) for the adsorption of solid-liquid systems. It is generally expressed as follows:

$$\frac{dQ_t}{dt} = k_1(Q_e - Q_t) \tag{1}$$

where, Q_e and Q_t are the adsorption capacity at equilibrium and at time t respectively (mg/g). K_1 is the rate constant of pseudo-first order adsorption (min^{-1}). After integration and applying boundary conditions, t = 0 to t = t and Q_t = 0 to Q_t = Q_t, the integrated form of equation (1) becomes:

$$\ln(Q_e - Q_t) = \ln(Q_e) - K_1 t \tag{2}$$

This equation is verified if the plot of in $(Q_e - Q_t)$ as a function of time gives a straight line. K_1 is deduced from the slope of the line and Q_t gives the vertical intercept.

Pseudo-second-order model

The pseudo second order adsorption kinetic rate equation is expressed as (Demirbas et al., 2007):

$$\frac{dQ_t}{dt} = k_2(Q_e - Q_t)^2$$

(3)

Where K_2 (mg^{-1} min^{-1}) is the rate constant of pseudo second order adsorption. From the boundary conditions $t = 0$ to $t = t$ and $Q_t = 0$ to $Q_t = Q_t$, the integrated form of equation (3) becomes:

$$\frac{1}{Q_e - Q_t} = \frac{1}{Q_e} + k_2 t$$

(4)

This is the integrated rate law for a pseudo second order reaction. Equation (4) can be rearranged to obtain equation (5) which has a linear form:

$$\frac{1}{Q_e} = \frac{1}{k_2 Q_e{}^2} + \frac{t}{Q_e}$$

(5)

If the initial adsorption rate ho (mg. g^{-1} min^{-1}) is:

$$h_o = K_2 Q_e{}^2$$

(6)

Then, equations (5) and (6) become:

$$\frac{1}{Q_e} = \frac{1}{h_0} + \frac{t}{Q_e}$$

(7)

Thus, from equation (7), plots of (t/Q_t) versus t give values of Q_e and K_2 from the slopes and intercepts respectively.

The Elovich model

The Elovich equation is generally expressed as:

$$\frac{dQ_t}{dt} = \alpha \varepsilon xp(-\beta Q_t)$$

(8)

Where, α- is the initial sorption rate (mg g^{-1}min) and β-is the desorption rate constant (g mg^{-1}) during any one experiment. To simplify the Elovich equation, Chien and Clayton, (1980) and Elkhatib et al., 2007) assumed αβt >> 1 and by applying the boundary conditions $Q_t = 0$ at $t = 0$ and $Q_t = Q_t$ at $t = t$, equation (8) becomes:

$$Q_e = \frac{1}{\beta} \ln(\alpha\beta) + \frac{1}{\beta} \ln t$$

(9)

Thus, a plot of Qt versus ln (t) should yield a linear relationship with a slope of (1/β) and an intercept of (1/β) ln (αβ) if the sorption

process fits the Elovich equation (Mirina and Mile, 2002).

Mass transfer model

The general equation for the mass transfer model is as follows:

$$C_t = D_e k_0 t$$

(10)

Where, C_o and C_t are the initial concentration and concentration of the solute (mg/L) at an instant t, while t is the agitation time (min), D is the mass transfer constant and K_0 the adsorption constant.
The linearization of equation (10) permits us to obtain the following expression:

$$\ln(C_0 - C_t) = \ln D + k_0 t$$

(11)

This equation is verified if the plot of ln $(C_o - C_t)$ as a function of time gives a straight line. K_o is deduced from the slope of the line and D is obtained from the vertical intercept (Pratik and Choksi, 2008).

The Intra-particle diffusion model

When the intra-particle mass transfer resistance is the rate limiting step, then the sorption process is described as being particle diffusion controlled (Igwe and Abia, 2007). The intra-particle diffusion model is expressed as equation (12)

$$R = k_{id} t^a$$

(12)

Where R is the percentage of lead (II) ions adsorbed, t the contact time, a, the adsorption mechanism, k_{id} the intra-particle diffusion rate constant (mg g^{-1} min$^{-1/2}$). A linear form of equation (12) is:

$$\ln R = \ln K_{id} + a \ln t$$

(13)

The plot of lnR versus lnt should give a linear relationship from where the constants a, and K_{id} can be determined from the slope and intercept of the plot respectively.
Higher values of K_{id} illustrate an enhancement in the rate of adsorption whereas larger a values illustrate a better adsorption mechanism. This is related to an improved bonding between lead (II) ions and the adsorbent particles.

Adsorption Isotherms

The equilibrium relationship between adsorbent and adsorbate are described by adsorption isotherm, usually the ratio between the quantity adsorbed and that remaining in solution at a fixed temperature at equilibrium. In this paper, the Langmuir Isotherm and the Freundlich Isotherm were investigated.

The Langmuir Isotherm

The Langmuir adsorption isotherm is often used for adsorption of a solute from a liquid solution. The Langmuir adsorption isotherm is perhaps the best known of all isotherms describing adsorption and is often expressed as:

$$Q_e = \frac{Q_m K C_e}{1 + K C_e} \qquad (14)$$

Where, Q_e (mg of adsorbate per g of adsorbent) is the adsorption density at the equilibrium solute concentration. C_e, C_e is the equilibrium concentration of adsorbate in solution (mg/L). Q_m (mg of solute adsorbed per g of adsorbent) is the maximum adsorption capacity corresponding to complete monolayer coverage. K is the Langmuir constant related to energy of the adsorption (L of adsorbate per mg of adsorbent).

The above equation can be rearranged to the following linear form:

$$\frac{C_e}{Q_e} = \frac{1}{Q_m K} + \frac{C_e}{Q_m} \qquad (15)$$

The linear form can be used for linearization of experimental data by plotting C_e/Q_e against C_e. The Langmuir constants Q_m and K can be evaluated from the slope and intercept of the linear equation.

This is the simplest physically possible isotherm. It is based on three assumptions:

1. Adsorption cannot proceed beyond monolayer coverage.
2. All surface sites are equivalent and can accommodate at most one adsorbed atom.
3. The ability of a molecule to adsorb at a given site is independent of the occupation of neighbouring sites (Lezeck et al., 2000).

Freundlich adsorption Isotherm

The Freundlich isotherm is the earliest known relationship describing the adsorption equation and is often expressed as (Najua et al., 2008):

$$Q_e = K_f C_e^{1/n} \qquad (16)$$

Where Q_e is the quantity of solute adsorbed at equilibrium (adsorption density: mg of adsorbate per g of adsorbent). C_e is the concentration of adsorbate at equilibrium. K_f and n are the empirical constants dependent on several factors and n is greater than one.

This equation is conveniently used in linear form by taking the logarithmic of both sides as:

$$\ln Q_e = \ln K_f + \frac{1}{n} \ln C_e \qquad (17)$$

A plot of $\ln Q_e$ against $\ln C_e$ yielding a straight line indicates the confirmation of the Freundlich isotherm for adsorption. The constants n and K can be determined from the slope and the intercept respectively.

These two models are widely used, the former assuming that maximum adsorption occurs when the surface is covered by one layer of adsorbate; and the latter being purely empirical.

RESULTS AND DISCUSSION

Effect of agitation time and initial lead (II) ions concentration on adsorption

The equilibrium adsorption capacity of kaolinite (MY22s)

and smectite (Sa01) was found to increase with increased in initial lead concentration (Figure 2), it was found out that on increasing the concentration of the lead (II) ions in solution from 50 to 130 ppm, the amount removed increases from 0.68 to 3.40 mg/g (MY22s), 1.52 to 4.68 (Sa01) as shown in Table 3. This indicates that there are plenty of adsorption sites on Sa01 than MY22s available for the adsorption of lead. Similar trends were observed for the other adsorption cases indicating that the adsorption capacity increases with increasing initial lead ion concentration within the experimental operating conditions (Reed and Matsumoto, 1993).

This is in agreement with the results obtained by other investigators (Hanafiah and Ngah, 2006) for the adsorption of lead (II) on rubber leaf powder. The increase of loading capacities of the sorbent with increasing initial lead (II) ion concentrations may also be due to a higher interaction between substituted lead (II) and sorbent (Ömer et al., 2005). Moreover, this can be explained by the fact that more adsorption sites were being covered as the metal ions concentration increased (Najua et al., 2008).

Effect of contact time

The results on Figure 2 shows that the enhanced adsorption of lead (II) ions with increase in agitation time may be due to the decrease in boundary layer resistance to mass transfer in the bulk solution and an increase in the kinetic energy of hydrated ions. By increasing the agitation time, the boundary layer resistance will be reduced and there will be an increase in the mobility of ions in the solution (Hanafiah and Ngah, 2006). For 50, 70, 90 and 130 ppm lead (II) ions initial concentration, the adsorption equilibrium time was obtained after 60 minutes for both MY22s and Sa01.

This result is interesting because equilibrium time is one of the important considerations for economical wastewater treatment applications (Francis, 2006). The maximum amount of lead ions was adsorbed within the first 60 min (92% of total metal ions adsorbed) and thereafter the adsorption proceeded at a slower rate until equilibrium was reached.

Kinetic studies

Kinetic equations have been developed to explain the transport of metals onto various adsorbents. To analyze the sorption rates of lead (II) metal ions onto smectite and kaolinite, five simple kinetic models were tested.

From these kinetic models, only the pseudo-second order model agreed well with the experimental data with Sa01 being the most following their correlation coefficients; Figure 3. These kinetic models are only concerned with the effect of the observable parameters on the over all rate of sorption (Cassey, 1997).

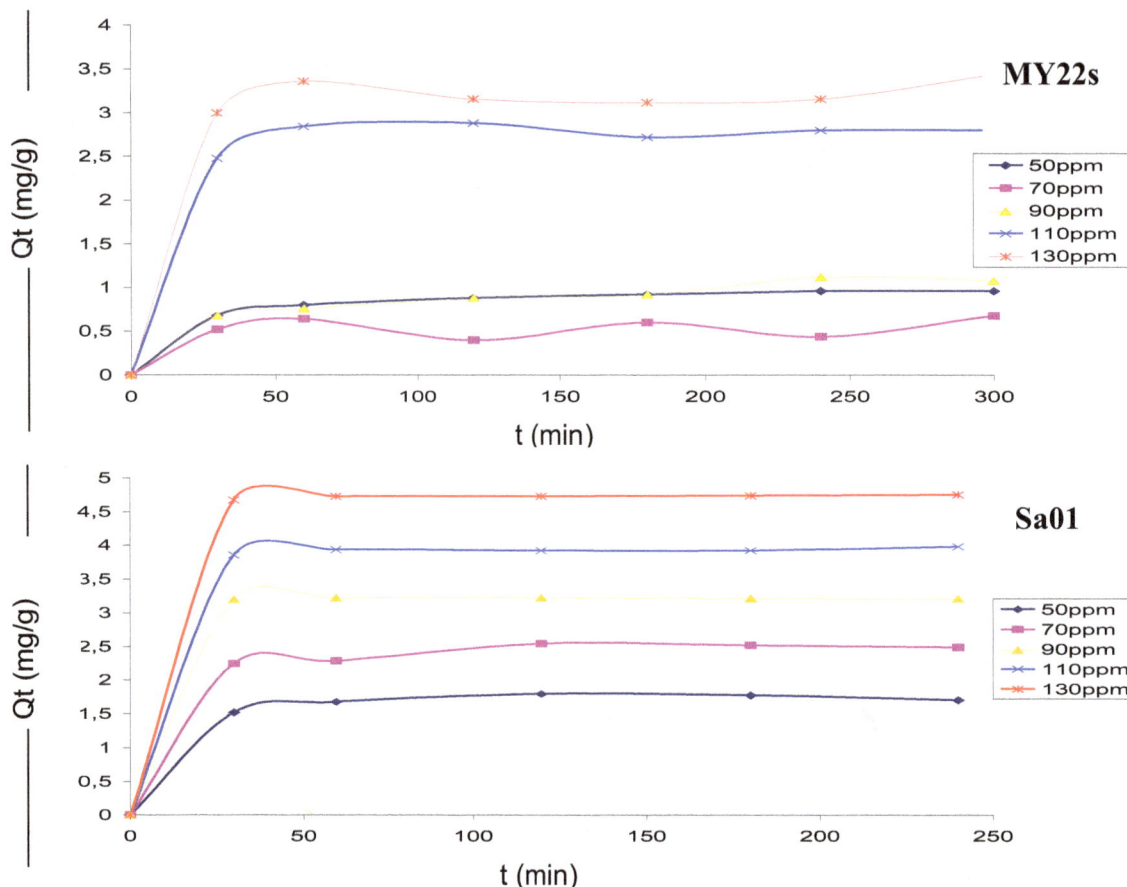

Figure 2. The adsorption capacity Q_t (mg/g) of lead (II) ion on MY22s and Sa01 at different initial concentrations C_0 (50-130 ppm) with time t (min).

Table 3. Langmuir and Freundlich equilibrium constants.

Parameter	Langmuir Isotherm				Freundlich Isotherm		
	$Q_{(exp)}$ (mg/g)	K_L (Lmg^{-1})	Q_m (mg/g)	R^2	K_f(L g^{-1})	1/n	R^2
MY22s 80 μm	3.4400	0.4760	0.9770	0.9770	24.7100	-0.2030	0.06220
Sa01 80 μm	4.7400	0.8170	3.1370	0.8190	0.4900	0.8480	0.9634

Pseudo-second order

The pseudo second order adsorption kinetics for the linearised rate equation is expressed as:

$$\frac{1}{Q_e} = \frac{1}{h_0} + \frac{t}{Q_e}$$

(7)

The plots of (t/Q_t) versus t are as shown in Figure 3. Absorption data at 25°C best fitted with the linearised Langmuir isotherm for MY22s, with the correlation coefficient constant (R^2) of 0.9770. For the Freundlich adsorption isotherm, only the smectite (Sa01) experimental data fitted the model with R^2 equal 0.9634.

Linear curves, obtained by plotting C_e/Q_e versus C_e, indicate the validity of Langmuir isotherm and lnQ_e versus lnC_e, indicate the validity of Freundlich isotherm (Mirina and Mile, 2002; Shanmuga et al., 2007).

The experimental data are better represented by the Langmuir isotherm as indicated by the high R^2 values in Table 3 and Table 4. The fact that the Langmuir isotherm fits the experimental data very well may be due to the homogenous distribution of active sites on the adsorbent surfaces; since the Langmuir equation assumes that the surface is homogenous. Thus the applicability of the Langmuir isotherm in the present system indicates the monolayer coverage of lead (II) ions on the outer surface of the adsorbents (Najua et al., 2008).

However, the Freundlich isotherm better describes

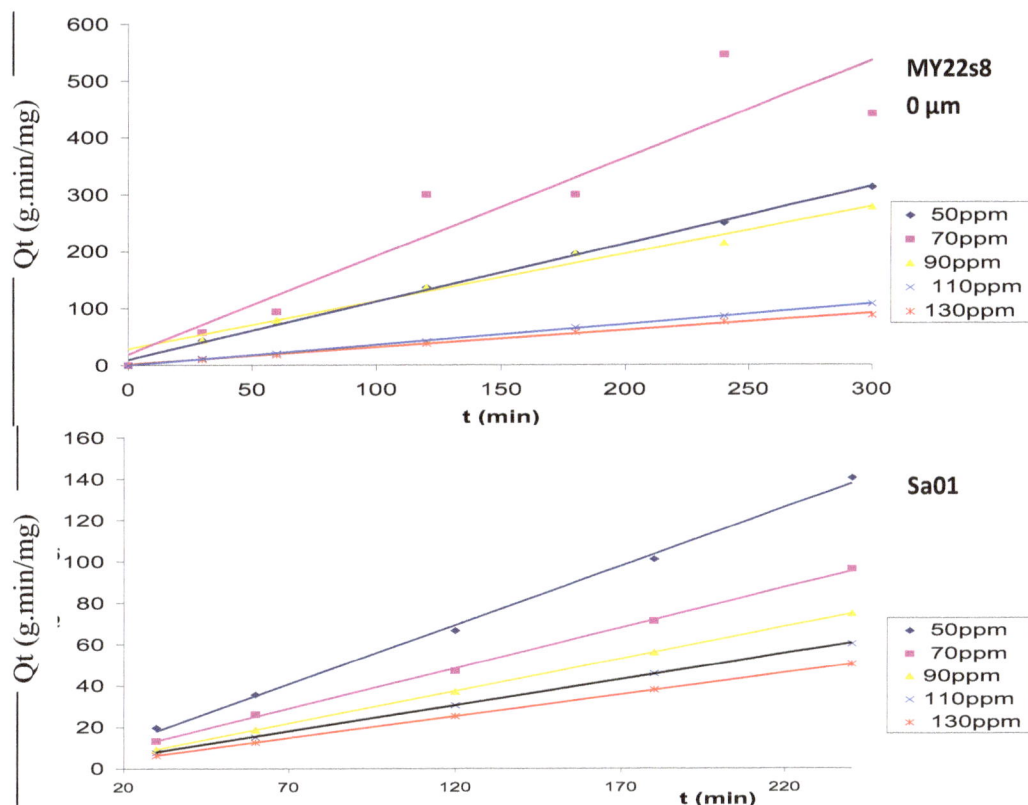

Figure 3. The linear plot from pseudo-second order kinetic model.

Table 4. Rate constants and correlation coefficients of kinetic models of MY22s, at different concentrations.

Models	Parameter	C$_o$ (ppm)				
		50	70	90	110	130
Pseudo-first order	R^2	0.9209	0.02902	0.9614	0.3628	0.5550
Pseudo-second order	R^2	0.9977	0.8835	0.9806	0.9993	0.9935
Elovich	R^2	0.9795	0.0073	0.9132	0.3590	0.2070
Mass diffusion	R^2	0.7934	0.0254	0.9189	0.1833	0.2299
Intra-particle diffusion	R^2	0.9659	0.0028	0.9460	0.3689	0.2181

adsorption with the adsorbent Sa01. The values of 1/n, ranging from 0 to 1, is a measure of adsorption intensity or surface heterogeneity that becomes more heterogeneous as its value gets closer to zero (Ketcha et al., 2007). A value for 1/n below 1 indicates a normal Langmuir adsorption isotherm, while 1/n above 1 is indicative of cooperative adsorption (Abia and Asuquo, 2006).

The value of 1/n = 0.8480 obtained with the adsorbent Sa01, supports the fact that, adsorption takes place homogenously on the surface.

Conclusion

Equilibrium and kinetic studies were determined for the

adsorption of lead (II) ions from aqueous solutions onto smectite (Sa01) and kaolinite (MY22s) in the concentration range from 50 to 130 ppm and the initial pH of 4.9 at an ambient temperature of 25°C.

Following the R^2 values, adsorption of lead (II) ions onto Sa01 followed both Langmuir and Freundlich isotherms models with that of Freundlich being higher than Langmuir. But with MY22s, the equilibrium data of adsorption are in good agreement only with Langmuir isotherms model at an ambient temperature (25°C).

The kinetic sorption model for pseudo-second other was in good agreement with the experimental equilibrium data. A contact time of one hour was enough for the system to reach equilibrium. Kinetic data would be useful for the fabrication and designing of wastewater treatment plants. The adsorption capacities of MY22s and Sa01 were compared for the uptake of lead (II) ions at optimum (maximum removal) conditions and the relative adsorption capacities are: MY22s, (3.36 mg/g) < Sa01, (4.73 mg/g).

The results of analysis of all the adsorbents clearly established that Sa01, is efficient in the adsorption of lead (II) ion from aqueous solution. Sa01 is not only economical, but easily available. Hence it is practically feasible for developing counties and will be useful for the economic treatment of wastewaters containing lead metal.

ACKNOWLEDGEMENT

We sincerely thank all the members of the Research Group: "Adsorption and Surfaces" of the Physical and Theoretical Chemistry Laboratory of the University of Yaoundé I, for their remarks and suggestions.

REFERENCES

Abia A, Asuquo ED (2006). "Lead (II) and NICKEL (II) Adsorption kinetics from Aqueous metal Solutions using Chemically modified and unmodified Agricultural Adsorbents." Afr. J. Biotechnol., 5(16): 1475-1482

Adesola BA, Oyebaniji B, Adebowale R (2006). "Biosorption of Lead ions from Aqueous Solution by Maize Leaf". Int. J. Phys. Sci., 1(1): 23-26.

Cassey TJ (1997). 'Unit Treatment Processes in Water and Wastewater Engineering.' John Wiley and Sons Ltd. England, pp.113-114,

Chien SH, Clayton WR (1980). 'Application of Elovich Equation to the Kinetics of Phosphate Release and Sorption in Soils.' Soil Sci. Soc. Am. J., 44 : 265-268.

Demirbas E, Kobya M, Senturk E, Ozkan T (2007). 'Adsorption Kinetics for the Removal of Chromium(VI) from Aqueous Solutions on the Activated Carbons prepared from Agricultural Wastes' Water. S.A., 30(4): 533-539.

Diaper DGM, Kuksis A (1957). 'Determination of lead by Dithizone in a single phase Water-Acetone system.' Can. J. Chem., 35: 1278-1284

Emmanuel OA, Olalekan AMB (2008). 'COD Removal from Industrial Wastewater using Activated Carbon Prepared from Animal Horns;' Afr. J. Biotech., 7(21): 3887-3891.

Elkhatib E, Mahdy A, Saleh M, Barakat N (2007). 'Kinetics of copper desorption from soils as affected by different organic Ligands 'Int. J. Environ. Sci. Tech., 4(3): 331-338.

Francis EN (2006). 'Heavy Metals Adsorption by Fiber Ash.' Master's Thesis, Chalmers University of Technology Göteborg, Sweden, p.28.

Hanafiah MAKM, Ngah WSW (2006). 'Kinetics and Thermodynamic study of lead Adsorption from Aqueous Solution onto rubber (*Hevea brasiliensis*) leaf powder' J. Appl. Sci., 6(13): 2762-2767.

Igwe JC, Abia A (2007). 'Adsorption kinetics and Intra-particulate Diffusivities for Bioremediation of Co(II), Fe(II) and Cu(II) ions from waste water using modified and unmodified maize cob.' Int. J. Phys. Sci., 2(5): 119-127.

Tonle KI, Ngameni E, Njopwouo D, Carteret C, Walcarius A (2003). 'Functionalization of natural Smectite-type Clays by Grafting with Organosilanes: Physico-chemical Characterisation and Application to Mercury (II) Uptake'. Phys. Chem. Chem. Phys., 5: 4951-4961.

Ishizaki C, Marti I (1983). "Surface oxide structure on commercial activated carbon" J. Anal. Chim. Acta. 19(6): 409.

Ketcha MJ, Ngomo MH, Kouotou D, Tchoua NP (2007). 'Kinetic and equilibrium studies of the adsorption of Nitrate ions in Aqueous Solutions by Activated Carbons and Zeolite'. Res. J. Chem. Environ., 11(3): 47-49.

Lezeck C, Mieczylaw R, Balys EK (2000). 'Some generalizationc of Langmuir adsorption Isotherm.' Int. J. Chem., 3(14): 1099-8292.

Mirina BS, Mile TK (2002). 'The Kinetics of Chromium (VI) adsorption from water on some Natural materials' APTEFF, 33: 1-174.

Nagarethinam K, Ananthakrishnam R (2002). "Comparative Study of Removal of Lead(II) Ions by Adsorption on Various Carbons." Fresenius Environ. Bull., 11(3): 160.

Najua DT, Luqman CA, Zawani Z, Suraya AR (2008). "Adsorption of Copper from Aqueous Solution by Elais Guineensis Kernel Activated Carbon." J. Eng. Sci. Technol., 2(180–189) : 186.

Njoya D, Elimbi A, Nkoumbou C, Njoya A, Njopwouo D, Lecomte G, Yvon J (2007). Contribution to study physico-chemical and clay mineralogy of Mayoum (Cameroon), Annals of Chemistry, Materials Sci., 32(1): 55-68.

Norotry K, Turzikova A, Komarek J (2000). 'Adsorption Kinetic Study for the Removal of Divalent and Trivalent ions from Aqueous Solution by Natural Adsorbents'. J. Anal. Chim. Acta. 411: 209-212.

Ömer Ş, Sevgi D, Mehmet FD (2005). "Removal of Lead(II) from Aqueous Solution by Antep Pistachio Shells" Frenius Environ. Bull., 11(14): 986-987.

Orumwense F (1996). 'Removal of lead from water by Adsorption on a Kaolinitic Clay'. J. Chem. Tech. Biotechnol., 65:363-369.

Pratik M, Choksi VY (2008). 'Adsorption kinetic Study for the Removal of Nickel(II) and Aluminium(III) from an Aqueous Solution by Natural Adsorbents.' National Sci. Tech. Lib., 1/3: 216-231.

Reed BE, Matsumoto MR (1993). 'Modeling Cadmium Adsorption by Activated Carbon Using Lagmuir and Freundlich Isotherm Expressions' Sep. Sci. Technol., 28: 2179.

Riaz Q, Sohail A (2005). "Kinetic Study of Lead Ion Adsorption on Active Carbon." Turk. J. Chem., 29: 95-99.

Shanmugavalli R, Syed SPS, Enchatesh R, Kadirvelu K, Madhavakrishnan S, Pattabhi S (2006). "Uptake of Lead (II) ion From Aqueous Solution Using Silk Cotton Hull Carbon: An Agricultural Waste Biomass" E-J. Chem., 13(3): 218-229.

Waid O, Hossam Al-I (2007). Removal of Pb^{+2} Ions from Aqueous Solution by Adsorption on Kaolinite Clay" Am. J. Appl. Sci., 4(7): 502-507.

Zamzow M, Eichbaum B, Sandgren K, Shanks D (1990). 'Removal of Heavy Metals and Other Cations from Wastewater Using Zeolites' Sep. Sci. Technol., 25(13-15): 1555-1569.

Use of Mediterranean plant as potential adsorbent for municipal and industrial wastewater treatment

Amina SOUDANI, Mohamed CHIBAN*, Mohamed ZERBET and Fouad SINAN

Department of Chemistry, Faculty of Science, Ibn Zohr University, Agadir, BP. 8106, Hay Dakhla, 80000, Morocco.

A batch adsorption system using dried *Carpobrotus edulis* plant as a new cheap adsorbent was investigated to remove some pollutants from raw wastewater. The results show that the uptake of nitrate, phosphate and some heavy metals ions from wastewater by *C. edulis* increased with increasing contact time. The percentage uptake of heavy metals from industrial wastewater by *C. edulis* particles was about ~94% for Cd(II), ~91% for Cu(II), ~99% for Pb(II) and ~98% for Zn(II). The removal percentage of phosphate and nitrate ions from municipal wastewaters by dried *C. edulis* was found to be ~96 and ~97%, respectively. The results indicate that the chemical oxygen demand values decreased after contact with micro-particles of dried *C. edulis* plant. The maximum uptake capacity was depending on the type of wastewater as well as the type of pollutant.

Key words: Dried *Carpobrotus edulis* plant, wastewater treatment, phosphate, nitrate, heavy metals.

INTRODUCTION

Municipal and industrial wastewaters frequently contain phosphate, nitrate and heavy metals ions. The industrial use of metals increases their concentrations in soil, air and water. The trace metals are widely spread in environment and may enter the food chain from the environment. It is well recognized that the presence of metals ions in the environment can be detrimental to a variety of living species (Benhima et al., 2008). Unlike organic pollutants, metals are non-biodegradable and because of this the removal of heavy metals becomes essential. Also, nitrate and phosphate ions are commonly found in various wastewaters. They can cause serious water pollution and threaten the environment (Barber and Stuckey, 2000). It is therefore, essential to control and prevent their unsystematic discharge in the environment. For this reason, increased attention is being focussed on the development of technical know how for their removal from nitrate, phosphate and metal bearing effluents before being discharged into water bodies and natural streams.

The conventional methods used to remove toxic cations and anions from wastewaters are membrane techniques (reverse osmosis, nanofiltration, etc.), oxidation/precipitation (hydroxides, sulfides, etc.), coagulation and flocculation, ion-exchange, adsorption by activated carbon (Zhao and Sengupta, 1998; Babel and Kurniawan, 2003; Igwe and Abia, 2003; Ozturk and Bektas, 2004; Mohan and Pittman Jr, 2007) etc.

However, these methods have certain disadvantages such as incomplete metal removal, high reagent and energy requirements, generation of toxic sludge or other waste products that require disposal (Özcan et al., 2005). The adsorption process is one of the most efficient methods of removing pollutants from wastewaters. It is the process that is used to collect soluble substances in solution on a suitable interface. The ability of adsorption to remove toxic chemicals without disturbing the quality of waters or leaving behind any toxic degraded products has augmented its usage in comparison to electro-chemical, biochemical or photochemical degradation processes (Mittal, 2006). Also, the adsorption process provides an attractive alternative treatment, especially if the adsorbent is inexpensive and readily available (Namasivayam et al., 2001). There are large numbers of studies in the literature in which various adsorbents are

*Corresponding author. E-mail: mmchiban@yahoo.fr.

used. Two recent reviews reported by (Mohan and Pitman, 2006; Kurniawan et al., 2006) can be referred for the other possible adsorbents for the removal of heavy metals. However, the adsorption capacity of the adsorbents is not very large. For the past few years, the focus of the research is to use cheap materials as potential adsorbents and the processes developed so far are based on exploring those natural adsorbent, which can prove economic and bring cost effectiveness (Benhima et al., 2008; Chiban et al., 2011).

The use of dried plants in the wastewaters treatment has been studied in recent years and the results of the laboratory investigations showed that dried plants are good adsorbents for the removal of arsenate, nitrate, phosphate, cadmium and lead ions from synthetic wastewaters (Benhima et al., 2008; Abdel-Halim et al., 2003; Chiban et al., 2005, 2009, 2011). In this study, the plant selected to be used as an environmentally friendly adsorbent of phosphate, nitrate and heavy metals ions from municipal and industrial wastewater samples was *C. edulis* plant from the Mediterranean area of Morocco. *Carpobrotus edulis* is a plant from the Aizoaceae family, the stems are spread over 2 m long, the leaves 4 to 8 mm long, 8 to 17 mm in width and color-bright green. The flowers are 10 to 20 mm in length. This plant has an effective antibacterial activity (Vander-Watt and Pretorius, 2001). The objective of our work was to investigate the possible use of a new cheap adsorbent obtained from the dried *C. edulis* plant as an alternate material for wastewater treatment using the batch equilibration technique.

MATERIALS AND METHODS

The material used in this study was obtained from *Carpobrotus edulis* plant as previously reported (Benhima et al., 2008; Chiban et al., 2011). A recent screening (Chiban et al., 2007) of the chemical composition and surface characterization of the plant points out that the major functional groups on *C. edulis* plant are polar hydroxyl, aldehydic and carboxylic groups. Due to this *C. edulis* has a great potential as adsorbent of anions and cations from aqueous solutions. The plant parts (leaves and stems) were dried under air during one week. The dried plants were chopped into small fragments, then again dried in oven at 35°C during 24 h and crushed with an electric grinder to get fine powders. The obtained micro-particles were used as adsorbent materials in batch experiments without any other pre-treatment to avoid extra costs.

Several raw wastewater samples were collected from two Agadir zones (Drarga and Anza regions) and stored in polyethylene bottles. Drarga and Anza wastewater samples are municipal and industrial wastewaters from Agadir zones, respectively. These samples were decanted and filtered on paper of 0.45 μm porosity. The pollutant charge of wastewater samples has been determined before using them for the batch experiments.

Batch experiments

The removal of pollutants from wastewaters was performed by batch technique at room temperature. About 1 g of dried *C. edulis* plant was accurately weighed and placed in Erlenmeyer glass flasks of 100 ml containing 40 ml of wastewater solution of known concentration and pH. The solutions were vigorously stirred by use of a magnetic stirrer at a constant temperature for a given time period to reach equilibrium. The agitation speed was kept constant for each run to ensure equal mixing. After different contact times (Tc), were centrifuged at 5000 rpm for 10 min and the supernatant was filtered. The concentrations of nitrate and phosphate ions of the filtrates were measured by Waters model capillary electrophoreses and HP model spectrophotometer, respectively.

The chemical oxygen demand (COD) was measured using HACH 8000 spectrophotometer and the total suspended solids (TSS) was determined by weighing filtered (Whatman filter paper of 0.45 μm) samples after drying for 24 h at 105°C. The instrument used for the determination of lead, cadmium, copper and zinc ions concentration was a Varian model 220FS atomic absorption spectrophotometer. The pH values of the wastewater samples were measured by a Mettler-Toledo meter (MP120) with a glass electrode. A mechanical shaker model Labotec was used for shaking the adsorption batches. The centrifuge Biofuge model (Heraeus Instruments) was used to separate dried plants from the solutions after complete adsorption experiments. An analytical balance Precisa model XT 220A was used for weighting the dried plants samples.

The pollutant concentration retained in the adsorbent phase (q_a, mg/g) and percentage uptakes (%) were calculated by the equations:

$$q_a = (C_i - C_t) \times \frac{V}{m}$$

$$\% \text{ uptake} = \left[\frac{C_i - C_t}{C_i} \right] \times 100,$$

where C_i (mg/l) is the initial concentration of pollutant in the feed solution, C_t (mg/l) is the concentration of pollutant in solution at a given time 't', V (ml) is the total volume of the feed solution; and m (g) is the weight of the material.

RESULTS AND DISCUSSION

The composition of wastewater samples is shown in Table 1. These results show that the raw wastewater samples are richer in heavy metals and phosphate and they also show the high COD (Chemical Oxygen Demand) and TSS (Total Suspended Solids) values. All the values obtained for pollutants, except nitrate ions, are superior to Moroccan norms and World Health Organization standard for drinking water (Standard, 1991; WHO, 2008). The average pH and temperature values of raw industrial wastewaters are 2.2 and ~24°C, respectively. These values show that the industrial wastewater samples are very acid.

These results indicate that municipal wastewater contain a low concentration of cations but high concentration of anions. It shows also the high values of COD (Chemical Oxygen Demand) and TSS (Total Suspended Solids). While, the industrial wastewater samples contain high heavy metals and phosphate concentration but low nitrate ions concentration. The average pH and temperature values of municipal and industrial

Table 1. Composition of municipal and industrial wastewater samples.

Pollutant	Unit	Municipal wastewater*	Industrial wastewater*	WHO standard for drinking water (WHO, 2008)
Pb^{2+}	mg/l	0.124	6.093	0.05
Cd^{2+}	mg/l	0.16×10^{-4}	0.068	0.005
Cu^{2+}	mg/l	0.185	2.131	1
Zn^{2+}	mg/l	0.32	17.35	5
NO_3^-	mg/l	92.9	2.976	50
PO_4^{3-}	mg/l	62.3	81.58	0.05
COD	mg O_2/l	1046	1575	30
TSS	mg/l	602	1034	25
pH	-	8.1	2.2	$6.5 < pH < 9.5$
T	°C	23	24	< 25

*: Indicate average pollutants concentration for several samples (mg/l).

wastewaters are 8.1, 2.2 and ~23, ~24°C, respectively. These values show that the industrial wastewater samples are very acid then municipal wastewaters. The variations of the residual concentration and % removal of both phosphate and nitrate ions from municipal and industrial wastewater samples versus the contact time are plotted in Figure 1. Based on previous work, the m/V ratio was chosen to 25 g/l (1g/40 ml) for this study in all batch experiments. A detailed study for determining the ratio of the weight of *C. edulis* adsorbent to volume of the aqueous phase (*m/V*) has been done by laboratory solution as it was described in our previous report (Benhima et al., 2008).

From these results, we note that the residual concentration of phosphate and nitrate ions in solution decreased with the increasing of the contact time. So, the % uptake of *C. edulis* particles increased with the increasing of the contact time. The adsorption process of both anions uptake by dried *C. edulis* plant appeared to follow a process in two phases characterized by an initial fast uptake step lasting at the maximum, less than 1 h, and corresponding to an uptake concentration of about ~53 to 77% and ~95 to 98% of the initial PO_4^{3-} and NO_3^- concentration ions of the both wastewater samples, followed by a slower step lasting for hours and tending to a steady state. The equilibrium time is attained in less than 3 hours for phosphate and 1 h for nitrate ions. This difference can be explained by the type of wastewater. The final nitrate ions concentrations are lower than 25 mg/l, which is the norm of drinking water. The maximum uptake capacities of phosphate and nitrate ions by unit of weight of dried plant were 2.40, 3.59 mg/g respectively. These values depend on the type of anions because the uptake of NO_3^- is higher than that for PO_4^{3-} ions.

This indicates some specificity of the interactions between anions and active sites of dried *C. edulis* plant responsible for the anions adsorption. The results of nitrate and phosphate ions uptake onto *C. edulis* with distilled water show that the negligible quantities of these

elements are release into solution by these micro-particles.

As shown in Tabe 1, the municipal wastewater samples contain a low concentrations of heavy metals including Cu(II), Cd(II) and Zn(II) comparing to the European norms. For this reason, it is not necessary to study the removal of these elements at different contact time. The variations of the residual concentration and the removal percentage of Pb(II) ions by dried *C. edulis* plant from municipal wastewaters versus the contact time are presented in Figure 2. These results show that the maximum uptake capacity of Pb(II) ions at equilibrium was found to be 3.36×10^{-3} mg/g. The percentage removal of Pb(II) ions from municipal wastewaters by dried plant was noticed at about 64% after 30 min of contact time. In the case of industrial wastewaters, the variations of the percentage removal of heavy metals ions by *C. edulis* micro-particles versus the contact time are plotted in Figure 3. These results show that the removal of heavy metals ions was very fast and the equilibrium was reached within 60 min. Equilibrium sorption efficiency for Pb(II) was achieved ~98 to 99% with an initial solution concentration of 6.09 mg/l. About 90% of the equilibrium Pb(II) uptake was removed rapidly within first 15 minutes. This indicates a high sorption rate for heavy metals ions from real wastewaters. It is also noticed that:

1. The uptake quantities of Cu(II), Cd(II), Pb(II) and Zn(II) ions by *C. edulis* particles are about 77.18, 2.53, 243.26 and 676.40 µg/g respectively and,
2. The percentage removed of heavy metals by *C. edulis* plant was found to be ~99% for Pb(II), ~93% for Cd(II), ~90% for Cu(II) and ~97% for Zn(II),
3. The treated wastewater on dried *C. edulis* plant fulfils the requirements of WHO (WHO, 2008).

It is clear that the maximum % removed was depending on the type of the metal ions. These values are much lower comparing to those obtained for laboratory

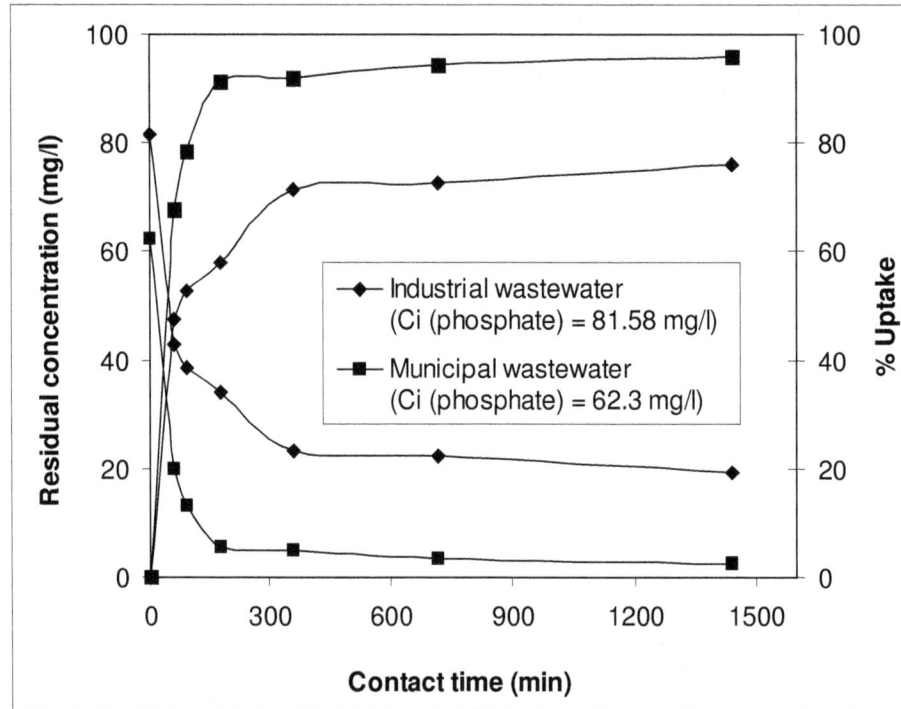

(a)

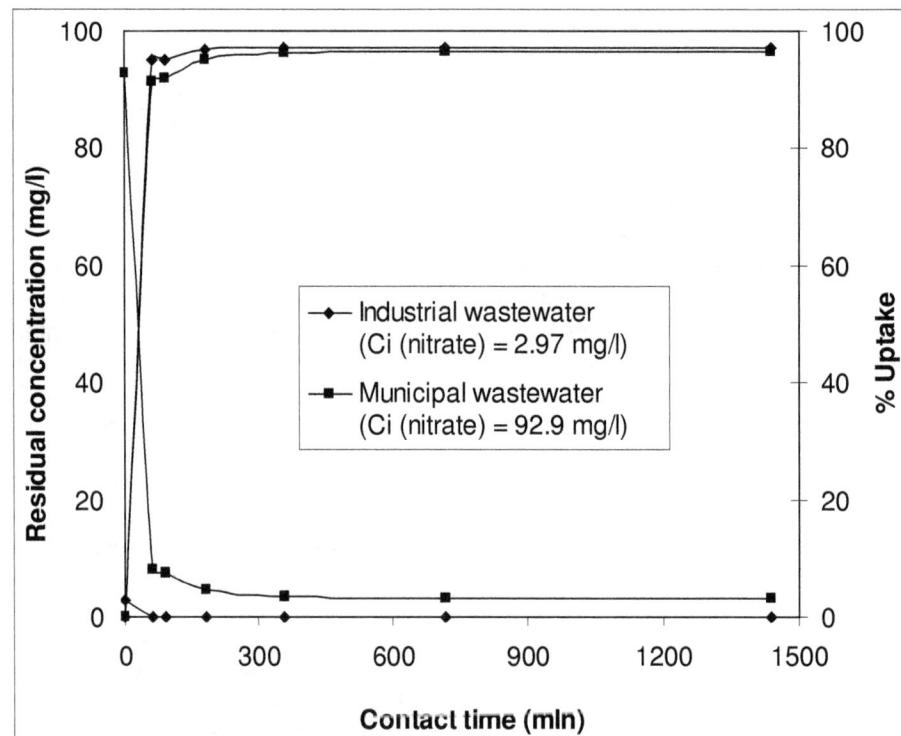

(b)

Figure 1. Residual concentration and % uptake of phosphate and nitrate from municipal and industrial wastewaters using *C. edulis:* (a) = phosphate ions; (b) = nitrate ions.

solutions at various concentrations (Benhima et al., 2008). In the studied conditions, the % removal of metal ions from industrial wastewater samples followed the order of Pb(II) > Cd(II) > Zn(II) > Cu(II).

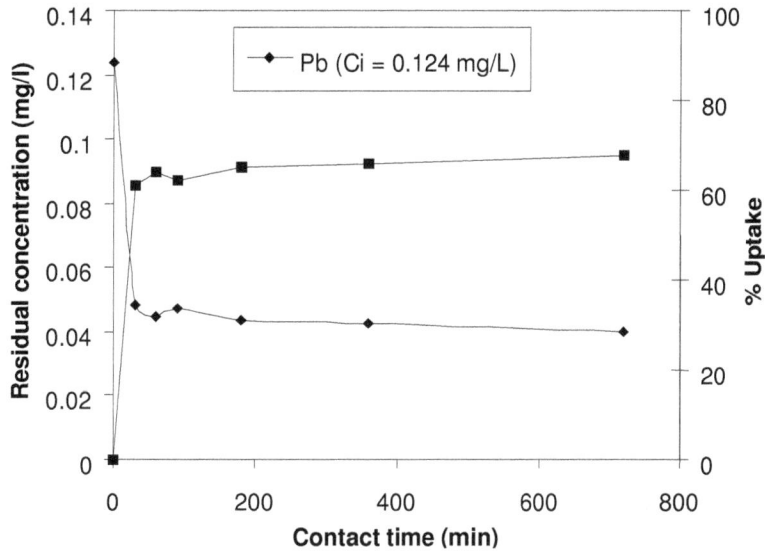

Figure 2. Residual concentration and uptake of Pb(II) ions from municipal wastewater by dried *C. edulis* plant : C_i(Pb)=0.124 mg/l, m/V=25 g/l, pH=8.1, T=23 °C.

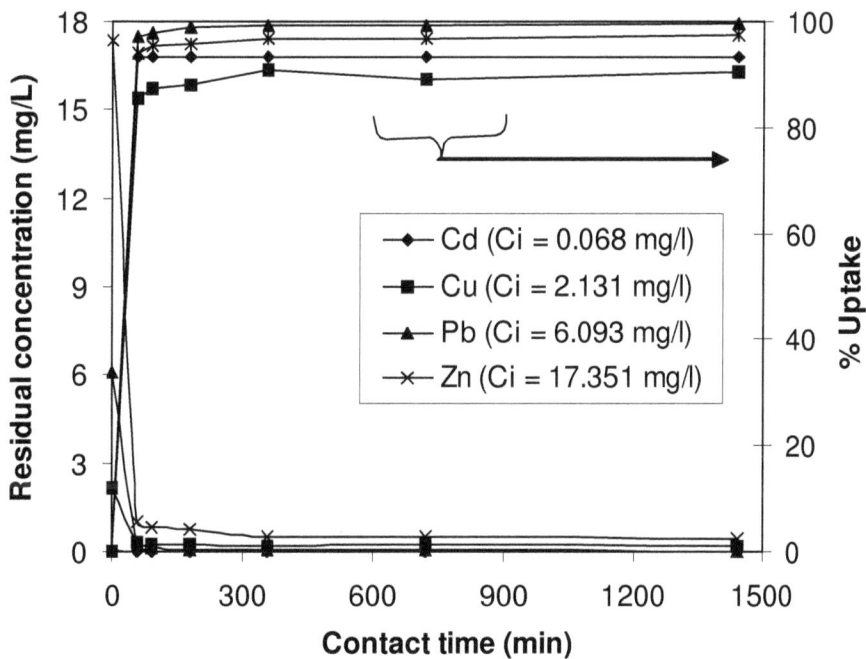

Figure 3. Residual concentration and uptake of heavy metals from industrial wastewater by dried *C. edulis* plant : C_i (Cd)=0.068 mg/l, C_i (Cu)=2.13 mg/l, C_i (Pb)=6.09 mg/l, Ci (Zn)=17.35 mg/l, pH=2.2, T=24 °C.

In order to evaluate any release of heavy metals ions by these micro-particles, the same amount of *C. edulis* plant was mixed with pure water. Released amounts of these ions in the order of 1.0×10^{-3} mg/l for Cd(II), 1.9×10^{-3} mg/l for Pb(II), 4.2×10^{-2} mg/l for Cu(II), 6.5×10^{-2} mg/l for Zn(II) have been measured after 12 h of

contact with dried *C. edulis* plant. These values are negligible.

The effect of contact time under agitation on the supplement COD from municipal and industrial wastewaters by *C. edulis* particles is illustrated in Figure 4. As shown in Figure 4, the release of organic matter in

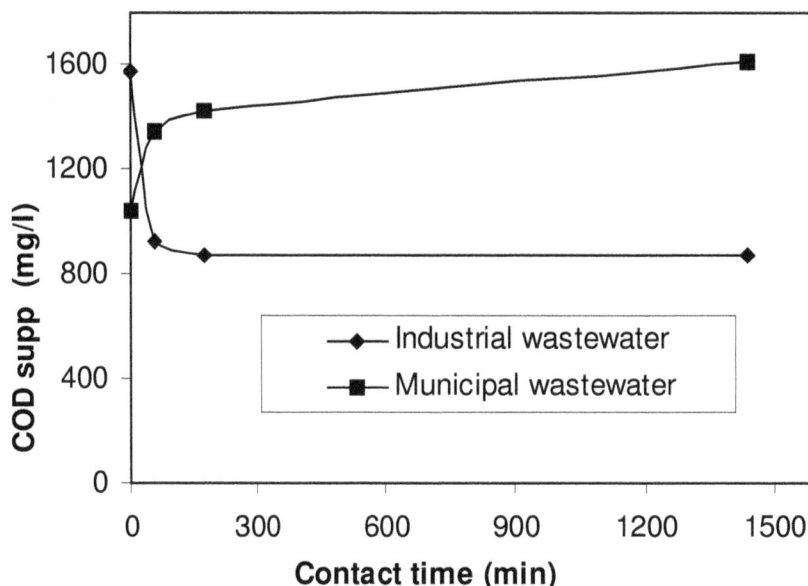

Figure 4. Effect of contact time on the COD supplement: T=25℃, *m/V*=25g/l and natural pH.

municipal wastewater samples by dried *C. edulis* increased with increasing of the contact time. In the case of industrial wastewater samples, these results show that the COD supplement decrease with contact time. This difference can be explained by the variation of the pH of wastewater samples (pH_{ind}=2.2, pH_{mun}=8.1). The tests of adsorption process with distilled water showed that the dried *C. edulis* plant can release important amounts of organic matter in solution, which suggests the necessity of a pre-wash before using these micro-particles of dried *C. edulis* plant.

The experimental results of this study can be used to design batch adsorption systems for the nitrate, phosphate and heavy metals ions removal. Such a batch system will be applicable to small industries which generate heavy metals-containing wastewaters. The adsorbent can be added to the wastewater collected in a tank and the mixture must be agitated for the equilibrium time found from this study. Then the liquid can be decanted and discharged. Because of low pH, the adjustment of pH is necessary before discharging to a sewer or water course. The used adsorbent has to be suitably disposed.

Conclusion

This study indicated that the dried *Carpobrotus edulis* plant could be used as an environmentally friendly material for municipal and industrial wastewater treatment. The maximum uptake percentage of *C. edulis* plant was found to be ~98% for NO_3^-, ~41% for PO_4^{3-}, ~94% for Cd(II), ~91% for Cu(II), ~99% for Pb(II) and ~98% for Zn(II). The maximum uptake capacities were depending

on the type of pollutant as well as on the type of wastewater.

The method presented in this work might be of interest for industrial and environmental applications for the removal of toxic metal ions from the environmental. Using dried plants for the removal of pollutants have the advantages of being available, cheap and efficient. The contaminated dried plants are expected to precipitate and became a part of the sediment. However, using this method for treatment of industrial effluents might require finding a way for a proper disposal of the dried plants.

ACKNOWLEDGEMENTS

This joint program was made possible by « Comité Mixte inter Universitaire Franco-Marocain (A.I. N°128/MA)». We gratefully acknowledge funding through European Membrane Institute (IEM) for technical support and analytical screening of samples.

REFERENCES

Abdel-Halim SH, Shehata AMA, El-Shahat MF (2003). Removal of lead ions from industrial waste water by different types of natural materials, Water Res., 37: 1678–1683.

Babel S, Kurniawan TA (2003). Low-cost adsorbents for heavy metals uptake from contaminated water: a review, J. Hazard. Mater, 97:219–243.

Barber WP, Stuckey DC (2000). Nitrogen removal in a modified anaerobic baffled reactor (ABR): 1, Denitrification, Water Res., 34: 2413–2422.

Benhima H, Chiban M, Sinan F, Seta P, Persin M (2008). Removal of lead and cadmium ions from aqueous solution by adsorption onto micro-particles of dry plants, Colloids Surf. B: Biointerfaces, 61: 10–16.

Chiban M, Amzeghal A, Benhima H, Sinan F, Tahrouch S, Seta P (2007). Phytochemical study of some inert plants from the South-Western region of Morocco (Etude phytochimique de certaines plantes inertes du sudmarocain), Rev. Biol. Biotechnol., 6: 40–43.

Chiban M, Benhima H, Saadi B, Nounah A, Sinan F (2005). Isotherms and kinetic study of dihydrogen and hydrogen phosphate ions ($H_2PO_4^-$ and HPO_4^{2-}) onto crushed plant matter of the semi-arid zones of Morocco: *Asphodelus microcarpus, Asparagus albus and Senecio anthophorbium*, J. Physique IV, 123: 393-399.

Chiban M, Lehutu G, Sinan F, Carja G (2009). Arsenate removal by *Withania frutescens* plant from the south–western Morocco. Environ. Eng. Manage. J., 8: 1377-1383.

Chiban M, Soudani A, Sinan F, Persin M (2011). Single, binary and multi-component adsorption of some anions and heavy metals on environmentally friendly *Carpobrotus edulis* plant, Colloids Surf. B: Biointerfaces, 82: 267–276.

Igwe JC, Abia AA (2003). Maize cob and husk as adsorbents for the removal of heavy metals from waste water, Phys. Scientist, 2: 83–92.

Kurniawan TA, Chan GYS, Lo W, Babel S (2006). Comparisons of low-cost adsorbents for treating wastewaters laden with heavy metals, Sci. Total Environ., 366: 409–426.

Mittal A (2006). Adsorption kinetics of removal of a toxic dye, Malachite Green, from wastewater by using hen feathers, J. Hazard. Mater., 133: 196–202.

Mohan D, Pitman Jr CU (2006). Activated carbons and low cost adsorbents for remediation of tri- and hexavalent chromium from water, J. Hazard. Mater., 138: 762–811.

Mohan D, Pittman Jr CU (2007). Arsenic removal from water/wastewater using adsorbents–a critical review, J. Hazard. Mater., 142: 1–53.

Namasivayam C, Radhika R, Suba S (2001). Uptake of dyes by a promising locally available agricultural solid waste: coir pith, Waste Manage., 21: 381-387.

Standard M (1991). Water quality for human consumption. DECREE No. 359 approval. 91 of 23 Rejab 1411.

Özcan A, Özcan AS, Tunali S, Akar T, Kiran I (2005). Determination of the equilibrium, kinetic and thermodynamic parameters of adsorption of copper(II) ions onto seeds of *Capsicum annuum*, J. Hazard. Mater., 124: 200–208.

Ozturk N, Bektas TE (2004). Nitrate removal from aqueous solution by adsorption onto various materials, J. Hazard. Mater., 112: 155–162.Vander-Watt E, Pretorius JC (2001). Purification and identification of active antibacterial components in *Carpobrotus edulis* L, J. Ethnopharmacol., 76: 87–91.

WHO, World Health Organization (2008). Guidelines for drinking Water Quality, Vol. 1, Geneva.

Zhao D, Sengupta AK (1998). Ultimate removal of phosphate from wastewater using a new class of polymeric ion exchangers, Water Res., 32: 1613–1625.

24

Occurrence of paraquat residues in some Nigerian crops, vegetables and fruits

Akinloye, O. A.[1]*, Adamson, I.[1], Ademuyiwa, O.[1] and Arowolo, T. A.[2]

[1]Department of Biochemistry, University of Agriculture, P. M. B. 2240, Abeokuta, Ogun-State, Nigeria.
[2]Department of Environmental Toxicology, University of Agriculture, P. M. B. 2240, Abeokuta, Ogun-State, Nigeria.

Pesticides constitute the major source of potential environmental hazard to man and animal, as they are present and concentrated in the food chain. Paraquat (1,1'-dimethyl-4,4'-bipyridylium dichloride) is one of the most highly toxic herbicide to be marketed over the last 60 years. Although, Paraquat (PQ) has been banned or severely restricted in most countries, its use continues in some, especially in Nigeria. Therefore, this study investigated possible occurrence of PQ in some commonly consumed vegetables, crops and fruits in Abeokuta using spectrophotometric method. This study was conducted on 150 samples of different kinds of crops, vegetables and fruits (harvested from Fadama farmland of the University of Agriculture, Abeokuta) to assess the presence and levels of PQ residues. PQ residues was found to present at 0.13±0.02, 0.27±0.02, 0.06±0.01, 0.10±0.03, 0.15±0.03, 0.09±0.02, 0.09±0.02, 0.04±0.01 and 0.05±0.01 ppm in Talinum triangulare, Corchorus olitorius, Amaranthus caudatus, Cratylia argentea, Capsicum frutescens, Lycopersicum esculentum, Raphanu sativus, Zea may and Dioscorea alata, respectively. Paraquat residues were not detectable in M. paradisicica and C. papaya using this method. The method achieves mean recovery of over 80% and is repeatable with overall coefficient of variation of 8.0% (n=10) at 0.05-1.0 ppm fortification level. All residue levels detected were within the PQ tolerance or maximum pesticide limits.

Key words: Occurrence, Paraquat residues, Nigerian, crops, vegetables, fruits.

INTRODUCTION

One of the major task facing control agencies in most of the developing and developed countries is monitoring for pesticide residues in foods. Pesticide residues monitoring program is important to enforce compliance with any legislated national and international maximum residue limits. Monitoring data are normally used to assess the dietary exposure of pesticide residues from food (Anderson and Poulsen, 2001; Camoni et al., 2001; European Commission, 2006).

Paraquat (PQ) was invented in England in 1956 and has been continuously used for agriculture all over the world because its splendid herbicidal effects have been producing large economic gains. It is used to control weeds and grasses in many agricultural and non-agricultural areas. For instance, it is used for pre-plant or

pre-emergence on vegetables, grains, potatoes and peanut areas; post emergence around fruit crops and soybeans during the dormant season on clover and other legumes. Paraquat products have been classified as restricted pesticides and even banned in some country (US EPA, 2001). More so, the occurrence of agro-chemical residues in foodstuffs, crops, vegetables, milk and milk products have been reported by different authors in many countries (Antonio and Manuela, 1995; Mohamed and Saad, 1995; Kinyamu et al., 1998; Rusibamayila et al., 1998; Osfor et al., 1998; PAN Europe, 2005; Codex Alimentarius, 2006).

On several occasions, incidents of poisoning after consuming these plants and animal products have been reported. These incidents were attributed to grower's misuse of the pesticides or to negligence in observing a safety interval after harvest. It has been reported that people were exposed to paraquat residues through diet (Selisker et al., 1995; Zeneca Agricochemicals, 1993). Determining the levels of herbicides present in plants and

*Corresponding author. E-mail: oaakin@yahoo.com

plant products has great relevance because these values are convenient indicators of environmental contamination by the herbicides. The present study was undertaken to evaluate quantitatively the paraquat residue levels in some vegetables, fruits and crops commonly consumed in Abeokuta, Nigeria, in order to assess the possible predisposition of human to this herbicide and its associated health risks such as Parkinson's disease, Alzheimer's disease and Amyotropic lateral sclerosis (Grandjean and Landrigan, 2006).

MATERIALS AND METHODS

Randomly selected samples of different kinds of vegetables, crops and fruits namely *Raphanus sativus* (Radish), *Zea mays* (Agbado, Maize), *Lycopersicon esculentum* (Tomato), *Talinum triagulare* (Gbure, Water leaf), *Chochorus olitorius* (Ewedu, Nalta jute), *Amaranthus caudatus* (Efo tete, Tassel flower), *Celocia argentea* (Efo soko,Cockscomb), *Capsicum frutescens* (Ata wewe, Chilli pepper), *Dioscorea alata* (Isu ewura, Water yam), *Musa paradisicica* (Ogede agbagba, Plantain) and *Carica papaya* (Ibepe, Pawpaw) used in this studies were harvested from the Fadama Farm of the University of Agriculture, Abeokuta, on which paraquat had been previously used (application rate of 2.2 Kg PQ/ha) as pre-emergence herbicides.

The control samples were obtained from the subsistence farmer's farmland on which there was no former history/record of PQ usage. PQ formulated as paraquat dichloride (Syngenta, Zwiterland) was purchased from one of the distributor (C.ZARD, Nig Ltd.). Sodium dithionite, calcium carbonate and sodium hydroxide were purchased from Aldrich Chemicals, sulphuric acid (analytical reagent grade) purchased from BDH Ltd. Spectrophotometer (Jenway, Model 6405 uv/vis) and centrifuge (JOL-802 Finlab, U.K, model TDL 80-2) were part of the apparatus used.

Preparations of samples

Randomly selected samples of each kinds of crops, fruits and vegetables aaforementioned, harvested from a 'homogeous lot' (same plantation, same paraquat treatment and same harvest date) on the Fadama farmland were thoroughly washed and rinsed with deionised water, copped, blended, homogenized and centrifuged at 5000 rpm for 30 min. The resulting supernatant was used for the analysis. Examination gloves (Top glove SDN BHD, Selangor D.E, Malaysia) were worn throughout the course of the experiment.

Extraction of paraquat residues

An extraction procedure adapted from Van Emon et al. (1987) as described by Selisker et al. (1995) was employed. Representative analytical sub-samples (2.5 g) were weighed into 25 ml calibrated polypropylene tubes followed by addition of concentrated tetraoxosulphate (VI) acid (2.5 ml 6 N H_2SO_4), closed and sonicated. The volume was made up to 5.0 ml with 6 N H_2SO_4, shaken for about 10 min in an orbital shaker. The mixture was then centrifuged at 5000 rpm for 10 min. The supernatant was carefully removed and kept at 4 °C until used.

Spectophotometric determination of PQ residues

The determination of PQ residues in the extracted samples was

carried out as described by Rai et al. (1997). 20 µl of the aforementioned supernatant was mixed with 4.9 ml of 0.01 M sodium acetate buffer, pH 5.0. This was followed by the addition of 2.0 ml 1% (w/v) aqueous sodium dithionite and 1 ml of 0.1 ml of 0.1 N sodium hydroxide. The mixture was allowed to stand for about 2 to 5 min and paraquat residues were quantified by measuring the absorbance at 600 nm. The amount of PQ residues was then estimated from a standard calibration curve plotted using comercially available PQ as standard, as seen in Figure 1. To examine the efficacy of extraction, three samples of each vegetable, crop and fruit were spiked with known concentration of PQ (0.01 to 0.5 ppm) and extraction was performed as described earlier and the mean percentage recovery level determined.

RESULTS AND DISCUSSION

The result of the levels of PQ residues determined from each sample analyzed is shown in Table 1. Among the samples analyzed, *Z. mays* and *C. olitorius* had the least (0.04 ppm) and highest (0.27 ppm) levels of PQ residues, respectively. The observed differences could be explained on the basis of different shape and morphology of these plants. The highest PQ level detected in *C. olitorius* could be attributed not only to the nature of it foliage system which could probably be responsible for the retention of this compound but also to the high fat content nature of the plant. PQ had been reported to be lipophobic, it is therefore not surprising to note that *Z. mays* whose seeds is majorly carbohydrate had the least amount of PQ residues. The highest level of PQ residues obtained in these studies was quite below that obtained in potatoes (0.5 ppm) in the work of Yuk et al. (1993). Also, the amount of PQ residues detected in *Z. mays* in this study (0.04 ppm) was not in agreement with the report of Plant Protection Ltd. (1986), where mean PQ level of 0.08 ppm was detected in maize. The differences may be attributed to the difference in the level of sensitivity of the method used. Codex online details for PQ reported that the maximum residue limits (MRLs) for PQ in leafy vegetables, maize, soya bean fodder, root and vegetables were 0.07, 0.05, 0.5 and 0.05 mg/kg, respectively (Codex Alimentarius, 2006). Costenla et al. (1990) also reported that PQ residues in coffee berries and bean in Costa Rica were at or below the limit detection of 0.02 mg/kg.

The European Union (EU) monitoring report on pesticide residues in food shows a frequency of samples with residues above MRLs of 5.5% and a frequency of samples with residues at or below the MRLs of 4.4%. Methomyl, methiocarb, dimethoate and bromopropylane were reported to be among substances found exceeding the MRLs. High level of exceedances were observed in grapes (5%), cucumber (3%) and aubergines (3%) (PAN Europe, 2005; European Commission, 2006). The percentage recovery of about 80, 85, 85, 86, 89, 85, 88, 85, 88 and 84% for *T. triangulare, C. olitorius, A. caudatus, C. argentea, C. frutescens, R. sativus, Z. mays, L. esculentum, D. alata, M. paradisicica* and *C. papaya* respectively was discovered in these samples. One could

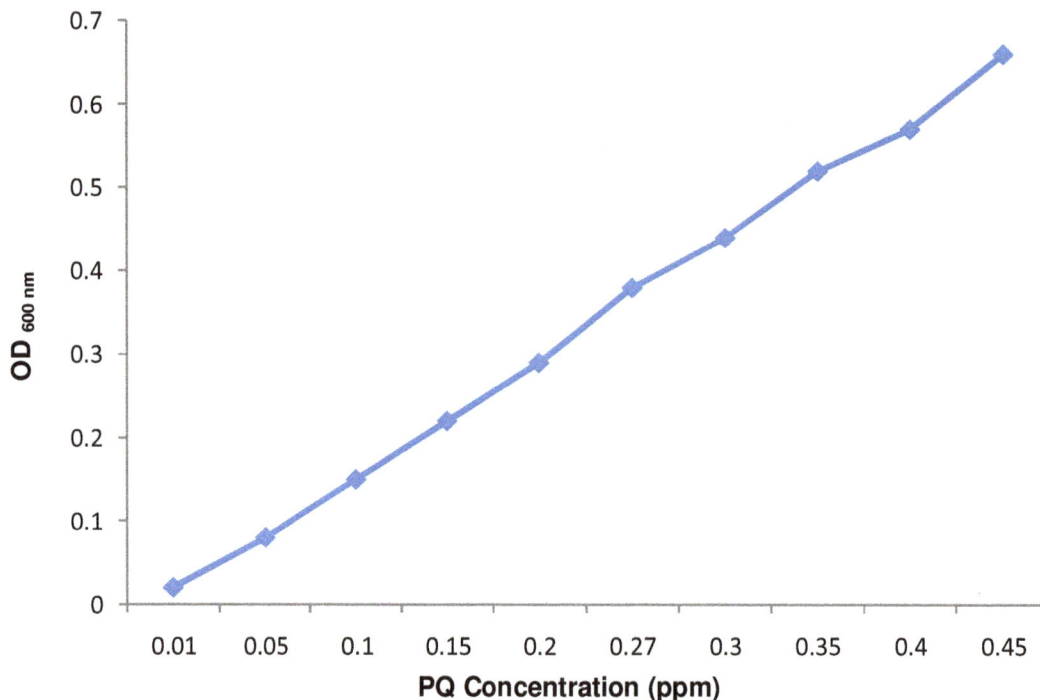

Figure 1. Calibration curve for paraquat residue concentration determination.

Table 1. Paraquat residues in some vegetables, crops and fruits.

Sample	Residues (ppm) mean±SEM
T. triangulare	0.13±0.02
C. olitorius	0.27±0.02
A. caudatus	0.06±0.01
C. argentea	0.10±0.03
C. frutescens	0.15±0.03
L. esculentum	0.09±0.02
R. sativus	0.09±0.02
Z. mays	0.04±0.01
D. alata	0.05±0.01
M. paradiscicica	ND
C. papaya	ND

ND, Not detected; n = 15 (where n is the number of each kind of individually randomly selected samples analyzed).

have thought that *C. papaya* and *M. paradisicica* would had the highest content of PQ residues, which could have arose from continuous accumulation due to their perennial nature. However, PQ residues in these plants were not detectable using the aforementioned methodology.

The present preliminary results revealed that in all the randomly selected samples tested, PQ residues detected were below the maximum residues limits (MRLs) of 0.5 mg/Kg set by the UK pesticides levels in crop foods and feedstuffs (Regulation, 1994; Statutory Instrument, 1985).

This is also in agreement with the report of Terry et al. (2002) that there was very little exposure to PQ for the consumers of treated crops as the vast majority of PQ uses do not result in detectable residues (>0.5 mg/kg) in food stuffs. Therefore, the consumption of these plant products does not seem to be a serious risk for consumer's health. On the other hand, we cannot rule out or forget the possible long-term effects of accumulation of this chemical in the body which may be deleterious. Thus, this calls for further investigations in this view.

To the best of our knowledge, this is the first reported notes on the occurrence of PQ residues in some Nigerian crops, vegetables and fruits using spectrophotometric method; work are still in progress on the use of high performance liquid chromatographic method which might offer faster, concise and possibly specific determinations.

REFERENCES

Anderson JH, Poulsen ME (2001). Results from monitoring of pesticides residues in fruits and vegetables in the Danish market, 1998-1999. Food Addit. Contam., 18: 906.

Antonio B, Manuela J (1995). Occurrence of organochlorine agrochemicals residues in Spanish cheese, Pestic. Chem., 44: 177-182.

Camoni I, Fabbrini R, Attias L, Muccio A, Cercere E, Consolino A, Roberti A (2001). Estimation of dietary intake of pesticide residue by the Italian population during 1997, Food Additives and Contaminants 18: 932. Codex Alimentarius Codex online details for paraquat-Pesticide residues in food and feed, 2006: 57.

Costenla, MAR, Kennedy H, Rojas CE, Mora LE, Stevens JEB (1990). Paraquat behavior in Costa Rican soils and residues in coffee. J. Agric. Food Chem., 38(10): 1985-1988.

European Commission (2006). Monitoring of Pesticide residues in products of plant origin in the European Union, Norway, Iceland and Liechtenstein, SEC (2004), 1416.

Grandjean P, Landrigan PJ (2006). Developmental neurotoxicity of industrial chemical. Lancet, 268: 2167-2178.

Kinyamu JK, Kanja, LW, Skaare JU, Maitho TE (1998). Level of organochlorine pesticides residues in milk of urban mothers in Kenya. Bull. Environ. Contam. Toxicol., 60(5): 732-738.

Mohamed TA, Saad MMI (1995). Residues of methomyl in strawberries, tomatoes and cucumbers. Pestic. Sci., 44: 197-199.

Osfor MM, Abdel Wahab AM, Dessonki SA (1998). Occurrence of pesticides in fish tissues, water and soil sediments from Manzala Lake and River Nile. Nahrung, 42(1): 39-41.

Pesticides Action Network (PAN) Europe (2005). Food residues in fruits. http://www. Pesticide residues.

Rai MK, Vanisha J, Gupta VK (1997). A sensitive determination of paraquat by spectrophotometry. Talanta, 45: 343-348.

Rusibamayila CS, Akhabuhaya JL, Lodenius M (1998). Determination of pesticide residues in some major food crops of Northern Tanzania, J. Environ. Sci. Health, 33(4): 399-409.

Selisker MY, Herzog DP, Erber RD, Fleeker JR, Itak AJ (1995). Determination of paraquat in fruits and vegetables by a magnetic particles based enzyme linked immunosorbent assay, J. Agric. Food Chem., 43: 544-547.

US EPA (2001). Paraquat dichloride. Registration eligibility decision (RED) U S Environmental Protection Agency, Washington DC, EPA 738-F-010-018.

Van Emon J, Seiber J, Hammock B (1987). Application of an enzyme linked immunosorbent assay (ELIZA) to determine paraquat residues in milk, beef and potatoes. Bull. Environ. Contam. Toxicol., 39: 490-497.

Yuk YK, Kathleen AM, Ralph G (1993). Simultaneous determination of residues of paraquat and diquat in potatoes using high Performance Capillary Electrophoresis with ultraviolet detection, J. Agric. Food Chem., 41: 2315-2318.

Zeneca Agrochemicals (1993). The determination of paraquat in crops- A spectrophometric method: ICI Plant Protection Division, Berkshire, England, p. 2.

25

Essentials of ecotoxicology in the tanning industry

Mwinyikione Mwinyihija

Kenya Leather Development Council, P. O. Box 14480, Nairobi, Kenya. E-mail: m_mwinyi@hotmail.co.uk.

Pollution within the tanning Industry has been a great concern with efforts recently directed towards mitigating the emerging challenges. The most inspiring is the development of novel ecotoxicological diagnostic tools which are considered rapid, cost effective and reliable. In the tanning industry, particular attention is directed towards use of chromium, salts, chlorinated phenols among others which are recognised as potential pollutants especially within the developing world where environmental restriction is still low. Therefore the review attempts to underscore the need to comprehend about these pollutants and the appropriate diagnostic approaches required. The application of biosensors and dehydrogenase is discussed to evaluate the toxicity levels and provide inherently the bioavailability status of the contaminants. Indeed while the diagnostic tools are crucial and opportune, other measures to ensure cleaner production within the tanning industry needs to be addressed to disalienate the notion that the tanning industry is still a major source of pollution.

Key words: Ecotoxicology, effluents, leather, pollutants.

INTRODUCTION

The world production of hides and skins is approximated at 1303 million pieces, while in Africa it is about 183 million pieces. Share wise therefore Africa has 14% of the world production. If compared to the world scenario it depicts 0.47% share to the world (Mwinyihija, 2011). The global leather industry is currently valued at more than $75 billion (Mwinyihija, 2011). The worldwide production of the leather is expected to have grown to 2.4 billion square metres by the end of 2000 (Komma, 1996). The leather and leather products industries have witnessed considerable shifts in the location of tanning and leather manufacturing to developing countries where production costs are lower and environmental regulations less stringent (Muchie, 2000). Indeed, while production of leather and fur products has decreased between 1975 and 1992 in developed countries, it has increased in the developing world from 4.4% (1975 – 1985) to 5.3% (1985 –1992) (UNIDO, 1992). Demand for leather and leather goods was also expected to rise, despite slowdowns as a result of the 1998 Asian financial crises (Muchie, 2000), while demand grows at 0.6% for the developed countries, the projected rise for developing countries is 1.1% (CCP, 1998), Mwinyihija, 2011.

When this performance is adjudged against the environmental impact the tanning industry also referred to as the leather sector exerts to the ecosystems, one wonders about the logical use of techniques that challenges negatively to the ecosystems. However emergence of sophisticated technologies in industrial, waste and water treatment techniques and new lifestyles means that the way the environmental media (terrestrial, aquatic and atmospheric systems) are imparted will differ at any point and place in time. This therefore poses a challenge in managing newer threats associated with stressors that are exposed to such environmental media. Of a major concern among the environmental media is the aquatic system that forms the basis of all living systems (Baumgartner, 1996; Sweeting, 1994, Mwinyihija et al., 2005, 2006a) and the terrestrial systems that supports such existence. For example the demand for clean water as a resource has risen with the increasing urbanisation (domestic supplies and recreation), industrialisation and agricultural intensification of the last century. Ironically, industrialisation and agricultural activities are the major contributors to pollution in the atmosphere, terrestrial and aquatic systems (SEPA, 1999; Mwinyihija et al., 2005). Soil contamination can be localised (e.g. related to industrial sites) or diffuse (the result of deposition of a pollutant over a wide area). More over the environmental threats could be related to physical or sensory attributes such as waste mound hills or resultant odour.

LEATHER SECTOR

To comprehend on the essentials of ecotoxicology it is important to review the basis of the leather sector and explore the production chain. Thus the transformation of animal hides and skins into leather primarily involves five main phases; flaying, preservation (whenever applicable) to arrest putrefaction; removal of hair and flesh, and tanning depending on the final leather products targeted through the application of chemical agents. The aim eventually is to render the raw material to be non-putrescible and enhance its durability (Mwinyihija, 2010).

Pollution from the leather processing industries has a negative long-term impact on both the ecosystem health and functionality; and economic growth potential of a country irrespective of the immediate profit accruals intended. Development of the industry in Africa for example does not match the technical know-how and capacity in protecting and predicting the environmental impacts related to the industry. Cleaning up of such environment will require expenditure of funds, which could have promoted positive and sustainable development. Tanning has thus been deemed as one of the largest polluters in the world.

Tanning pollutants and its effect to the ecosystem

Tanning Industry involves the following processes in brief, linked closely with pollution:

Soaking

The objective in carrying out soaking at this stage is to wash the rawstock from debri, dirt and other related physically bound materials (e.g. mostly insecticides, salts ($NaCl_2$) and other preservatives). Primarily and very important (other than washing) the purpose also at this stage is to rehydrate the hides and skins to replace lost moisture during curing to allow for the next level's process involved with application of various chemical to allow for permeation of such intended chemicals, during subsequent processes.

Liming

After the hide and skins is rehydrated, the material is ready to move to the next stage where the use of alkaline medium is undertaken e.g. lime. The aim at this stage is to remove the hair, flesh and splitting up of the structural protein e.g. fibre bundles by chemical and physical means (Ramasami et al., 1991). In this process Na_2S is added to facilitate in dehairing (Flaherty, 1959). It is estimated that for processing 1ton of raw skins weight of skins before soaking the input in a typical input audit

processing(kg) of lime is 100 with an output of 12.3, while Na_2S has an input of 35 with an output of 18.3 (Thanikaivelan, 2000).

Deliming, bating and pickling

For these processes weak organic acids, digestive enzymes and inorganic acids respectively are used to remove the lime, digest and remove the non-structural proteins and eventually bring the pH to a level that will enhance the tanning process.

Chrome tanning

Nearly 90% of all leather produced is tanned using Cr salts (Stein, 1994) especially in the developed world. 8% of the basic chromium sulphate salt is used for conventional tanning. It binds with the collagenous protein to convert to leather. This conversion renders the material non putrescible.

To understand these fundamental processes which are, but a few, of the total that are found in processing of leather, form the vital point of intervention in pollution control. Main pollutants found during leather processing include, $NaCl_2$ and pesticides, strong alkalines and sulphides, inorganic residual compounds, dissolved matter and chromium salts. Chlorinated phenols are important compounds to be investigated due to the various mixtures used in the tanning industry and their ecotoxicity potential.

In another approach of tanning, vegetable tannin are used mostly to retan leather to impart certain specific properties desired or could be used alone in producing leather especially at the rural tanning level by the use of plant material (this could be tree barks and pods which are commonly used in Africa). Mostly according to the tanning sciences, the tannin materials are derived from plants and consist of condensed or hydrolysable tannins (Zywicki et al., 2002). In the East African region, wattle and certain species of acacia (e.g. in arid and semi arid areas) is extensively used as a tannin material. Previous efforts to study the polyphenolic structures of condensed tannins have been hampered by the fact that the structure rapidly transforms during the tanning process to yet unknown products (Zywicki et al., 2002).

Other related ecotoxicological approaches

Biosensors: There is growing need for quick, cheap and reliable bioassays for toxicity testing of compounds. The demand for such a necessity emanates from the continued escalation of pollution load related to elevated anthropogenic activities (i.e. agricultural, urbanisation and industrial activities). Although there is zeal to use cleaner technologies in various processing techniques in the manufacturing sector, contamination of various eco-systems in the world continue to be conspicuous. There is need at this juncture to understand two terms used frequently in environmental sciences that is *pollution* and *contamination* more clearly. Pollution is a term that defines the state of the anthropogenic introduction of substances into the environment causing hazard to human, living resource, damage to structure, and eco-systems. Contamination while used synonymously with pollution demonstrates the concentration leverage of a substance when it exceeds its natural occurrence but not with the set standard threshold. Thus pollutants are chemicals that cause environmental harm (Harrion, 2001) in contrast to contaminant which does not necessarily impact an evident harm.

Therefore the study of these contaminants and pollution in the environment require novel ecotoxi-cological techniques in understanding and predicting the impact of such chemicals on terrestrial and aquatic systems. Manly, 2000 defines ecotoxicology as the study of toxic effects of substances on the biotic and abiotic components of the biosphere, especially on populations and communities within defined ecosystems. However it is imperative to understand that ecotoxicology is a discipline within a wider field of environmental toxicology. Thus bioassays that integrate toxicity tests developed to examine the effects or impacts of chemicals in a broad range of ecosystems using various species are presently the modern preferred ecotoxicological techniques. Indeed bioassay may be used to determine both short term (acute) and long term (chronic) impact of pollutants, as well as being used to study the mode of action and routes of transport through the ecosystems. Bioassays are therefore described as tests determining or estimating the effects of biologically active substances under standardised and reproducible conditions.

To that effect, emerging novel techniques for ecoto-xicological analysis has included biosensors. Mwinyihija (2010) found out that efficiency, accuracy, rapidity, convenience and on-line monitoring are some of the advantages conferred by the use of biosensors over other forms of biomonitoring. Vo-Dinh and Cullum, 2000 defined biosensors as a combination of a bioreceptor, biological component, and a transducer as the detector. The interaction of the analyte with the bioreceptor has the functional characteristic of producing an effect which is measured with a transducer. The transducer then con-verts the information into measurable effect such as an electrical or optical signal also referred to as bio-luminescence. Moreover biosensors have been reported to provide reliable technique of measuring biological effects (e.g. acute and chronic physiological toxicity, genotoxicity, immunotoxicity and endocrine toxicity) and the concentration of specific analytes which are difficult to detect and are important contaminants of water, waste, soil or air (e.g. surfactants, chlorinated hydrocarbons,

sulphophenyls carboxylates, sulphonated dyes, fluorescent whitening agents, naphthalensulphonates, carboxylic acids, dioxins, pesticides and metabolites). The protocols for the biosensor (e.g. *E. coli* HB101 pUCD 607) in flow chart as indicated in Figure 1.

During the research period *lux*-marked bacteria biosensors (Mwinyihija, 2010) (Figure 2) offered a powerful way forward for rapid screening of the effect of environmental variables on the fate and toxicity of the pollutants.

Dehydrogenase: Mwinyihija (2010) provides an in depth use of Dehydrogenase test as a tool in an Ecotoxicological diagnosis especially for the river sediments impacted by the tanning industry for the first time. The technique is a simple and inexpensive test used to determine the degree of toxicity by measuring the microbial activity through the production of the enzyme dehydrogenase. Dehydrogenase activity (DHA) tends to be a common and suitable test made to quantify the impact of heavy metals on soil micro-organism. The mea-surement of enzyme concentrations may be used as an indirect measure of the soil microbial activity. This may serve as a supplement to biomass measurement. This is because DHA is assumed to be proportional to microbial respiration (Stevenson, 1959; Thalman, 1968; Skajins, 1973; Frankenberger and Dick, 1983). The sensitivity of the particular enzyme to the pollutant depends on the success of such a related study. Dehydrogenase is an enzyme which is unspecific in its activity unlike urease. Zn and Cu additions were found to decrease the activity of acid phosphatase and urease activity (Tyler, 1976) and amylase (Ebregt and Boldewijn, 1977). Dehydrogenase functions in living cells while phosphatase is an extracellular enzyme. Adverse effects on the metabolic functioning of the cell would result in reduction of Dehydrogenase activity (DHA). Compara-tively, reductions in phosphate activity would not be as noticeable as those of Dehydrogenase when the microbial metabolic function reduces.

The mode of action for the dehydrogenase encompasses a theorem where an endocellular enzyme Dehydrogenase facilitates in the transfer of hydrogen and electrons from organic compounds to appropriate electron acceptors during the initial oxidation of the substrate (Skujins, 1978).Normally when electrons pass along a chain of intermediate carriers, Oxygen act as the final electron acceptor. In the dehydrogenase test tetrazolium salts (e.g. triphenyltetrazolium chloride (TTC) and 2-(4-iodophenyl)-3-(4-nitrophenyl)-5-phenyl tetrazolium chloride) act as the terminal electron acceptors in the anaerobic environments (Trevors et al., 1981). Banefield et al. (1977) suggested that in O_2 depleted conditions aerobic as well as anaerobic dehydrogenase use TTC as an electron acceptor. In effect TTC is reduced to TPF (triphenyltetrazolium

Stage one:
Construction of *E. coli* HB101 puce 607

Stage two:
Freeze drying of *lux* marked *E.coli* which will involve ;-
i. Preparation of glycerol stock
ii. Resuscitation of freeze dried vials

Stage three:
Growth curve of *lux* marked *E.coli*

Stage four:
lux marked bacterial assay;-
~ To involve the preparation of standard solutions for the chlorinated phenols

Stage five:
Data Analysis

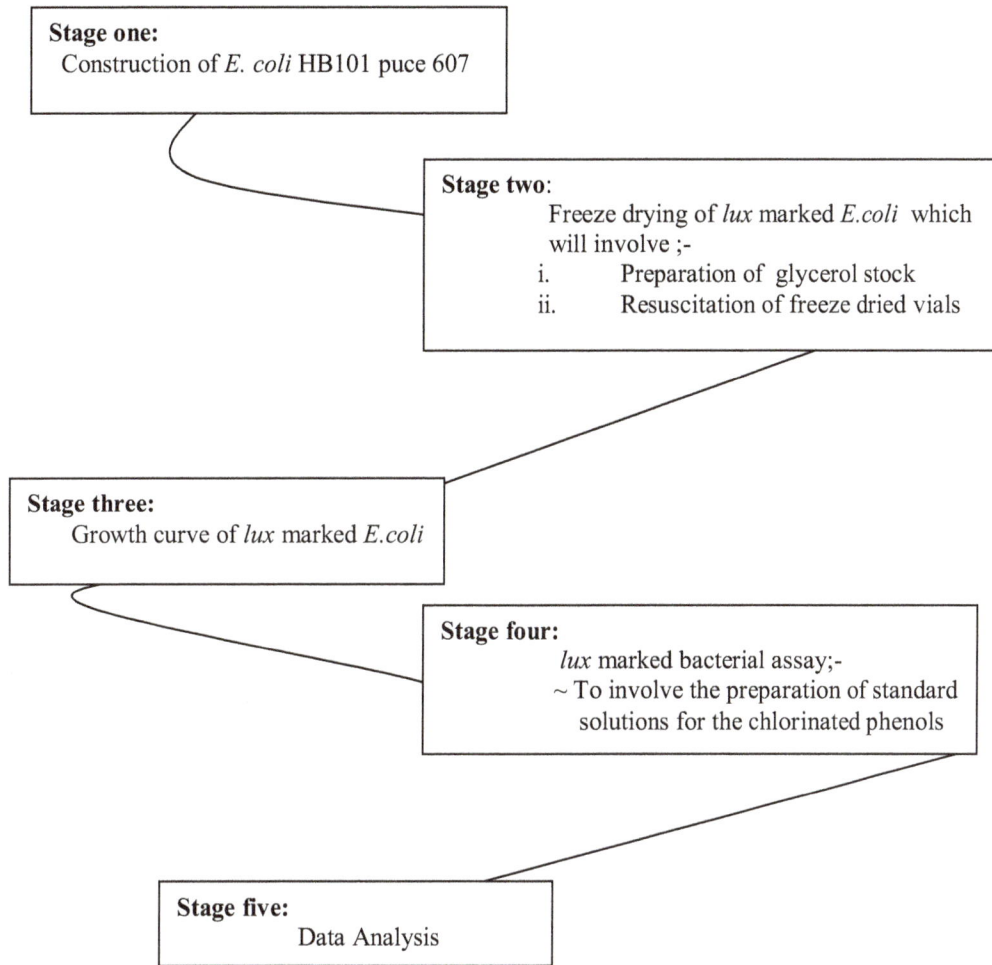

Figure 1. Protocols To assess lux-marked *E. coli*'s response in the form of toxic response and examine the effect of exposure time on toxicity of pollutants.

How do lux toxicity-based biosensors (e.g. E.coli HB101 (pUCD607) work?

The presence of the target analyte induces the expression of the specific gene sequence and consequently of the reporter gene with synthesis of the luciferase enzyme , and luciferin/luciferase-mediated light output occurs.[2]

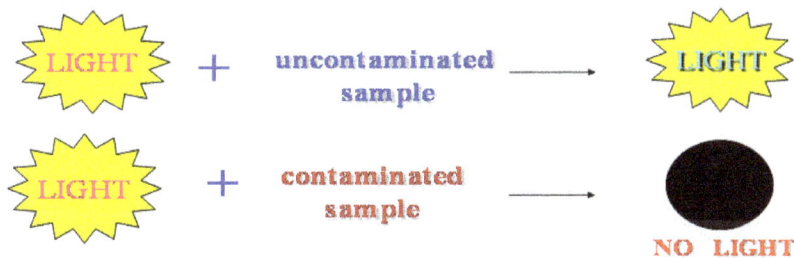

LIGHT + uncontaminated sample → LIGHT

LIGHT + contaminated sample → NO LIGHT

(Specific, "lights on biosensors" can target individual contaminants but care must be taken in their use as they usually detect families of contaminants)

Figure 2. Illustration of how *lux* toxicity based biosensors work.

formazon) a red insoluble compound. Ethanol is used to dissolve the precipitate and calorimetrically measured by a spectrometer.

i. Application of the DHA as an ecotoxicological tool

The Dehydrogenase test has been used to determine the effects of heavy metals present in sewage sludge (during the study a reduction of the DHA in metal contaminated soils was observed suggesting a decline in microbial activity in polluted soils) on the soil microbial activity (Cenci and Morozzi, 1979; Ruhling and Tyler, 1973; Doelman and Haanstra, 1979; Schinner et al., 1980; Brookes et al., 1984). Similarly DHA was used to determine the microbial biomass of the soil (Malkommes, 1988; Wilke and Keuffel, 1988). A correlation between the DHA and the biomass of the soil microflora was established (Beck, 1984) and also other processes involving nitrification (Skujins, 1973). Recently it was used as a tool to ascertain river health impacted by the tanning Industry (Mwinyihija et al., 2006a; Mwinyihija, 2010).

However there are limitations through use of Dehydrogenase Tests. For example Dehydrogenase in the presence of Cu (unlike other enzymes such as urease, invertase and cellulase which) causes a reduction in TPF (triphenyltetrazolium formazon) absorbance (Chander and Brookes, 1991). The presence of alternate electron acceptors in the soil may cause a problem in comparing the DHA of different soil types due to a large proportion of O_2 uptake which remains unaccounted for (Burns,1978) (Somerville et al., 1987). DHA inhibition may also be caused by humic acid (Pflug and Ziechman, 1981) or soils with high inorganic nitrogen (Trevors, 1984) and long incubation periods.

In retrospect ways of addressing these shortcomings have been developed. For example competition of DHA with other electron acceptors in the soil may be addressed by using INT (2-p-iodophenyl-3-p-nitrophenyl-5-phenyl-tetrazolium chloride) instead of TTC (Benefield et al., 1977; Mwinyihija, 2006b). INT has a rapid response, high sensitivity and compete well with O_2 for liberated electrons. INT as an electron acceptor is pH sensitive with optimum result at pH 4.8 (Benefield et al., 1977). DHA as a test may further be enhanced by shortening the incubation period from 24 to 6 hours and by adding yeast extract (where a 53% increase in DHA was observed) (Rossel and Tarradellas,1991), good correlation between short term respiration and DHA (Beck ,1984) and lack of correlation occurring between DHA and microbial numbers especially in nutrient starved soils ,may be due to the presence of low metabolism, dormant or dwarf forms of organisms(Marshall, 1980, Roszack and Colwell, 1987). DHA as a test estimates the total potential activity of the microflora rather than its total effective activity.

Fundamentals of applying ecotoxicological tools for diagnosis

There is increasing movement towards the use of biological techniques for the monitoring of environmental pollution by industry and regulators a like (McHenry, 1995; Mwinyihija, 2010; 2009; 2006a). This will further strengthen capacity in quality control where detrimental and beneficial aspects of the environment influenced by the tanning industry will be monitored more efficiently and effectively.

The fundamentals of the study by Mwinyihija (2010) related to the tanning industry and mostly centred on; characterising the chemistry of effluents of Kenyan tanning industry, developed toxicity testing procedures for risk assessment and also undertook field based assessment of environmental impact of chromium. The manipulation linked to toxicity for the purpose of remediation will include additive techniques such as; purging, ion exchange resin extraction, pH adjustment and filtration through activated charcoal. Subsequently the use of whole cell microbial biosensors as a way forward for monitoring environmental contamination by heavy metals (e.g. Cr for this study) (Knight et al., 1999) and organic contaminants (e.g. chlorinated phenolics) as well as toxicity in soils and water contaminated by industrial effluents (Brown et al., 1996) including river sediments and tannery dust (Mwinyihija et al., 2005, 2006b; Mwinyihija, 2010)

Selected potential pollutants of the tanning industry

The following areas form important part of this review to identify impact areas of the tanning industry effluent on soils and water;

NaCl₂: Salt as a curing agent in the primary level of preservation is an inorganic chemical that strongly has been identified as a pollutant. In a study in Egypt, $NaCl_2$ concentration varied between 40, 000 to 50, 000 mg/l in the effluent discharge in the tannery under study (Hafez et al., 2002). Salinity or ionic strength can cause a small decrease in the solubility of non-polar organic compounds (e.g. Naphthalene, Benzene, toluene etc) through a process known as salting-out effect (Pepper et al., 1996). This is a very critical point of interest in the tanning industry and its one of the most important pollutant that need to be addressed urgently.

Organic matter: Most of these will include biodegradable organic matter e.g. Proteins and Carbohydrates. Their main problem is the depression of the dissolved oxygen content of stream waters caused by their microbial decomposition. Their impacts are primarily the loss of

Table 1. Summary of various Cr oxidation numbers, type and environmental behaviour.

Valency	Environment behaviour	Remarks
Cr	Unstable	
Cr^{1+}	Unstable	
Cr^{2+}	Readily oxidised to Cr_3 but stable only in the absence of any oxidant	Active under anaerobic condition
Cr^{3+}	Most stable	Considerable energy required to convert to lower or higher states
Cr^{4+}	Forms Unstable intermediate reactions to trivalent and oxidation states	Exhibits this phase during oxidation and reduction
Cr^{5+}	Unstable intermediate	Observed during oxidation and reduction
Cr^{6+}	In acidic conditions demonstrates very high positive redox potential and unstable in the presence of electron donors	Strongly oxidizing

dissolved oxygen, which is detrimental to aquatic organisms. Secondly, their effect is on dissolved oxygen that is consumed by aerobic microbial oxidation of the waste; anaerobic decomposition becomes prominent and releasing noxious gases (Pepper et al., 1996).

Chromium salts as important tanning industry ingredients: Chromium basic sulphate is the most widely used tanning substance today. The exhausted bath coming from the chromium tannage contains about 30% of the initial salt and is normally sent for cleaning up (Cassano et al., 2000). Here chromium salts are entrained in the sludge creating serious problems for their disposal (Gauglhofer, 1986). Chromium can exist in several chemical forms displaying oxidation numbers from 0 to VI. Trivalent and hexavalent forms are most stable in the environment exhibiting different environmental behaviour as depicted in Table 1.

In nature, Cr originates from weathering or rock, wet precipitation and dry fallout from the atmosphere and run off from the terrestrial systems (Kotas and Stasicka, 2000). In rivers and lakes the Cr concentration is usually limited to 0.5 – 100 nM (Handa, 1988; Kaczynski and Kieber, 1992), in seawaters it varies from 0.1- 16 nM. Tanning industry can contribute highly to the increase of Cr concentration if located near the water systems. In nature Cr exists in its two stable oxidation states, Cr^{3+} and Cr^{6+}. The presence and ratio between these two forms depend on various processes; Chemical and photochemical, redox transformation, precipitation/dissolution. Campanella et al. (1996) described the Cr^{3+} in oxygenated aqueous solution as predicted by thermodynamic calculations on the stable species at pH ≤ 6 whereas pH ≥ 7 the CrO_4^{2-} ion as predominate under anoxic and suboxic condition, trivalent Cr should be the

only form (Kotas and Stasicka, 2000).

The nature and behaviour of various Cr forms found in wastewater can be quite different from those present in natural water because of altered physico-chemical condition of the effluents originating from various industrial sources (Kotas and Stasicka, 2000). The presence and concentration of Cr forms in effluents depends on Cr compounds applied during processing, pH, and organic and / or inorganic waste coming from the material processing ((Kotas and Stasicka, 2000).

Cr^6 will dominate in wastewater from metallurgical industry, metal finishing industry (Cr hard plating), refractory industry and production or application of pigments (chromate colour pigments and corrosion inhibition pigments). Cr_3 dominates from tannery, textile (printing, dyeing), decorative plating industry waste. In reference to the tanning industry Cr_3 in the effluents is the most expected form but with redox reactions occurring in the sludge, an increase in hexavalent form can occur.

Under slightly acidic or neutral pH, condition in this type of wastewater the poorly soluble $Cr(OH)_3$.aq should be the preferred form, but a high content of organic matter originated from hide/skin material processing is effective in forming soluble Cr^3 complexes (Stein et al., 1994; Walsh and Halloran, 1996).

In soils an increase in local Cr concentration originates from fallout and wash out of atmospheric Cr containing particles as well as from the chrome bearing sludge and refuse from industrial activity (Kotas and Stasicka, 2000). Cr^3 adsorption into humic acids renders it insoluble, immobile and unreactive. This process is the most effective within the pH range of 2.7 – 4.5 (Walsh, 1996). Other macromolecular ligands behave similarly. In contrast mobile ligands such as citric acid, diethylene triamine pentaacetic acid (DTPA) and fulvic acid form soluble Cr^{3+} complexes, which mediate its relocation and oxidation to Cr^{6+} in soils (James and Bartlett, 1983; James, 1996). Dechromification is thought as being of vital importance because without it, theoretically all atmospheric oxygen could be a threat to life on earth (James and Bartlett, 1983).

Whilst in the atmosphere Cr present in the atmosphere originates from anthropogenic sources, which account for 60-70%, as well as from natural sources, which account for the remaining 30-40%. Industrial activities still remain the major source of pollution to the atmospheric systems. Other could be natural sources like volcano eruptions and erosions of soil and rocks ((Kotas and Stasicka, 2000). Sea salt particles and forest wild fires do not seem to be important sources of Cr. Average atmospheric concentrations of this metal are, 1 ng/m^3 in rural to 10 ng/m^3 in polluted urban areas. The amount of Cr at any particular time depends on the intensity of industrial processes, proximity to the sources, the amount of Cr released and meteorological factors (Kotas and Stasicka, 2000).

Cr forms mostly ionic compounds, their vapour could be neglected and one may assume that gaseous Cr species do not exist at ambient atmospheric temperatures and Cr is present in the atmosphere in form of particles and droplet aerosols. On the other hand Cr from sources releasing the element in lager particles (particle diameter varies: 0.2-50 μm (Kotas and Stasicka, 2000) is deposited locally and can migrate through individual particular environmental media. The size of particles is of importance for consideration of Cr toxicity: Friess et al. (1989) found that only the particles of diameters from 0.2 to 10 μm are respirable, and that their retention in the lung can pose carcinogenic risk.

Chlorinated phenols related to the tanning industry

The curing and storage phase of the hides and skins utilises various types of insecticides and antimould chemical compounds. All these including the bacteriostats used in the tanning industry are eventually spewed out into the tannery effluents. Most of these compounds belong to the chlorophenols, which enter the environment through several pathways (Steirt and Crawford, 1985). It was reported by Escher et al. (1996) that phenolic compounds exert toxic effect on microorganisms disrupting energy transduction either by uncoupling oxidative phosphorylation or inhibiting electron transfer. Substituted phenols act by destroying proton gradient by transporting protons back across the membranes and/or inhibiting electron flow by binding to specific components of the transport chain Escher et al. (1996).

Chlorophenols are routinely determined by chemical analysis. The disadvantage of this method is the inability to determine the concentration of the chlorophenols present, in effect providing no information on availability of the compounds to the biomass. Further, chemical analysis does not provide information in the toxicity of the compounds and factors that affect toxicity. In retrospect biological assessment provide sensors that can identify the toxicity of chlorophenols in a variety of environmental matrices. It assesses the impact of any environmental factor on the toxicity of these compounds much quickly and easily.

During previous study by Mwinyihija (2010) *lux* marked *E. coli* was used as a biosensor. It was responsive to a wide variety of chlorophenols depending on the number and position of chlorine substitution. The advantages associated with the use of this biosensor as an appropriate Ecotoxicological diagnostic tool, was that they are rapid, reliable tools for the toxicity testing and easy in assessing environmental factors associated with the pollutants (Mwinyihija et al., 2006b; Mwinyihija, 2009)

Perspective of ecotoxicology on public health and ecology

Adopting of this perspective of study assisted in investigating the occupational hazards within the tanning industry e.g. understanding the impact levels of contaminated tannery dust. Indeed an in-depth qualitative assessment of the risk of human exposure to dust was made throughout a commercial Kenyan tannery using this technique and reported upon (Mwinyihija et al., 2006c). The success of applying this novel technique also for the first time, was obtaining information which demonstrated the high-risk points in the processing line and identifying the dust sampling regimes developed. A support study by Mwinyihija et al. (2005) using an optical set-up using microscopy and digital imaging techniques was used to determine dust particle numbers and size distributions. The results showed that chemical handling was the most hazardous (12 mg m^{-3}). A Monte Carlo method was used to estimate the concentration of the dust in the air throughout the tannery during a working day. Thus toxicity studies related to the ecology could also engage other methodologies apart from bioassays. This method also proved beyond reasonable doubts that Ecotoxicological diagnosis could also provide mitigation of occupational risk within the tannery workers environment.

The effect on pollution of terrestrial and aquatic ecosystem carries a positive correlation to human health as a component of the environmental matrix.

Soil and effluent samples characterisation as an important ecotoxicological approach

Characterisation of soil and effluent samples from the tanning sector is fundamental in understanding the primary sources of toxicity in both terrestrial and aquatic environment as caused by the sector. This process will develop the appropriate diagnostic and remediation strategies to be deployed subsequently. Various systems of characterization have been used recently including x-ray diffraction, photo acoustic spectroscopy and electron paramagnetic resonance (Wararatananurak, 2000).

It has been reported by investigations undertaken by Ritchie et al. (1972), that the humus–clay complexes which are formed with smectite clays form bonding not easily destroyed by even microorganism. Therefore with high amounts of organic matter the shrink/swell propensity is reduced (Ritchie et al., 1972). Many organism such as fungi's, diatoms and earthworms are found in such soils irrespective of discomforts due to the shrinking (due to dry conditions) and swelling(due to wet conditions). According to Killham (2001), he denotes that soil being heterogeneous in nature it is imperative during its study to investigate and eventually conceptualise all the soil parameters to aquire an integrated understanding of the role of microsites in soil ecology. These parameters include minerals, organic matter, living biomass, structure, water, atmosphere, redox, pH, temperature and light. Thus in our current research proposal forming the basis of the proposed thesis, the study of microsites impacted by the effluent form the curing sites will form the main basis of investigation, enveloping the understanding of interaction between different components of the soil biota (Killham, 2001). In view of that, during the field studies, modern molecular techniques involving the marking and *in situ* detection of components of the soil biota coupled with the traditional soil micromophological methods would be considered to better characterise the soil at the microsite level.

To provide the best Ecotoxicological diagnosis the investigator intends to obtained effluents that are raw; this will be for the purpose of identifying the predominant toxicants and establishing their levels of toxicity when spewed out to the environment. The traditional method of soil and effluent analysis were undertaken following *in situ* sample collection. This included (a). sample preparation, moisture content determination, particle size analysis, and measurement of soil pH, organic carbon, cation exchange capacity, and total nitrogen and (b) determination of major constituents dissolved in riverwater within the effluent discharge areas and outwith the discharge areas.

Conclusion

Ecotoxicology as an emergent science has all the advantages of finding the causes and effects to pollutants and their associated impact to various ecosystems. The current study leading to the proposed thesis will investigate the influence of effluents emanating from the curing premises at a selected site in Kenya is opportune and appropriate so as to provide a cost effective and easy to apply technique in the proposed area. Indeed previous study by the author has had a fundamental

influence and scientific backing to the proposed area of investigation. Thus the focus on areas such as the study of complexity of effluents form the curing of hides and skins will benefit the zeal to expand the knowledge horizon in ecotoxicology as an emerging science. This will not only provide the essentials in ecotoxicology but provide a an in-depth scope to this field of stud as it has never been attempted before therefore depicting the efforts as a potential to very interesting results to be shared with the whole world.

ACKNOWLEDGEMENT

All Staff at the International Atlantic University (IAU) are acknowledged for their support during the post doctorate programme that leads to this work.

REFERENCES

Baumgartner HL (1996). Surface water pollution, chapter 13, pp 189 – 208. In: Pepper, L.I., Gerba, P.C., Brusseau, M.L Pollution Science, Academic Press (Elsevier Science, USA), San Diego, California, USA.
Beck TH (1984). Mikrobiologische und biochemische charakterisierung lanwirtschaftich genutzer Boden I. Mittelung: Die Ermittlung einer Boden Mikrobiologischen kennzahl. Z. Pflanzenernahr. Bodenkd, 147: 456-466.
Benefield CB, Haward PJA, Howard DM (1977). The estimation of dehydrogenase activity in soil. Soil Biol. Biochem., 9: 67 – 70.
Brookes PC, Mcgrath SP, Klein DA, Elliot ET (1984). Effects of heavy metals on microbial activity and biomass in field soils treated with sewage sludge. In: Environmental Contamination (International Conference, London, July (1984), pp. 574-583, CEP, Edinburgh.
Brown JS, Rattray EAS, Paton GI, Reid G, Caffoor I, Killham K (1996). Comparative assessment of the toxicity of a aper mill effluent by respirometry and a luminescence-based bacterial assay. Chemosphere, 32: 1553 –1561.
Burns RG (1978). Soil Enzymes. Academic Press, New York, (1978).
Campanella L (1996). Problems of speciation of elements in Selenium. In: Caroli,S.,(Ed) Element speciation in Bioinorganic chemistry. Wiley Interscience,New York, pp. 419 – 444.
Cassano A, Molinari, A., Romano M., Drioli, E (2000). Treatment of the aqueous effluents of the leather Industry by membrane process A review, J. Membr. Sci., 181: 111 – 126.
Cenci H, Morozzi, G (1979). The validity of the TTC-test for dehydrogenase activity of activated sludge in the presence of chemical inhibitors. Zentralblatt fur Bacteriologie and Hygiene Abstract B, 169: 320-330.
Chander K, Brookes PC (1991). Is the dehydrogenase assay invalid as a method to estimate microbial activity in copper contaminated soils? Soil Biol. Biochem., 23(10): 909-915.
Committee on Commodity Problems (CCP) (1998). Intergovermental group on meat; sub-group (on Hides and Skins, sixth session, (CCP:ME/HS 98/3) –Capetown, South Africa, September.
Doleman P, Haanstra L (1979). Effect of lead on soil respiration and dehydrogenase activity. Soil Biol Biochem, 11: 475 – 479.
Ebregt A, Boldewijn JMAM (1977). Influence of heavy metals in spruce forest on amylase activity, CO_2 evolution from starch and soil respiration. Plant and Soil, 47: 137-148.
Escher BI, Snozzi M, Schwarzenbach RP (1996). Uptake speciation and uncoupling activity of substituted phenols in energy transducing membranes. Environ. Sci. Technol., 30, 3071 – 3079.
Flaherty O, Roddy W, Lollar TRM (1959). The chemistry and Technology of Leather, Vol. 1 E. Robert Krieger Publishing company, NewYork.
Frankenberger WT Jr, Dick WA (1983). Relationship between enzyme

activities and microbial growth and activities indices in soil. Soil Sci. Soc. Am. J., 47: 945-951.

Friess SL (1989). Carcinogenic risk assessment criteria associated with inhalation of air borne particulates containing chromium (VI/III). Sci. Tot. Environ., 86: 109 – 112.

Gauglhofer J (1986). Environmental aspects of Tanning with chromium, J. Society of leather Technol. Chem., 70(1)(11).

Hafez AI, El-Manharawy MS, Khedr MA (2002). RO membrane removal of untreated chromium from spent tanning effluent. A pilot scale study, part 2., Desalination, 144: 237 – 242.

Handa BK (1988). Occurance and distribution of chromium in natural waters of India. In: Nriagu, J.O., Nieboer, E. (Eds), Chromium in Natural and Human Environment. Wiley Interscience, New York, pp. 189 – 215.

James BR, Bartlett RJ (1983). Behaviour of chromium in soils: VI. Interactions between oxidation-reduction and organic complexation J. Environ. Qual, 12: 173 – 176.

James BR (1996). The challenge of remediating chromium contaminated soils. Environ. Sci. Technol., 30: 248 –251.

Kaczynski SE, Kieber RJ (1993). Aqueous trivalent chromium photoproduction in natural waters. Environ. Sci. Technol., 27, 1572 – 1576.

Killham K (2001). Soil Ecology, Cambridge University press, 1994 (reprinted 2001), pg. 1 – 32.

Knight B, McGrath SP, Killham K, Preston S, Paton GI (1999). Assessment of the toxicity of heavy metals in soils amended with sewage sludge using a chemical speciation techniques and a *lux* biosensors. Environ. Toxicol. Chem., 18: 659 –663.

Komma consultants (1996). Eco trade manual. Environmental challenges for exporters to the European union. February 1996, p. 115

Kotas J, Stasicka Z (2000). Chromium occurrence in the environment and its speciation. Environ. Pollut., 107: 263 –283.

Malkommes HP (1988). Einfluss einmaliger und wiederholter Herbizia-Gaben auf Mikrobielle Pozesse in Bodenproben unter laborbedingungen. Pedobiologia, 31: 323-338.

Manly R (2000). Ecotoxicology In: Environmental Analytical Chemistry (Eds. Fifield F. W. And Haines P .J.). 2nd ed. Blackwell Science Ltd., Oxford, pp. 425-470.

Marshall KC (1980). Adsorption of microorganisms to soils and sediments, pp. 317-329. In: G.Bitton and K.C Marshall (eds). Wiley-Interscience, New York.

McHenery JG (1995). Ecotoxicity testing-effective use in field monitoring, In Richardson M, ed, Environmental toxicology Assessment. Taylor & Francis, Rickmansworth, U.K, pp. 55 – 64.

Muchie M (2000) Leather processing in Ethiopia and Kenya: Lessons from India, Tech. Soc., 22: 537 – 555.

Mwinyihija M (2011). Emerging World Leather trends and continental shifts on leather and Leathergoods production. World leather congress proceedings, Rio de-janeiro, Brazil.

Mwinyihija M, Killham K (2006a). Is the Kenyan tanning industry integral to prioritised environmental sustainability targets set in the quest to industrialisation by 2020 AD? J. Environ. Sci., 3(2): 113-134.

Mwinyihija M, Meharg A, Strachan NJC, Killham K (2006b). An ecotoxicological approach to assessing the impact of tanning Industry effluent on river health. Arch Environ. Contam Toxicol., 50(3): 316.

Mwinyihija M, Strachan N, Rotariu O, Meharg A, Killham K (2006c). Ecotoxicological screening of Kenyan tannery dust using a luminescent-based bacterial biosensor. Int. J. Environ. Health Res., 16(1): 47-58.

Mwinyihija M (2010). Ecotoxicological Diagnosis in the tanning Industry – Springer Publisher (New York, USA).

Mwinyihija M (2009). An Overview of Selected *lux*-marked Biosensors and Its Application As A Tool To Ecotoxicological Analysis on Book titled 'Biosensors, Uses and Application' (Eds; Rafael Comeaux and Pablo Novotny), pp 199 - 214 published by Nova publishers (USA).

Mwinyihija M, Killham K, Rotariu O, Meharg A, Strachan N (2005). Assessing the occupational risk of dust particles using rapid image processing and microscopy techniques. Int. J. Environ. Health Res., 15: 53- 62.

Mwinyihija M, Meharg A, Strachan NJC, Killham K (2005a). *Biosensor based toxicity dissection of tannery and associated environmental samples,* J AM LEATHER CHEM AS., 100 (12):481- 490.

Mwinyihija M, Meharg A, Strachan NJC, Killham K (2005b). *Ecological Risk Assessment of the Kenyan tannery industry.* J AM LEATHER CHEM AS., 100(11): 380-395.

Pepper IL, Gerba CP, Brussean ML (1996). Pollution Science, Academic press Inc., pp. 194.

Pflug W, Ziechman W (1981). Inhibition of malate dehydrogenase by humic acids. Soil Biol. Biochem .,13: 293-297.

Ramasami T, Prasad BGS (1991). Proceedings of the Lexpo – XV, pg 43.

Ritchie, J.T., Kissel, D.E., Burnett, E., Water movement in undisturbed swelling clay. Soil Sci. Am. Proc., 36: 874-879.

Ritchie GSP, Sposito G (1972). Speciation in soil. In: Ure,A.M., Davidson,C.M (Eds), chemical speciation in the Environment.Blackie Academic and Professional , Glasgow, pp. 201 – 233.

Rossel D, Tarradellas J (1991). Dehydrogenase activity of soil microflora: Significance in ecotoxicological test. Environ. Toxicol. Water Qual., 6:17-33.

Roszack DB, Colwell RR (1987). Survival strategies of bacteria in natural environment. Microbiol Rev., 51: 365-379.

Ruhling A, Tyler G (1973). Heavy metal pollution and decomposition of spruce needle litter. *Oikos,* 24: 402 – 416.

Schinner FA, Niederbacher R, Neuwinger I (1980). Influence of compound fertiliser and cupric sulphate on soil enzymes and CO_2 evolution. Plant Soil, 57: 85 – 93.

SEPA. (1999). Improving Scotland's Water Environment. Scottish Environmental Protection Agency, Stirling,

Skujins J (1978). History of abiotic soil enzyme research, pp 1- 49. In: Soil Enzymes (Burns, R.G (ed.)). Academic Press, New York.

Stein K, Schwed TG (1994). Fresenius J., Anal. Chem., (38) 350.

Steirt JG, Crawford RL (1985). (Microbial degradation of chlorinated phenols. Trends Biotechnol., 3: 300 – 305.

Stevenson IL (1959). Dehydrogenase activity in soil. Can J. Microbiol., 5: 229-235.

Sweeting RA (1994). River pollution, pp. 23 – 33. In: The Rivers Handbook. Vol. 2 Hydrological and Ecological Principles, (Calow P. & Petts, G.E. (Eds.)). Blackwell Scientific, Oxford.

Thalmann A (1968). Zur bestimmung der deydrogenase activitat im boden mittels Triphenytetrazoliumchlorid (TTC). Landwirt. Forsch, 21: 249-258.

Thanikaivelan P, Raghava Rao J, Nair BU, Ramasami T (2000). Approach towards zero discharge tanning: role of concentration on the development of eco- friendly liming-reliming process, J. Clean Prod., 11:79-90.

Trevors JT, Mayfield CI, Innis WE (1981). A rapid toxicty test using *Pseudomonas fluorescens.* Bull Environ. Contam. Toxicol.,, 26: 433-437.

Tyler G (1976). Heavy metal pollution, phosphatase activity and mineralisation of organic phosphorous in forest soils. Soil Biol. Biochem., 8: 327-332.

UNIDO (1992). Industry and development report. UNIDO-Vienna: UN Publications, pg. 10.

Vo-Dinh T, Cullum BM (2000). Biosensors and biochips for bioanalysis. Fresen J. Anal. Chem., 366: 540-551.

Walsh AR, O'Halloran J (1996). Chromium speciation in Tanery effluents – II Speciation in effluent and in a receiving estuary water Res., 30: 2401 – 2412.

Wararatananurak P, (2000). Fractionation of chromium toxicity in water. PhD Thesis, University of Aberdeen, U.K.

Wilke BM, Keuffel AB (1988). Short term experiments for the estimation of lopng term effects of inorganic pollutants on soil microbiol activity. Z. Pflanzenernahr Bodenkd, 151: 399-403.

Zywicki B, Reemtsma T, Jekel M (2002). Analysis of commercial vegetable tanning agents by reversed- phase liquid chromatography – electrospray ionisation- tandem mass spectrometry and its application to the waste water, J. Chrom., 970 (1-2): 191-200.

26

Physicochemical dynamics of the impact of paper mill effluents on Owerrinta River, eastern Nigeria

Ihejirika Chinedu Emeka[1]*, Emereibeole Enos Ihediohamma[1], Nwaogu Linus[2], Uzoka Christopher Ndubuisi[1] and Amaku Grace Ebele[1]

[1]Department of Environmental Technology, Federal University of Technology, Owerri, Nigeria.
[2]Department of Biochemistry, Federal University of Technology, Owerri, Nigeria.

Industrial effluent discharge constitute major source of water pollution. Effects of effluent discharge from three paper mill industries on recipient Owerrinta River was determined by subjecting samples to standard physicochemical analysis. All values were within standard excepting the pH value of Effluent-II sample (3.92) and Total Suspended Solids values of all the effluent samples (84, 496, and 165 mg/L) respectively. There were significant variations ($P < 0.05$) between effluents and river samples and within effluents and river samples respectively, for all the parameters. The values varied as follows: temperature (24.70 – 24.12); pH (6.68 – 3.92); conductivity (64.67 – 0.02); Turbidities (259.00 – 16.00); Total Dissolved Solids (29.50 – 1.50); Total Suspended Solids (496.00 – 2.02); nitrate (NO_3^-) (19.10 – 0.08); phosphate (PO_4^{2-}) (0.81 – 0.02); sulphate (SO_4^{2-}) (34.00 – 0.06); Biochemical Oxygen Demand (1.09 – 0.41); Chemical Oxygen Demand (8.25 – 0.72) and Oil and Grease (4.01 – 1.92). Variations from River samples indicated impact from effluents discharge, while variations in effluents values implied the contributory pattern of the effluents to River quality. The River recovered from some parameter. Treatment of effluents to insignificant values will reduce the impact on River quality.

Key words: Paper mill, effluents, Owerrinta River, water quality, sustainable, development.

INTRODUCTION

The introductions of contaminants through effluent and sludge to different environments can often over whelm the self-cleaning capacity of recipient ecosystems and thus result in the accumulation of pollutants to problematic or even harmful levels. An awareness of environmental problems and potential hazards caused by industrial wastewaters has prompted many countries to limit the discharge of certain toxic effluents. The raw wastewaters from pulp, paper and board mills can be potentially very polluting. Wastewaters have prompted many countries to limit the discharge of certain toxic effluents. Indeed, a survey within the UK industry has found that their Chemical Oxygen Demand (COD) can be as high as 11,000 mg/L (Thompson et al., 2001). The amount of pollutant in pulp and paper mill effluent is measured in terms of two key parameters, Total Suspended Solids (TSS) and Biochemical Oxygen Demand (BOD) (OFIA, 2005).

Full access to safe drinking water to citizens living in developing countries was the decision of the United Nations Assembly of 10th November 1980. However, almost two decades after, over two billion people especially in the developing countries, lack safe water and sanitation (ODA, 1997). In Nigeria, especially in the Eastern region, the large scale pollution of streams and rivers is not only a major public health problem but also constitute a principal obstacle to socio-economic advancement and fight against poverty and malnutrition (Okpokwasili and Ogbulie, 1993). This problem has had its toll on aquatic species extinction and fish diseases of various consequences (Okpokwasili et al., 1995; Ogbulie and Okpokwasili, 1998).

Community based studies (Izuagba and Ogbulie, 1997) revealed that the use of natural water bodies for industrial and domestic waste disposal is expected to worsen in the

---
*Corresponding author. E-mail: ceihejirika@yahoo.com.

Table 1. Comparison of physicochemical parameters of effluents and stream samples with FEPA standards at Owerrinta Point of Imo River.

Parameter	Stream samples			Effluent samples			
	A	B	C	I	II	III	FEPA STD
Temperature (°C)	24.12	24.14	24.15	24.70	24.4	24.70	<40
pH	6.68	6.35	6.67	6.22	3.92	6.13	6-9
Conductivity (µS/cm)	0.01	0.02	0.01	59.00	48.00	53.00	NA
Turbidity (NTU)	16.00	28.00	6.80	51.00	259	121	NA
TDS (mg/L)	2.10	6.90	1.50	29.50	24.00	26.50	2000
TSS (mg/L)	4.80	17.00	2.02	84.00	496	165	30
NO_3^- (mg/L)	0.09	2.00	0.08	3.40	8.50	19.10	20
PO_4^{2-} (mg/L)	0.02	0.16	0.08	0.04	0.81	0.26	5
SO_4^{2-} (mg/L)	0.33	1.00	0.06	19.00	34.00	30.00	500
BOD_5 (mg/L)	0.41	0.66	0.78	1.09	1.02	0.48	50
COD (mg/L)	0.72	1.24	1.22	4.77	8.25	6.97	NA
Oil and grease (mg/L)	1.92	3.22	2.01	4.01	3.91	3.77	10

NA = Not available.

and domestic waste disposal is expected to worsen in the nearest future. Previous studies have revealed that our sources of water are not only polluted by sewage but also by toxic discharge and emission from industrial and other sources (Okpokwasili et al., 1997). This is a serious source of concern considering the rapid population growth in the developing countries.

Owerrinta River provides water for domestic, industrial, and small scale agricultural irrigation practices in addition to fishery and recreational activities. This work therefore was aimed at determining the physicochemical characteristics of the paper mill industrial effluents and the recipient River samples and comparing the values to understand the contributions of individual paper mill industries in Owerrinta River quality and the River's ability to recover from the impact, to assist environmental regulatory agencies and other stake holders in controlling discharges from individual industries into the River for a sustainable environment and development.

MATERIALS AND METHODS

Study area

Owerrinta River is located within longitude 7°17'E and Latitude 5°18'N and serves as a recipient of effluents from three paper mill industries (Effluent I - Star paper mill, Effluent II- Apex paper mill, and Effluent III- Industrial paper mill) closely sited together, and provides sand for excavators, source of fishes and water for domestic uses.

Sample collection

Samples were collected in triplicates with the aid of clean 1 liter water sampling cans. Collected samples were transported to the laboratory for analysis. Effluent and River samples were collected

from discharge points before discharge into Owerrinta Point of Imo River for two years (2008 and 2009). River samples were collected thus: upstream (100 m) before the first discharge point; discharge point (20 m) after the third discharge point; and downstream (100 m) after the third discharge point.

METHODOLOGY

The temperature, pH, conductivity, and turbidity were determined using digital meters. Total Dissolved Solids and Total Suspended Solids measurements were carried out by using the conductivity/total dissolved solids meter (HACH DR/2010 Spectrophotometer Hand Book, 1997). Nitrate, phosphate, and sulphate were determined by using a spectrophotometer with Nitra var 5 nitrate, Phosphor var 5 phosphate, and Sulfa var 4 sulphate reagents as described in the HACH Water Analysis Handbook (HACH, 1981). Biochemical Oxygen Demand (BOD_5), Chemical Oxygen Demand (COD), and Oil and Grease were determined as described by the *Standard Methods for the Examination of Water and Wastewater* (APHA/AWWA/WPCF, 1985).

Statistical analysis

The result was subjected to different statistical analyses and presentations by using tools ranging from T-test, Analysis of Variance (ANOVA), and Tukey Grouping, by the method of Statistical Package for Social Sciences (SPSS 16.0).

RESULTS

Table 1 shows the comparison of physicochemical parameters of effluents and stream samples with FEPA standards at Owerrinta Point of Imo River. All the values of the parameters were within FEPA standard excepting the value of pH of Effluent - II sample (3.92) that was acidic and the values of TSS of all the effluent samples

Table 2a. Physicochemical variations between paper mill effluents and river water samples.

Sample	Temperature (°C)	pH	Conductivity (µS/cm)	Turbidity (NTU)	TDS (mg/L)	TSS (mg/L)
Upstream	24.12±0.026^A	6.68±0.017^E	0.13±0.06^A	16.00±1.00^A	2.10±0.10^A	1.80±0.17^A
Discharge point	24.14±0.026^A	6.35±0.01^D	0.02±0.01^A	28.00±1.00^B	6.90±0.20^B	17.00±1.00^B
Down stream	24.13±0.025^A	6.67±0.01^E	0.02±0.01^A	16.00±1.00^A	2.10±0.10^A	2.02±0.01^A
Effluent I	24.70±0.100^C	6.22±0.01^C	64.67±6.66^D	50.67±1.53^C	29.50±0.17^E	84.00±1.73^C
Effluent II	24.40±0.173^B	3.92±0.01^A	47.00±1.00^B	259.00±1.73^E	24.00±0.50^C	496.00±1.00^E
Effluent III	24.70±0.100^C	6.13±0.02^B	52.00±1.00^C	120.67±1.53^D	26.50±0.50^D	165.00±1.00^D

At P > 0.05, Tukey grouping with same letters are not significantly different. At P < 0.05, Tukey grouping with different letters are significantly different.

Table 2b. Physicochemical variations between paper mill effluents and river water samples.

Sample	Nitrate (mg/L)	Phosphate (mg/L)	Sulphate (mg/L)	BOD$_5$ (mg/L)	COD (mg/L)	Oil and Greas (mg/L)
Upstream	0.09±0.01^A	0.02±0.01^A	0.33±0.01^A	0.41±0.01^A	0.72±0.01^A	1.92±0.01^A
Discharge point	1.67±0.66^B	0.16±0.01^D	1.00±0.10^A	0.66±0.01^B	1.24±0.01^C	3.22±0.01^C
Down Stream	0.08±0.10^A	0.08±0.01^C	0.33±0.01^A	0.78±0.01^C	1.22±0.01^B	2.01±0.01^B
Effluent I	3.40±0.10^C	0.40±0.01^B	19.00±1.00^B	1.09±0.01^F	4.77±0.01^D	4.01±0.01^F
Effluent II	8.50±0.10^D	0.81±0.01^F	34.00±1.00^D	1.02±0.01^E	8.25±0.01^F	3.91±0.01^E
Effluent III	19.10±0.10^E	0.26±0.01^E	30.00±1.00^C	0.97±0.01^D	6.97±0.01^E	3.77±0.01^D

At P < 0.05, Tukey grouping with different letters are significantly different. At P > 0.05, Tukey grouping with same letters are not significantly different.

(84, 496 and 165 mg/l) respectively.

Table 2 shows the temperature values between paper mill effluents and river water samples. There was significant variation in temperature (P<0.05) between the river and effluent samples. Upstream, Discharge point and downstream samples (Turkey group–A) were not significantly different because they fell within the same domain of mean values. While group–A was significantly different from group-B (Effluent II) and group–C (Effluent - I and III); and group–B (Effluent - II) was significantly different from group-C (Effluent - I and III) and vice versa, because they fell with different domains of values.

There was significant variation in pH (P<0.05) between the river and effluent samples. Group-A (Effluent II) was very acidic (pH= 3.92) and was significantly different from other groups. Group-B (Effluent - III) was significantly different from group-C, D and E (Effluent - I, Discharge point, and downstream samples) respectively and vice versa because they fell with different domains of values.

There was significant variation in Conductivity (P<0.05) between the river and effluent samples. Upstream, discharge point and downstream samples (Tukey group-A) were not significantly different because they fell within the same domain of values. Group-A was significantly different from group-B (Effluent - II), group-C (Effluent - III) and group-D (Effluent - I), while group-B was significantly different from group-C and group-D and vice

versa. Conductivity was lowest in group-A (0.02 µS/cm) and highest in group-D (64.67 µS/cm).

There was significant variation in Turbidity (P<0.05) between the river and effluent samples. Upstream and downstream (group-A) were significantly different because they fell within the same domain of mean values. Group-A varied significantly from groups-B, C, D and E (Discharge point, Effluent - I, Effluent - III and Effluent - II) respectively, while group-B varied significantly from groups-C, D, and E respectively and vice versa. Turbidity was lowest in group-A (16.00 NTU) and highest in group-E (259.00 NTU).

There was significant variation in TDS (P<0.05) between the river and effluent samples. The upstream and downstream samples (Tukey group-A) were not significantly different because they fell within the same domain of mean values. Group-A was significantly different from group-B (Discharge point), group-C (Effluent - II), group-D (Effluent - III) and group-E (Effluent - I), and vice versa. TDS was lowest in group in group-A (2.10 mg/l) and highest in group-E (29.50 mg/L).

There was significant variation in TSS (P<0.05) between the river and effluent samples. Upstream and downstream samples (Tukey group-A) were not significantly different because they fell within the same domain of mean values. Group A was significantly different from group-B (Discharge point), group-C

(Effluent - I), group-D (Effluent -III), and group-E (Effluent -II), vice versa. TSS was lowest in group-A (1.8 mg/L) and highest in group-E (496.0 mg/L).

There was significant variation in nitrate (P<0.05) between the river and effluent samples. Upstream and downstream samples (Tukey group-A) were not significantly different because they fell within the same domain of mean values. While group-A was significantly different from group-B (Discharge point), group-C (Effluent - I), group-D (Effluent - II), and group-E (Effluent - III), group-B was significantly different from group-C, D, and E and vice versa. Nitrate was lowest in group-A (0.08 mg/L) and highest in group-E (19.10 mg/L).

There was significant variation in phosphate (PO_4^{2-}) (P<0.05) between the river and effluent samples. Group-A varied significantly from groups-B, C, D, E and F and vice versa. Phosphate was lowest in group-A (0.02 mg/L) and highest in group-F (Effluent - II) (0.8 mg/L).

There was significant variation in sulphate (P<0.05) between the river and effluent samples. Upstream, discharge and downstream samples (Tukey group-A) were not significantly different because they fell within the same domain of mean values. Group-A was significantly different from group-B (Effluent - I), Group-C (Effluent - III), and group-D (Effluent - II), and vice versa. Sulphate was lowest in group-A (0.33 mg/L) and highest in group-D (34.0 mg/L).

There was significant variation in BOD_5 (P<0.05) between the river and effluent samples. Group-A (upstream) varied significantly from groups-B, C, D, E, and F (Discharge point, Downstream, Effluent III, Effluent II and Effluent I) respectively. BOD_5 was lowest in group-A (0.41 mg/L) and highest in group–F (1.09 mg/L).

There was significant variation in COD (P<0.05) between the river and effluent samples. Group-A (upstream) varied significantly from groups-B, C, D, E and F (Downstream, Discharge point, Effluent - I, Effluent - II and Effluent - III) respectively. COD was lowest in group-A (1.92 mg/L) and highest in group-F (4.01 mg/L).

There was significant variation in Oil and Grease (P<0.05) between the river and effluent samples. Group-A (upstream) varied significantly from groups-B, C, D, E, and F (Downstream, Discharge point, Effluent - III, Effluent - II and Effluent - I) respectively. Oil and grease was lowest in group-A (1.92 mg/l) and highest in group-F (4.01 mg/L).

Discussion

The parameters as determined and shown in Table 1 are those of effluents from industries and stream samples, and were compared with the guidelines for effluent discharge limitations (FEPA, 1991). The temperatures of the water samples and the effluent samples were within the <40°C standard of FEPA. There was significant variation (P <0.05) between the mean values of the temperatures of effluent and water samples.

That temperatures of the upstream, discharge point and downstream water samples did not vary from each other and were lower than that of the effluent implied that the impact of the effluent could not elevate the temperature of the stream and the stream recovered quickly from the impact of the varied temperature of the effluents. This is in accordance with the reports of Sharples and Evans (1998) and Nwaedozie (2000).

There was significant variation between the effluent pH values and those of the water samples. The individual pH of each effluent (I, II and III) varied significantly which might be caused by the different chemical compositions of the effluents. The pH of the upstream water sample did not vary from the pH of the downstream butt both varied from the pH of the discharge point. The variation recorded with the discharge point might be due to the impact of the effluents discharged into the river at that point while the similarity in with the pH values of the upstream and downstream might be due to possible recovery of the stream from the impact of the effluents. This corroborates the reports of Odoemelam (1999).

The conductivity recorded significant variations between the stream samples and the effluent samples, though there was no variation within the stream samples while the effluent recorded variation in conductivity within the samples. This implied that the effluents contained higher levels of ionized salts from industrial activities than the stream samples, though the stream samples recovered from the impact of the effluent discharge. This corroborates the reports of Oluwande et al. (1983).

The turbidity measurement and analysis of stream samples and effluent samples showed significant variation. There was similarity between upstream and downstream samples while the samples varied from discharge point sample. This showed that the stream might have been recovered from turbidity levels of the different effluent samples. This corroborated the report of Sharples and Evans (1998).

The TDS that was recorded between the mean values of the effluents and the mean values of stream water samples indicated higher dissolved solutes in paper mill effluents than the water samples. The mean values of the upstream and downstream were the same but varied from the value of the discharge point sample which might be probably due to the impact of the effluent discharge at the point and possible recovery of the river at downstream. This is in line with the work of Odoemelam (1999), Colodey and Wells (1992).

The TSS values of the effluents recorded variations with the stream water samples similar to the TDS and might inferences. This is in line with the report of Colodey and Wells (1992).

The sulphate values of the effluents varied significantly with the mean values of the stream water samples. This might imply higher dissolved sulphate solutes in the effluent than the stream samples. These high values of sulphate in effluent did not influence the value of the discharge point and subsequently the value of the downstream probably due to natural ability of the river to recover from the impact. This is supported by the report of Anyam (1990).

The values of nitrate of effluents and the mean values of river water samples showed significant variation between effluents and river water samples, variations within effluent samples and variations within River samples, though the value of the upstream sample did not vary from the value of the downstream samples. These variations might imply that there were higher values of nitrate in effluents than the stream sample, and that the higher value indicated in discharge point over the upstream and downstream values might be due to the impact of effluent discharge from the paper mill industry. While the similarity in the values of upstream and downstream indicated possible recovery of the river from the impact of the effluent discharge. The values of the effluent varied in this pattern: Effluent III >Effluent II > Effluent I, implying that this was the contributory pattern of nitrate to the river. This report is in accordance with that of Beecroft and Oladimeji (1987).

The values of phosphate of effluent and the values of river samples showed significant variation between effluent and river samples. The value of phosphate in Effluent - I was less than the values at discharge point and downstream. This implied the major phosphate in river were Effluent II an Effluent III. The value of downstream was higher than the upstream value indicating possible non-recovery of the river at the point from impact of phosphate discharges from effluents. The values of the effluent varied in this pattern: Effluent II > Effluent > Effluent I, implying that this was the contributory pattern of phosphate to the river. This is in conformity with the report of Odoemelam (1999).

The five – day Biochemical Oxygen Demand (BOD_5) indicated that there were significant variations in BOD_5 of effluents and river samples, variations within the river samples and variations within the effluent samples. The values in the river showed the following trend; downstream > discharge point > upstream. This implied possible rise in BOD_5 due to the effluent discharges at the discharge point and further increase recorded at downstream probably due to discharges from drainages that empty into the river and other human activities. The trend in the values of BOD_5 effluents showed: Effluent I > Effluent II > Effluent III which might imply that it was the contributory pattern of organic materials into the river. This corroborates the works of Sharples and Evans (1998).

The records of Chemical Oxygen Demand (COD) indicated that there was a significant variation in COD of effluent and river samples, variations within the river effluents and variation within the effluent samples. The values in the river showed the following trend: discharge point > downstream > upstream. This implied that high

COD recorded at discharge point might be due to high chemical discharge at the point and the rivers gradual recovery at the downstream. The trend in the values of COD of effluents showed: Effluent II > Effluent III > Effluent I which may mean that it was the contributory pattern of chemicals into the River. This corroborates the works of Sharples and Evans (1998), and Sial et al. (2006).

The values of oil and grease analysis shared significant variations in oil and grease values of effluents and River samples, variations within the River samples and variations within the effluent samples. The values in the River showed the following trend: discharge point > downstream > upstream. This implied that high oil and grease recorded at discharge point might be due to high oil spill, leakages, and discharge of spent oil from generator engines, machines, vehicles, and tanks.

This is in conformity with the report of Otokunefor and Obiukwu (2005). The trend in the values of oil and grease of effluents showed: Effluent I > Effluent II > Effluent III which might imply that it was the contributory pattern of oil and grease into the River. This corroborates the work of Sial et al. (2006).

Conclusion

The research work revealed the impact of the paper mill effluents from the three paper mill industries on the Owerrinta River. Though the impact might not be conclusive by comparison of values of parameters with local and international water and effluent regulatory standards, the dynamism of the River values exposed the impact and the River's natural ability to contain with the impact from the industries. Treatment of effluent has not reduced the impact of the discharges on the empirical quality of water bodies which might expose organisms to toxic effects.

REFERENCES

Anyam RW (1990). Pollution in River Kaduna. A preliminary report. The Nigerian Fields. 55:129-132.

APHA/AWWA/W/WPCF (1985) Standard Methods for the Examination of Water and Wastewater. 16th ed. American Public Health Association Washington D.C. pp. 76 – 538.

Beecroft GA, Olademeji AA (1987). Pollution monitoring of the Kaduna River (Feb-April). Report submitted to the Environmental Planning and Protection Division, Federal Ministry of Works and Housing, Abuja.

Colodey AG, Wells PG (1992). Effects of pulp and paper mill effluents on estuarine and marine ecosystems in Canada: a review. J. Aquat. Ecosyst. Health 1:201-226.

FEPA (1991). Effluent limitation guidelines in Nigeria for all categories of industries. Federal Republic of Nigeria Official Gazette.

HACH (1981). HACH Water Anaysis Handbook. Hach Company, USA. pp 16-19.

HACH (1997). HACH DR/2010 Sepctrophotometic Handbook. Hach Company, USA. pp.147-303.

Izuagba AC, Ogbulie JN (1997). Women and Environmental Management in Africa. In: Environment and Citizenship Education C. N. Ndoh (ed) CRC Publishers Nigeria. pp. 34-56.

Nwaedozie, JM (2000). Environmental pollution: a case study of waste water effluent parameters of some industries in Kaduna. Afr. J. Envir. Stud., pp.84-89.

ODA (1997). Manual of Environmental Appraisal Overseas Development Administration London .

Odoemelam SA (1999). Effects of industrialization on water quality: A case study of effluents from the three industries in Aba River. Env. Anal., 2: 120-126.

OFIA (Ontario Forest Research Association) (2005). Pulp and paper mill effluent on the environment.

Ogbulie JN, Okpokwasili GC (1998). Efficacy of chemotherapeutic agents in controlling bacterial diseases of cultured fish. J. Aquaculture Trop., 13:285-292.

Okpokwasili GC, Ogbulie JN (1993). Bacterial and metal quality of Tilapia (Oreochromis nolitica) Aquaculture systems. Inter. J. Environ. Health Res., 3:190-202.

Okpokwasili GC, Ogbulie JN, Eleke FN, Okpokwasili NR (1997). Substrate specificity of bacterial isolates from Nigerian fish culture systems. International J. Agri. Trop., 13:269-276.

Oluwande PA, Sridhar MKC, Bammeke AO, Okubadejo J (1983). Pollution levels in some Nigeria river. Water Res., 17(9): 957-963.

Otokunefor TV, Obiukwu C (2005). Impact of refinery effluent on the physicochemical properties of a water body in the Niger Delta. Apll. Ecol. Environ. Res., 3(1): 61-72.

Sharples AE, Evans CW (1998). Impact of pulp and paper mill effluents on water quality and fauna in a New Zealand hydro-electric lake. New Zealand J. Mar. Freshw. Res., 32: 31-53.

Sial RA, Chaudlary MF, Abbas ST, Latif MI, Khan AG (2006). Quality of Effluents from Hattar Industrial Estate. J. Zhejiang Univ. Sci. B., 7(12): 974-980.

Thompson G, Swain J, Key M, Foster CF (2001). The treatment of pulp and paper mill effluent: A Review. Biores. Technol., 77(3): 275-286.

Permissions

List of Contributors

K. J. Alagoa
Department of Agric-Technology, Bayelsa State College of Arts and Science, P. M. B. 74, Agudama-Epie, Yenagoa, Bayelsa State, Nigeria

Olusegun Peter Abiola
Department of Science Laboratory Technology, Faculty of Science, The Polytechnic, Ibadan, Oyo State, Nigeria

H. M. Zakir
Department of Agricultural Chemistry, Faculty of Agriculture, Bangladesh Agricultural University, Mymensingh- 2202, Bangladesh

N. Shikazono
Laboratory of Geochemistry, School of Science for Open and Environmental Systems, Faculty of Science and Technology, Keio University, Hiyoshi 3-14-1, Yokohama 223-8522, Japan

Ezemonye Lawrence
Department of Animal and Environmental Biology (AEB) University of Benin, P. M. B. 1154, Benin City, Edo State, Nigeria

Tongo Isioma
Department of Animal and Environmental Biology (AEB) University of Benin, P. M. B. 1154, Benin City, Edo State, Nigeria

Lichao DING
School of Resource and Environmental Engineering, Jiangxi University of Science and Technology, Ganzhou, China

Yunnen CHEN
School of Resource and Environmental Engineering, Jiangxi University of Science and Technology, Ganzhou, China

Jingbiao FAN
School of Resource and Environmental Engineering, Jiangxi University of Science and Technology, Ganzhou, China

Asma Ben Ghnaya
Université Européenne de Bretagne, France
Laboratoire de Biotechnologie et Physiologie Végétales, ESMISAB, Université de Bretagne Occidentale Technopôle Brest-Iroise, 29280 Plouzané, France

Annick Hourmant
Université Européenne de Bretagne, France
Laboratoire de Biotechnologie et Physiologie Végétales, ESMISAB, Université de Bretagne Occidentale Technopôle Brest-Iroise, 29280 Plouzané, France
Laboratoire de Toxicologie Alimentaire et Cellulaire, Université de Bretagne Occidentale, UFR Sciences et Techniques, C.S. 93 837, 29238 Brest Cedex 3, France

Michel Couderchet
Laboratoire Plantes Pesticides et Développement Durable, Université de Reims Champagne-Ardenne, URVVC-SE, BP 1039, 51687 Reims, France

Michel Branchard
Université Européenne de Bretagne, France
Laboratoire de Biotechnologie et Physiologie Végétales, ESMISAB, Université de Bretagne Occidentale Technopôle Brest-Iroise, 29280 Plouzané, France

Gilbert Charles
Université Européenne de Bretagne, France
Laboratoire de Biotechnologie et Physiologie Végétales, ESMISAB, Université de Bretagne Occidentale Technopôle Brest-Iroise, 29280 Plouzané, France
Laboratoire d'Ecophysiologie et Biotechnologie des Halophytes et Algues Marines (LEBHAM), Université de Bretagne Occidentale, IUEM, Technopole Brest Iroise, Place Nicolas Copernic, 29280 Plouzané, France

A. Mishra
Aquatic Toxicology Division, IITR Lucknow, Uttar Pradesh, India

C. P. M. Tripathi
D. D. U. Gorakhpur University, Gorakhpur, India

A. K. Dwivedi
National Bureau of Fish Genetic Resources, Canal Ring Road, PO, Dilkusha, Lucknow- 26002, Uttar Pradesh, India

V. K. Dubey
National Bureau of Fish Genetic Resources, Canal Ring Road, PO, Dilkusha, Lucknow- 26002, Uttar Pradesh, India

David Adeyemi
Department of Pharmaceutical Chemistry. Faculty of Pharmacy, University of Lagos, Nigeria

Chimezie Anyakora
Department of Pharmaceutical Chemistry. Faculty of Pharmacy, University of Lagos, Nigeria

Grace Ukpo
Department of Pharmaceutical Chemistry. Faculty of Pharmacy, University of Lagos, Nigeria

Adeleye Adedayo
Nigerian Institutes of Oceanography and Marine Research, Victoria Island, Lagos, Nigeria

Godfred Darko
Department of Chemistry, Rhodes University, Grahamstown 6140, South Africa

Salem Fathallah
Laboratoire d'Aquaculture – Institut National des Sciences et Technologies de la Mer BP59, route de Khniss 5000 Monastir, Tunisie

Mohamed Néjib Medhioub
Laboratoire d'Aquaculture – Institut National des Sciences et Technologies de la Mer BP59, route de Khniss 5000 Monastir, Tunisie

Amel Medhioub
Laboratoire d'Aquaculture – Institut National des Sciences et Technologies de la Mer BP59, route de Khniss 5000 Monastir, Tunisie

Mohamed Mejdeddine Kraiem
Laboratoire d'Aquaculture – Institut National des Sciences et Technologies de la Mer BP59, route de Khniss 5000 Monastir, Tunisie

M. O. Aremu
Department of Chemistry, Nasarawa State University, P. M. B. 1022, Keffi, Nigeria

B. O. Atolaiye
Department of Chemistry, Nasarawa State University, P. M. B. 1022, Keffi, Nigeria

B. L. Gav
Department of Chemistry, Nasarawa State University, P. M. B. 1022, Keffi, Nigeria

O. D. Opaluwa
Department of Chemistry, Nasarawa State University, P. M. B. 1022, Keffi, Nigeria

D. U. Sangari
Department of Geography, Nasarawa State University, P. M. B. 1022, Keffi, Nigeria

P. C. Madu
Department of Chemistry, Nasarawa State University, P. M. B. 1022, Keffi, Nigeria

Moêz Smiri
Bio-physiologie cellulaires, faculté des sciences de Bizerte, 7021 zarzouna, Tunisia

Rohan Dasika
Department of Chemistry, Osmania University, Hyderabad, India

Siddharth Tangirala
Department of Chemistry, Osmania University, Hyderabad, India

Padmaja Naishadham
Department of Chemistry, Osmania University, Hyderabad, India

D. F. OGELEKA
Department of Chemistry, Western Delta University, Oghara, Delta State, Nigeria

L. E. TUDARARO-AHEROBO
Department of Environmental Sciences, Federal University of Petroleum Resources, Effurun, Delta State, Nigeria

Ismail I. El-Fakharany
Pesticides Department, Faculty of Agriculture kafr-EL-Shiekh University, 33516 Egypt

Ahmed H. Massoud
Pesticides Department, Faculty of Agriculture kafr-EL-Shiekh University, 33516 Egypt

Aly S. Derbalah
Pesticides Department, Faculty of Agriculture kafr-EL-Shiekh University, 33516 Egypt

Mostafa S. Saad Allah
Pesticides Department, Faculty of Agriculture kafr-EL-Shiekh University, 33516 Egypt

Opio Alfonse
Department of Biology, Faculty of Science, Gulu University, Gulu-Uganda

Julius Otutu Oseji
Department of Physics, Delta State University, Abraka, Delta State, Nigeria

D. Zogo
Société Nationale des Eaux du Bénin (SONEB), Republic of Bénin

L. M. Bawa
Laboratoire de Chimie de l'Eau, Faculté Des Sciences, Université de Lomé, BP 1515, Lomé, Togo

H. H. Soclo
Unité de Recherche en Ecotoxicologie et Etude de Qualité (UREEQ), Université d'Abomey-Calavi, Republic of Bénin

D. Atchekpe
Société Nationale des Eaux du Bénin (SONEB), Republic of Bénin

Abd El-Moneim M. R. Afify
Biochemistry Department, Faculty of Agriculture, Cairo University, Giza, Egypt, 12613

Sayed A. Fayed
Biochemistry Department, Faculty of Agriculture, Cairo University, Giza, Egypt, 12613

Emad A. Shalaby
Biochemistry Department, Faculty of Agriculture, Cairo University, Giza, Egypt, 12613

Aderonke A. Okoya
Institute of Ecology and Environmental Studies, Obafemi Awolowo University, Ile-Ife, Nigeria

Olabode I. Asubiojo
Department of Chemistry, Obafemi Awolowo University, Ile-Ife, Nigeria

Adeagbo A. Amusan
Department of Soil Science, Obafemi Awolowo University, Ile - Ife, Nigeria

I. J. Alinnor
Department of Pure and Industrial Chemistry, Federal University of Technology, P. M. B. 1526, Owerri, Imo State, Nigeria

M. A. Nwachukwu
Department of Earth and Environmental Studies, Montclair State University, New Jersey, USA

Alvarez A. Anthon
Faculty of Medicine, University A. of Sinaloa, Mexico

A. D. Campaña-Salcido
Faculty of Medicine, University A. of Sinaloa, Mexico

Joseph KETCHA MBADCAM
Physical and Theoretical Chemistry Laboratory, Faculty of Science, University of Yaoundé , Yaoundé – Cameroon

Solomon Gabche ANAGHO
Department of Chemistry, University of Dschang, Dschang – Cameroon

Julius NDI NSAMI
Physical and Theoretical Chemistry Laboratory, Faculty of Science, University of Yaoundé , Yaoundé – Cameroon

Adélaïde Maguie KAMMEGNE
Physical and Theoretical Chemistry Laboratory, Faculty of Science, University of Yaoundé , Yaoundé – Cameroon

Amina SOUDANI
Department of Chemistry, Faculty of Science, Ibn Zohr University, Agadir, BP. 8106, Hay Dakhla, 80000, Morocco

Mohamed CHIBAN
Department of Chemistry, Faculty of Science, Ibn Zohr University, Agadir, BP. 8106, Hay Dakhla, 80000, Morocco

Mohamed ZERBET
Department of Chemistry, Faculty of Science, Ibn Zohr University, Agadir, BP. 8106, Hay Dakhla, 80000, Morocco

Fouad SINAN
Department of Chemistry, Faculty of Science, Ibn Zohr University, Agadir, BP. 8106, Hay Dakhla, 80000, Morocco

O. A. Akinloye
Department of Biochemistry, University of Agriculture, P. M. B. 2240, Abeokuta, Ogun-State, Nigeria

I. Adamson
Department of Biochemistry, University of Agriculture, P. M. B. 2240, Abeokuta, Ogun-State, Nigeria

O. Ademuyiwa
Department of Biochemistry, University of Agriculture, P. M. B. 2240, Abeokuta, Ogun-State, Nigeria

T. A. Arowolo
Department of Environmental Toxicology, University of Agriculture, P. M. B. 2240, Abeokuta, Ogun-State, Nigeria

Mwinyikione Mwinyihija
Kenya Leather Development Council, P. O. Box 14480, Nairobi, Kenya

Ihejirika Chinedu Emeka
Department of Environmental Technology, Federal University of Technology, Owerri, Nigeria

Emereibeole Enos Ihediohamma
Department of Environmental Technology, Federal University of Technology, Owerri, Nigeria

Nwaogu Linus
Department of Biochemistry, Federal University of Technology, Owerri, Nigeria

Uzoka Christopher Ndubuisi
Department of Environmental Technology, Federal University of Technology, Owerri, Nigeria

Amaku Grace Ebele
Department of Environmental Technology, Federal University of Technology, Owerri, Nigeria

www.ingramcontent.com/pod-product-compliance
Lightning Source LLC
Chambersburg PA
CBHW050452200326
41458CB00014B/5148